ESSENTIAL BIOINFORMATICS

Essential Bioinformatics is a concise yet comprehensive textbook of bioinformatics that provides a broad introduction to the entire field. Written specifically for a life science audience, the basics of bioinformatics are explained, followed by discussions of the state-of-the-art computational tools available to solve biological research problems. All key areas of bioinformatics are covered including biological databases, sequence alignment, gene and promoter prediction, molecular phylogenetics, structural bioinformatics, genomics, and proteomics. The book emphasizes how computational methods work and compares the strengths and weaknesses of different methods. This balanced yet easily accessible text will be invaluable to students who do not have sophisticated computational backgrounds. Technical details of computational algorithms are explained with a minimum use of mathematical formulas; graphical illustrations are used in their place to aid understanding. The effective synthesis of existing literature as well as in-depth and up-to-date coverage of all key topics in bioinformatics make this an ideal textbook for all bioinformatics courses taken by life science students and for researchers wishing to develop their knowledge of bioinformatics to facilitate their own research.

Jin Xiong is an assistant professor of biology at Texas A&M University, where he has taught bioinformatics to graduate and undergraduate students for several years. His main research interest is in the experimental and bioinformatics analysis of photosystems.

Essential Bioinformatics

JIN XIONG

Texas A&M University

CAMBRIDGE
UNIVERSITY PRESS

CAMBRIDGE
UNIVERSITY PRESS

University Printing House, Cambridge CB2 8BS, United Kingdom

One Liberty Plaza, 20th Floor, New York, NY 10006, USA

477 Williamstown Road, Port Melbourne, VIC 3207, Australia

4843/24, 2nd Floor, Ansari Road, Daryaganj, Delhi - 110002, India

79 Anson Road, #06-04/06, Singapore 079906

Cambridge University Press is part of the University of Cambridge.

It furthers the University's mission by disseminating knowledge in the pursuit of education, learning and research at the highest international levels of excellence.

www.cambridge.org
Information on this title: www.cambridge.org/9780521600828

First published 2006

A catalogue record for this publication is available from the British Library

Library of Congress Cataloging in Publication data
Xiong, Jin. Essential bioinformatics / Jin Xiong.
p. ; cm.
Includes bibliographic references and index.
ISBN-13: 978-0-521-84098-9 (hardback)
ISBN-10: 0-521-84098-8 (hardback)
ISBN-13: 978-0-521-60082-8 (pbk.)
ISBN-10: 0-521-60082-0 (pbk.)
1. Bioinformatics. I. Title.
[DNLM: 1. Computational Biology. QU 26.5 X6e 2006]
QH324.2.X56   2006
572.80285 – dc22   2005029453

ISBN 978-0-521-84098-9 Hardback
ISBN 978-0-521-60082-8 Paperback

Contents

Preface

With a large number of prokaryotic and eukaryotic genomes completely sequenced and more forthcoming, access to the genomic information and synthesizing it for the discovery of new knowledge have become central themes of modern biological research. Mining the genomic information requires the use of sophisticated computational tools. It therefore becomes imperative for the new generation of biologists to be familiar with many bioinformatics programs and databases to tackle the new challenges in the genomic era. To meet this goal, institutions in the United States and around the world are now offering graduate and undergraduate students bioinformatics-related courses to introduce them to relevant computational tools necessary for the genomic research. To support this important task, this text was written to provide comprehensive coverage on the state-of-the-art of bioinformatics in a clear and concise manner.

The idea of writing a bioinformatics textbook originated from my experience of teaching bioinformatics at Texas A&M University. I needed a text that was comprehensive enough to cover all major aspects in the field, technical enough for a college-level course, and sufficiently up to date to include most current algorithms while at the same time being logical and easy to understand. The lack of such a comprehensive text at that time motivated me to write extensive lecture notes that attempted to alleviate the problem. The notes turned out to be very popular among the students and were in great demand from those who did not even take the class. To benefit a larger audience, I decided to assemble my lecture notes, as well as my experience and interpretation of bioinformatics, into a book.

This book is aimed at graduate and undergraduate students in biology, or any practicing molecular biologist, who has no background in computer algorithms but wishes to understand the fundamental principles of bioinformatics and use this knowledge to tackle his or her own research problems. It covers major databases and software programs for genomic data analysis, with an emphasis on the theoretical basis and practical applications of these computational tools. By reading this book, the reader will become familiar with various computational possibilities for modern molecular biological research and also become aware of the strengths and weaknesses of each of the software tools.

The reader is assumed to have a basic understanding of molecular biology and biochemistry. Therefore, many biological terms, such as *nucleic acids*, *amino acids*, *genes*, *transcription*, and *translation*, are used without further explanation. One exception is *protein structure*, for which a chapter about fundamental concepts is included so that algorithms and rationales for protein structural bioinformatics can be better understood. Prior knowledge of advanced statistics, probability theories, and calculus is of course preferable but not essential.

This book is organized into six sections: biological databases, sequence alignment, genes and promoter prediction, molecular phylogenetics, structural bioinformatics, and genomics and proteomics. There are nineteen chapters in total, each of which is relatively independent. When information from one chapter is needed for understanding another, cross-references are provided. Each chapter includes definitions and key concepts as well as solutions to related computational problems. Occasionally there are boxes that show worked examples for certain types of calculations. Since this book is primarily for molecular biologists, very few mathematical formulas are used. A small number of carefully chosen formulas are used where they are absolutely necessary to understand a particular concept. The background discussion of a computational problem is often followed by an introduction to related computer programs that are available online. A summary is also provided at the end of each chapter.

Most of the programs described in this book are online tools that are freely available and do not require special expertise to use them. Most of them are rather straightforward to use in that the user only needs to supply sequences or structures as input, and the results are returned automatically. In many cases, knowing which programs are available for which purposes is sufficient, though occasionally skills of interpreting the results are needed. However, in a number of instances, knowing the names of the programs and their applications is only half the journey. The user also has to make special efforts to learn the intricacies of using the programs. These programs are considered to be on the other extreme of user-friendliness. However, it would be impractical for this book to try to be a computer manual for every available software program. That is not my goal in writing the book. Nonetheless, having realized the difficulties of beginners who are often unaware of or, more precisely, intimidated by the numerous software programs available, I have designed a number of practical Web exercises with detailed step-by-step procedures that aim to serve as examples of the correct use of a combined set of bioinformatics tools for solving a particular problem. The exercises were originally written for use on a UNIX workstation. However, they can be used, with slight modifications, on any operating systems with Internet access.

In the course of preparing this book, I consulted numerous original articles and books related to certain topics of bioinformatics. I apologize for not being able to acknowledge all of these sources because of space limitations in such an introductory text. However, a small number of articles (mainly recent review articles) and books related to the topics of each chapter are listed as "Further Reading" for those who wish to seek more specialized information on the topics. Regarding the inclusion of computational programs, there are often a large number of programs available for a particular task. I apologize for any personal bias in the selection of the software programs in the book.

One of the challenges in writing this text was to cover sufficient technical background of computational methods without extensive display of mathematical formulas. I strived to maintain a balance between explaining algorithms and not getting into too much mathematical detail, which may be intimidating for beginning students and nonexperts in computational biology. This sometimes proved to be a tough balance for me because I risk either sacrificing some of the original content or losing the reader. To alleviate this problem, I chose in many instances to use graphics instead of formulas to illustrate a concept and to aid understanding.

I would like to thank the Department of Biology at Texas A&M University for the opportunity of letting me teach a bioinformatics class, which is what made this book possible. I thank all my friends and colleagues in the Department of Biology and the Department of Biochemistry for their friendship. Some of my colleagues were kind enough to let me participate in their research projects, which provided me with diverse research problems with which I could hone my bioinformatics analysis skills. I am especially grateful to Lisa Peres of the Molecular Simulation Laboratory at Texas A&M, who was instrumental in helping me set up and run the laboratory section of my bioinformatics course. I am also indebted to my former postdoctoral mentor, Carl Bauer of Indiana University, who gave me the wonderful opportunity to learn evolution and phylogenetics in great depth, which essentially launched my career in bioinformatics. Also importantly, I would like to thank Katrina Halliday, my editor at Cambridge University Press, for accepting the manuscript and providing numerous suggestions for polishing the early draft. It was a great pleasure working with her. Thanks also go to Cindy Fullerton and Marielle Poss for their diligent efforts in overseeing the copyediting of the book to ensure a quality final product.

Jin Xiong

Introduction and Biological Databases

CHAPTER ONE

Introduction

Quantitation and quantitative tools are indispensable in modern biology. Most biological research involves application of some type of mathematical, statistical, or computational tools to help synthesize recorded data and integrate various types of information in the process of answering a particular biological question. For example, enumeration and statistics are required for assessing everyday laboratory experiments, such as making serial dilutions of a solution or counting bacterial colonies, phage plaques, or trees and animals in the natural environment. A classic example in the history of genetics is by Gregor Mendel and Thomas Morgan, who, by simply counting genetic variations of plants and fruit flies, were able to discover the principles of genetic inheritance. More dedicated use of quantitative tools may involve using calculus to predict the growth rate of a human population or to establish a kinetic model for enzyme catalysis. For very sophisticated uses of quantitative tools, one may find application of the "game theory" to model animal behavior and evolution, or the use of millions of nonlinear partial differential equations to model cardiac blood flow. Whether the application is simple or complex, subtle or explicit, it is clear that mathematical and computational tools have become an integral part of modern-day biological research. However, none of these examples of quantitative tool use in biology could be considered to be part of bioinformatics, which is also quantitative in nature. To help the reader understand the difference between bioinformatics and other elements of quantitative biology, we provide a detailed explanation of what is bioinformatics in the following sections.

Bioinformatics, which will be more clearly defined below, is the discipline of quantitative analysis of information relating to biological macromolecules with the aid of computers. The development of bioinformatics as a field is the result of advances in both molecular biology and computer science over the past 30–40 years. Although these developments are not described in detail here, understanding the history of this discipline is helpful in obtaining a broader insight into current bioinformatics research. A succinct chronological summary of the landmark events that have had major impacts on the development of bioinformatics is presented here to provide context.

The earliest bioinformatics efforts can be traced back to the 1960s, although the word *bioinformatics* did not exist then. Probably, the first major bioinformatics project was undertaken by Margaret Dayhoff in 1965, who developed a first protein sequence database called *Atlas of Protein Sequence and Structure*. Subsequently, in the early 1970s, the Brookhaven National Laboratory established the Protein Data Bank for archiving three-dimensional protein structures. At its onset, the database stored less than a dozen protein structures, compared to more than 30,000 structures today. The first sequence alignment algorithm was developed by Needleman and Wunsch in 1970. This was a fundamental step in the development of the field of bioinformatics, which paved the way for the routine sequence comparisons and database searching practiced by modern biologists. The first protein structure prediction algorithm was developed by Chou and Fasman in 1974. Though it is rather rudimentary by today's standard, it pioneered a series of developments in protein structure prediction. The 1980s saw the establishment of GenBank and the development of fast database searching algorithms such as FASTA by William Pearson and BLAST by Stephen Altschul and coworkers. The start of the human genome project in the late 1980s provided a major boost for the development of bioinformatics. The development and the increasingly widespread use of the Internet in the 1990s made instant access to, and exchange and dissemination of, biological data possible.

These are only the major milestones in the establishment of this new field. The fundamental reason that bioinformatics gained prominence as a discipline was the advancement of genome studies that produced unprecedented amounts of biological data. The explosion of genomic sequence information generated a sudden demand for efficient computational tools to manage and analyze the data. The development of these computational tools depended on knowledge generated from a wide range of disciplines including mathematics, statistics, computer science, information technology, and molecular biology. The merger of these disciplines created an information-oriented field in biology, which is now known as *bioinformatics*.

WHAT IS BIOINFORMATICS?

Bioinformatics is an interdisciplinary research area at the interface between computer science and biological science. A variety of definitions exist in the literature and on the world wide web; some are more inclusive than others. Here, we adopt the definition proposed by Luscombe et al. in defining bioinformatics as a union of biology and informatics: *bioinformatics* involves the technology that uses computers for storage, retrieval, manipulation, and distribution of information related to biological macromolecules such as DNA, RNA, and proteins. The emphasis here is on the use of computers because most of the tasks in genomic data analysis are highly repetitive or mathematically complex. The use of computers is absolutely indispensable in mining genomes for information gathering and knowledge building.

Bioinformatics differs from a related field known as *computational biology*. Bioinformatics is limited to sequence, structural, and functional analysis of genes and genomes and their corresponding products and is often considered *computational molecular biology*. However, computational biology encompasses all biological areas that involve computation. For example, mathematical modeling of ecosystems, population dynamics, application of the game theory in behavioral studies, and phylogenetic construction using fossil records all employ computational tools, but do not necessarily involve biological macromolecules.

Beside this distinction, it is worth noting that there are other views of how the two terms relate. For example, one version defines *bioinformatics* as the development and application of computational tools in managing *all kinds* of biological data, whereas *computational biology* is more confined to the theoretical development of algorithms used for bioinformatics. The confusion at present over definition may partly reflect the nature of this vibrant and quickly evolving new field.

GOALS

The ultimate goal of bioinformatics is to better understand a living cell and how it functions at the molecular level. By analyzing raw molecular sequence and structural data, bioinformatics research can generate new insights and provide a "global" perspective of the cell. The reason that the functions of a cell can be better understood by analyzing sequence data is ultimately because the flow of genetic information is dictated by the "central dogma" of biology in which DNA is transcribed to RNA, which is translated to proteins. Cellular functions are mainly performed by proteins whose capabilities are ultimately determined by their sequences. Therefore, solving functional problems using sequence and sometimes structural approaches has proved to be a fruitful endeavor.

SCOPE

Bioinformatics consists of two subfields: the development of computational tools and databases and the application of these tools and databases in generating biological knowledge to better understand living systems. These two subfields are complementary to each other. The tool development includes writing software for sequence, structural, and functional analysis, as well as the construction and curating of biological databases. These tools are used in three areas of genomic and molecular biological research: molecular sequence analysis, molecular structural analysis, and molecular functional analysis. The analyses of biological data often generate new problems and challenges that in turn spur the development of new and better computational tools.

The areas of sequence analysis include sequence alignment, sequence database searching, motif and pattern discovery, gene and promoter finding, reconstruction of evolutionary relationships, and genome assembly and comparison. Structural analyses include protein and nucleic acid structure analysis, comparison, classification, and prediction. The functional analyses include gene expression profiling, protein–protein interaction prediction, protein subcellular localization prediction, metabolic pathway reconstruction, and simulation (Fig. 1.1).

The three aspects of bioinformatics analysis are not isolated but often interact to produce integrated results (see Fig. 1.1). For example, protein structure prediction depends on sequence alignment data; clustering of gene expression profiles requires the use of phylogenetic tree construction methods derived in sequence analysis. Sequence-based promoter prediction is related to functional analysis of

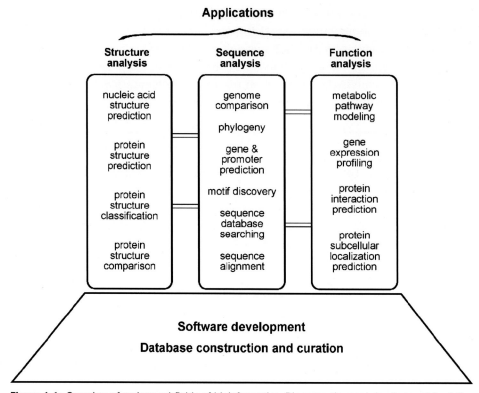

Figure 1.1: Overview of various subfields of bioinformatics. Biocomputing tool development is at the foundation of all bioinformatics analysis. The applications of the tools fall into three areas: sequence analysis, structure analysis, and function analysis. There are intrinsic connections between different areas of analyses represented by bars between the boxes.

coexpressed genes. Gene annotation involves a number of activities, which include distinction between coding and noncoding sequences, identification of translated protein sequences, and determination of the gene's evolutionary relationship with other known genes; prediction of its cellular functions employs tools from all three groups of the analyses.

APPLICATIONS

Bioinformatics has not only become essential for basic genomic and molecular biology research, but is having a major impact on many areas of biotechnology and biomedical sciences. It has applications, for example, in knowledge-based drug design, forensic DNA analysis, and agricultural biotechnology. Computational studies of protein–ligand interactions provide a rational basis for the rapid identification of novel leads for synthetic drugs. Knowledge of the three-dimensional structures of proteins allows molecules to be designed that are capable of binding to the receptor site of a target protein with great affinity and specificity. This informatics-based approach significantly reduces the time and cost necessary to develop drugs with higher potency, fewer side effects, and less toxicity than using the traditional trial-and-error approach. In forensics, results from molecular phylogenetic analysis have been accepted as evidence in criminal courts. Some sophisticated Bayesian statistics and likelihood-based methods for analysis of DNA have been applied in the analysis of forensic identity. It is worth mentioning that genomics and bioinformtics are now poised to revolutionize our healthcare system by developing personalized and customized medicine. The high speed genomic sequencing coupled with sophisticated informatics technology will allow a doctor in a clinic to quickly sequence a patient's genome and easily detect potential harmful mutations and to engage in early diagnosis and effective treatment of diseases. Bioinformatics tools are being used in agriculture as well. Plant genome databases and gene expression profile analyses have played an important role in the development of new crop varieties that have higher productivity and more resistance to disease.

LIMITATIONS

Having recognized the power of bioinformatics, it is also important to realize its limitations and avoid over-reliance on and over-expectation of bioinformatics output. In fact, bioinformatics has a number of inherent limitations. In many ways, the role of bioinformatics in genomics and molecular biology research can be likened to the role of intelligence gathering in battlefields. Intelligence is clearly very important in leading to victory in a battlefield. Fighting a battle without intelligence is inefficient and dangerous. Having superior information and correct intelligence helps to identify the enemy's weaknesses and reveal the enemy's strategy and intentions. The gathered information can then be used in directing the forces to engage the enemy and win the battle. However, completely relying on intelligence can also be dangerous if the intelligence is of limited accuracy. Overreliance on poor-quality intelligence can yield costly mistakes if not complete failures.

It is no stretch in analogy that fighting diseases or other biological problems using bioinformatics is like fighting battles with intelligence. Bioinformatics and experimental biology are independent, but complementary, activities. Bioinformatics depends on experimental science to produce raw data for analysis. It, in turn, provides useful interpretation of experimental data and important leads for further experimental research. Bioinformatics predictions are not formal proofs of any concepts. They do not replace the traditional experimental research methods of actually testing hypotheses. In addition, the quality of bioinformatics predictions depends on the quality of data and the sophistication of the algorithms being used. Sequence data from high throughput analysis often contain errors. If the sequences are wrong or annotations incorrect, the results from the downstream analysis are misleading as well. That is why it is so important to maintain a realistic perspective of the role of bioinformatics.

Bioinformatics is by no means a mature field. Most algorithms lack the capability and sophistication to truly reflect reality. They often make incorrect predictions that make no sense when placed in a biological context. Errors in sequence alignment, for example, can affect the outcome of structural or phylogenetic analysis. The outcome of computation also depends on the computing power available. Many accurate but exhaustive algorithms cannot be used because of the slow rate of computation. Instead, less accurate but faster algorithms have to be used. This is a necessary trade-off between accuracy and computational feasibility. Therefore, it is important to keep in mind the potential for errors produced by bioinformatics programs. Caution should always be exercised when interpreting prediction results. It is a good practice to use multiple programs, if they are available, and perform multiple evaluations. A more accurate prediction can often be obtained if one draws a consensus by comparing results from different algorithms.

NEW THEMES

Despite the pitfalls, there is no doubt that bioinformatics is a field that holds great potential for revolutionizing biological research in the coming decades. Currently, the field is undergoing major expansion. In addition to providing more reliable and more rigorous computational tools for sequence, structural, and functional analysis, the major challenge for future bioinformatics development is to develop tools for elucidation of the functions and interactions of all gene products in a cell. This presents a tremendous challenge because it requires integration of disparate fields of biological knowledge and a variety of complex mathematical and statistical tools. To gain a deeper understanding of cellular functions, mathematical models are needed to simulate a wide variety of intracellular reactions and interactions at the whole cell level. This molecular simulation of all the cellular processes is termed *systems biology*. Achieving this goal will represent a major leap toward fully understanding a living system. That is why the system-level simulation and integration are considered the future of bioinformatics. Modeling such complex networks and making predictions about their behavior present tremendous challenges and opportunities for bioinformaticians. The ultimate goal of this endeavor is to transform biology from a qualitative science to a quantitative and predictive science. This is truly an exciting time for bioinformatics.

FURTHER READING

Attwood, T. K., and Miller, C. J. 2002. Progress in bioinformatics and the importance of being earnest. *Biotechnol. Annu. Rev.* 8:1–54.

Golding, G. B. 2003. DNA and the revolution of molecular evolution, computational biology, and bioinformatics. *Genome* 46:930–5.

Goodman, N. 2002. Biological data becomes computer literature: New advances in bioinformatics. *Curr. Opin. Biotechnol.* 13:68–71.

Hagen. J. B. 2000. The origin of bioinformatics. *Nat. Rev. Genetics* 1:231–6.

Kanehisa, M., and Bork, P. 2003. Bioinformatics in the post-sequence era. *Nat. Genet.* 33 Suppl:305–10.

Kim, J. H. 2002. Bioinformatics and genomic medicine. *Genet. Med.* 4 Suppl:62S–5S.

Luscombe, N. M., Greenbaum, D., and Gerstein, M. 2001. What is bioinformatics? A proposed definition and overview of the field. *Methods Inf. Med.* 40:346–58.

Ouzounis, C. A., and Valencia, A. 2003. Early bioinformatics: The birth of a discipline – A personal view. *Bioinformatics* 19:2176–90.

CHAPTER TWO

Introduction to Biological Databases

One of the hallmarks of modern genomic research is the generation of enormous amounts of raw sequence data. As the volume of genomic data grows, sophisticated computational methodologies are required to manage the data deluge. Thus, the very first challenge in the genomics era is to store and handle the staggering volume of information through the establishment and use of computer databases. The development of databases to handle the vast amount of molecular biological data is thus a fundamental task of bioinformatics. This chapter introduces some basic concepts related to databases, in particular, the types, designs, and architectures of biological databases. Emphasis is on retrieving data from the main biological databases such as GenBank.

WHAT IS A DATABASE?

A *database* is a computerized archive used to store and organize data in such a way that information can be retrieved easily via a variety of search criteria. Databases are composed of computer hardware and software for data management. The chief objective of the development of a database is to organize data in a set of structured records to enable easy retrieval of information. Each record, also called an *entry*, should contain a number of fields that hold the actual data items, for example, fields for names, phone numbers, addresses, dates. To retrieve a particular record from the database, a user can specify a particular piece of information, called *value*, to be found in a particular field and expect the computer to retrieve the whole data record. This process is called *making a query*.

Although data retrieval is the main purpose of all databases, biological databases often have a higher level of requirement, known as *knowledge discovery*, which refers to the identification of connections between pieces of information that were not known when the information was first entered. For example, databases containing raw sequence information can perform extra computational tasks to identify sequence homology or conserved motifs. These features facilitate the discovery of new biological insights from raw data.

TYPES OF DATABASES

Originally, databases all used a flat file format, which is a long text file that contains many entries separated by a *delimiter*, a special character such as a vertical bar (|). Within each entry are a number of fields separated by tabs or commas. Except for the raw values in each field, the entire text file does not contain any hidden instructions for computers to search for specific information or to create reports based on certain fields from each record. The text file can be considered a single table. Thus, to search a flat file for a particular piece of information, a computer has to read through the entire file, an obviously inefficient process. This is manageable for a small database, but as database size increases or data types become more complex, this database style can become very difficult for information retrieval. Indeed, searches through such files often cause crashes of the entire computer system because of the memory-intensive nature of the operation.

To facilitate the access and retrieval of data, sophisticated computer software programs for organizing, searching, and accessing data have been developed. They are called *database management systems.* These systems contain not only raw data records but also operational instructions to help identify hidden connections among data records. The purpose of establishing a data structure is for easy execution of the searches and to combine different records to form final search reports. Depending on the types of data structures, these database management systems can be classified into two types: *relational database management systems* and *object-oriented database management systems.* Consequently, databases employing these management systems are known as *relational databases* or *object-oriented databases*, respectively.

Relational Databases

Instead of using a single table as in a flat file database, relational databases use a set of tables to organize data. Each table, also called a *relation*, is made up of columns and rows. Columns represent individual fields. Rows represent values in the fields of records. The columns in a table are indexed according to a common feature called an *attribute*, so they can be cross-referenced in other tables. To execute a query in a relational database, the system selects linked data items from different tables and combines the information into one report. Therefore, specific information can be found more quickly from a relational database than from a flat file database.

Relational databases can be created using a special programming language called *structured query language* (SQL). The creation of this type of databases can take a great deal of planning during the design phase. After creation of the original database, a new data category can be easily added without requiring all existing tables to be modified. The subsequent database searching and data gathering for reports are relatively straightforward.

Here is a simple example of student course information expressed in a flat file which contains records of five students from four different states, each taking a different course (Fig. 2.1). Each data record, separated by a vertical bar, contains four fields describing the name, state, course number and title. A relational database is also created to store the same information, in which the data are structured as a number of tables. Figure 2.1 shows how the relational database works. In each table, data that fit a particular criterion are grouped together. Different tables can be linked by common data categories, which facilitate finding of specific information.

Flat File	Name, States, Course number, Course name|John Smith, Texas, Biol 689, Bioinformatics|Jane Doe, Kansas, Bich 441, Biochemistry|William Brown, Illinois, Chem 289, Organic Chemistry|Jennifer Taylor, New York, Hort 201, Horticulture|Howard Douglas, Texas, Math 172, Calculus

Table A

Student #	Name	State
1	John Smith	Texas
2	Jane Doe	Kansas
3	William Brown	Illinois
4	Jennifer Taylor	New York
5	Howard Douglas	Texas

Table B

Student #	Course #
1	Biol 689
2	Bich 441
3	Chem 289
4	Hort 201
5	Math 172

Table C

Course #	Course name
Biol 689	Bioinformatics
Bich 441	Biochemistry
Chem 289	Organic chemistry
Hort 201	Horticulture
Math 172	Calculus

Figure 2.1: Example of constructing a relational database for five students' course information originally expressed in a flat file. By creating three different tables linked by common fields, data can be easily accessed and reassembled.

For example, if one is to ask the question, which courses are students from Texas taking? The database will first find the field for "State" in Table A and look up for Texas. This returns students 1 and 5. The student numbers are colisted in Table B, in which students 1 and 5 correspond to Biol 689 and Math 172, respectively. The course names listed by course numbers are found in Table C. By going to Table C, exact course names corresponding to the course numbers can be retrieved. A final report is then given showing that the Texans are taking the courses Bioinformatics and Calculus. However, executing the same query through the flat file requires the computer to read through the entire text file word by word and to store the information in a temporary memory space and later mark up the data records containing the word *Texas*. This is easily accomplishable for a small database. To perform queries in a large database using flat files obviously becomes an onerous task for the computer system.

Object-Oriented Databases

One of the problems with relational databases is that the tables used do not describe complex hierarchical relationships between data items. To overcome the problem, object-oriented databases have been developed that store data as objects. In an object-oriented programming language, an object can be considered as a unit that combines data and mathematical routines that act on the data. The database is structured such that the objects are linked by a set of pointers defining predetermined relationships between the objects. Searching the database involves navigating through the objects with the aid of the pointers linking different objects. Programming languages like C++ are used to create object-oriented databases.

The object-oriented database system is more flexible; data can be structured based on hierarchical relationships. By doing so, programming tasks can be simplified for data that are known to have complex relationships, such as multimedia data. However,

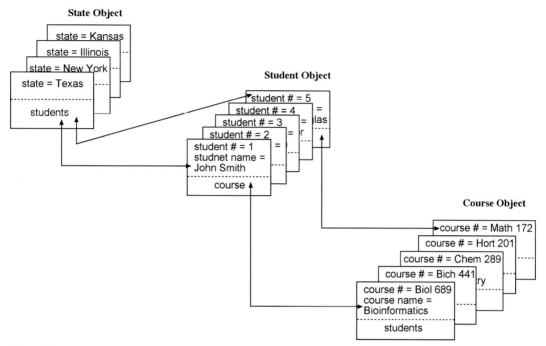

Figure 2.2: Example of construction and query of an object-oriented database using the same student information as shown in Figure 2.1. Three objects are constructed and are linked by pointers shown as arrows. Finding specific information relies on navigating through the objects by way of pointers. For simplicity, some of the pointers are omitted.

this type of database system lacks the rigorous mathematical foundation of the relational databases. There is also a risk that some of the relationships between objects may be misrepresented. Some current databases have therefore incorporated features of both types of database programming, creating the *object–relational database management system.*

The above students' course information (Fig. 2.1) can be used to construct an object-oriented database. Three different objects can be designed: student object, course object, and state object. Their interrelations are indicated by lines with arrows (Fig. 2.2). To answer the same question – which courses are students from Texas taking – one simply needs to start from Texas in the state object, which has pointers that lead to students 1 and 5 in the student object. Further pointers in the student object point to the course each of the two students is taking. Therefore, a simple navigation through the linked objects provides a final report.

BIOLOGICAL DATABASES

Current biological databases use all three types of database structures: flat files, relational, and object oriented. Despite the obvious drawbacks of using flat files in database management, many biological databases still use this format. The justification for this is that this system involves minimum amount of database design and the search output can be easily understood by working biologists.

Based on their contents, biological databases can be roughly divided into three categories: primary databases, secondary databases, and specialized databases. *Primary databases* contain original biological data. They are archives of raw sequence or structural data submitted by the scientific community. GenBank and Protein Data Bank (PDB) are examples of primary databases. *Secondary databases* contain computationally processed or manually curated information, based on original information from primary databases. Translated protein sequence databases containing functional annotation belong to this category. Examples are SWISS-Prot and Protein Information Resources (PIR) (successor of Margaret Dayhoff's Atlas of Protein Sequence and Structure [see Chapter 1]). Specialized databases are those that cater to a particular research interest. For example, Flybase, HIV sequence database, and Ribosomal Database Project are databases that specialize in a particular organism or a particular type of data. A list of some frequently used databases is provided in Table 2.1.

Primary Databases

There are three major public sequence databases that store raw nucleic acid sequence data produced and submitted by researchers worldwide: GenBank, the European Molecular Biology Laboratory (EMBL) database and the DNA Data Bank of Japan (DDBJ), which are all freely available on the Internet. Most of the data in the databases are contributed directly by authors with a minimal level of annotation. A small number of sequences, especially those published in the 1980s, were entered manually from published literature by database management staff.

Presently, sequence submission to either GenBank, EMBL, or DDBJ is a precondition for publication in most scientific journals to ensure the fundamental molecular data to be made freely available. These three public databases closely collaborate and exchange new data daily. They together constitute the International Nucleotide Sequence Database Collaboration. This means that by connecting to any one of the three databases, one should have access to the same nucleotide sequence data. Although the three databases all contain the same sets of raw data, each of the individual databases has a slightly different kind of format to represent the data.

Fortunately, for the three-dimensional structures of biological macromolecules, there is only one centralized database, the PDB. This database archives atomic coordinates of macromolecules (both proteins and nucleic acids) determined by x-ray crystallography and NMR. It uses a flat file format to represent protein name, authors, experimental details, secondary structure, cofactors, and atomic coordinates. The web interface of PDB also provides viewing tools for simple image manipulation. More details of this database and its format are provided in Chapter 12.

Secondary Databases

Sequence annotation information in the primary database is often minimal. To turn the raw sequence information into more sophisticated biological knowledge, much postprocessing of the sequence information is needed. This begs the need for

TABLE 2.1. Major Biological Databases Available Via the World Wide Web

Databases and Retrieval Systems	Brief Summary of Content	URL
AceDB	Genome database for *Caenorhabditis elegans*	www.acedb.org
DDBJ	Primary nucleotide sequence database in Japan	www.ddbj.nig.ac.jp
EMBL	Primary nucleotide sequence database in Europe	www.ebi.ac.uk/embl/index.html
Entrez	NCBI portal for a variety of biological databases	www.ncbi.nlm.nih.gov/gquery/gquery.fcgi
ExPASY	Proteomics database	http://us.expasy.org/
FlyBase	A database of the *Drosophila* genome	http://flybase.bio.indiana.edu/
FSSP	Protein secondary structures	www.bioinfo.biocenter.helsinki.fi:8080/dali/index.html
GenBank	Primary nucleotide sequence database in NCBI	www.ncbi.nlm.nih.gov/Genbank
HIV databases	HIV sequence data and related immunologic information	www.hiv.lanl.gov/content/index
Microarray gene expression database	DNA microarray data and analysis tools	www.ebi.ac.uk/microarray
OMIM	Genetic information of human diseases	www.ncbi.nlm.nih.gov/entrez/query.fcgi?db=OMIM
PIR	Annotated protein sequences	http://pir.georgetown.edu/pirwww/pirhome3.shtml
PubMed	Biomedical literature information	www.ncbi.nlm.nih.gov/PubMed
Ribosomal database project	Ribosomal RNA sequences and phylogenetic trees derived from the sequences	http://rdp.cme.msu.edu/html
SRS	General sequence retrieval system	http://srs6.ebi.ac.uk
SWISS-Prot	Curated protein sequence database	www.ebi.ac.uk/swissprot/access.html
TAIR	Arabidopsis information database	www.arabidopsis.org

secondary databases, which contain computationally processed sequence information derived from the primary databases. The amount of computational processing work varies greatly among the secondary databases; some are simple archives of translated sequence data from identified open reading frames in DNA, whereas others provide additional annotation and information related to higher levels of information regarding structure and functions.

A prominent example of secondary databases is SWISS-PROT, which provides detailed sequence annotation that includes structure, function, and protein family assignment. The sequence data are mainly derived from TrEMBL, a database of translated nucleic acid sequences stored in the EMBL database. The annotation of each entry is carefully curated by human experts and thus is of good quality. The protein annotation includes function, domain structure, catalytic sites, cofactor binding, posttranslational modification, metabolic pathway information, disease association, and similarity with other sequences. Much of this information is obtained from scientific literature and entered by database curators. The annotation provides significant added value to each original sequence record. The data record also provides cross-referencing links to other online resources of interest. Other features such as very low redundancy and high level of integration with other primary and secondary databases make SWISS-PROT very popular among biologists.

A recent effort to combine SWISS-PROT, TrEMBL, and PIR led to the creation of the UniProt database, which has larger coverage than any one of the three databases while at the same time maintaining the original SWISS-PROT feature of low redundancy, cross-references, and a high quality of annotation.

There are also secondary databases that relate to protein family classification according to functions or structures. The Pfam and Blocks databases (to be described in Chapter 7) contain aligned protein sequence information as well as derived motifs and patterns, which can be used for classification of protein families and inference of protein functions. The DALI database (to be described in Chapter 13) is a protein secondary structure database that is vital for protein structure classification and threading analysis (to be described in Chapter 15) to identify distant evolutionary relationships among proteins.

Specialized Databases

Specialized databases normally serve a specific research community or focus on a particular organism. The content of these databases may be sequences or other types of information. The sequences in these databases may overlap with a primary database, but may also have new data submitted directly by authors. Because they are often curated by experts in the field, they may have unique organizations and additional annotations associated with the sequences. Many genome databases that are taxonomic specific fall within this category. Examples include Flybase, WormBase, AceDB, and TAIR (Table 2.1). In addition, there are also specialized databases that contain original data derived from functional analysis. For example, GenBank EST database and Microarray Gene Expression Database at the European Bioinformatics Institute (EBI) are some of the gene expression databases available.

Interconnection between Biological Databases

As mentioned, primary databases are central repositories and distributors of raw sequence and structure information. They support nearly all other types of biological databases in a way akin to the Associated Press providing news feeds to local news media, which then tailor the news to suit their own particular needs. Therefore, in the biological community, there is a frequent need for the secondary and specialized databases to connect to the primary databases and to keep uploading sequence information. In addition, a user often needs to get information from both primary and secondary databases to complete a task because the information in a single database is often insufficient. Instead of letting users visiting multiple databases, it is convenient for entries in a database to be cross-referenced and linked to related entries in other databases that contain additional information. All these create a demand for linking different databases.

The main barrier to linking different biological databases is format incompatibility current biological databases utilize all three types of database structures – flat files, relational, and object oriented. The heterogeneous database structures limit communication between databases. One solution to networking the databases is to use a specification language called Common Object Request Broker Architecture (COBRA), which allows database programs at different locations to communicate in a network through an "interface broker" without having to understand each other's database structure. It works in a way similar to HyperText Markup Language (HTML) for web pages, labeling database entries using a set of common tags.

A similar protocol called eXtensible Markup Language (XML) also helps in bridging databases. In this format, each biological record is broken down into small, basic components that are labeled with a hierarchical nesting of tags. This database structure significantly improves the distribution and exchange of complex sequence annotations between databases. Recently, a specialized protocol for bioinformatics data exchange has been developed. It is the distributed annotation system, which allows one computer to contact multiple servers and retrieve dispersed sequence annotation information related to a particular sequence and integrate the results into a single combined report.

PITFALLS OF BIOLOGICAL DATABASES

One of the problems associated with biological databases is overreliance on sequence information and related annotations, without understanding the reliability of the information. What is often ignored is the fact that there are many errors in sequence databases. There are also high levels of redundancy in the primary sequence databases. Annotations of genes can also occasionally be false or incomplete. All these types of errors can be passed on to other databases, causing propagation of errors.

Most errors in nucleotide sequences are caused by sequencing errors. Some of these errors cause frameshifts that make whole gene identification difficult or protein translation impossible. Sometimes, gene sequences are contaminated with sequences from cloning vectors. Generally speaking, errors are more common for sequences produced before the 1990s; sequence quality has been greatly improved since. Therefore, exceptional care should be taken when dealing with more dated sequences.

Redundancy is another major problem affecting primary databases. There is tremendous duplication of information in the databases, for various reasons. The causes of redundancy include repeated submission of identical or overlapping sequences by the same or different authors, revision of annotations, dumping of expressed sequence tags (EST) data (see Chapter 18), and poor database management that fails to detect the redundancy. This makes some primary databases excessively large and unwieldy for information retrieval.

Steps have been taken to reduce the redundancy. The National Center for Biotechnology Information (NCBI) has now created a *nonredundant* database, called RefSeq, in which identical sequences from the same organism and associated sequence fragments are merged into a single entry. Proteins sequences derived from the same DNA sequences are explicitly linked as related entries. Sequence variants from the same organism with very minor differences, which may well be caused by sequencing errors, are treated as distinctly related entries. This carefully curated database can be considered a secondary database.

As mentioned, the SWISS-PROT database also has minimal redundancy for protein sequences compared to most other databases. Another way to address the redundancy problem is to create sequence-cluster databases such as UniGene (see Chapter 18) that coalesce EST sequences that are derived from the same gene.

The other common problem is erroneous annotations. Often, the same gene sequence is found under different names resulting in multiple entries and confusion about the data. Or conversely, unrelated genes bearing the same name are found in the databases. To alleviate the problem of naming genes, reannotation of genes and proteins using a set of common, controlled vocabulary to describe a gene or protein is necessary. The goal is to provide a consistent and unambiguous naming system for all genes and proteins. A prominent example of such systems is *Gene Ontology* (see Chapter 17).

Some of the inconsistencies in annotation could be caused by genuine disagreement between researchers in the field; others may result from imprudent assignment of protein functions by sequence submitters. There are also some errors that are simply caused by omissions or mistakes in typing. Errors in annotation can be particularly damaging because the large majority of new sequences are assigned functions based on similarity with sequences in the databases that are already annotated. Therefore, a wrong annotation can be easily transferred to all similar genes in the entire database. It is possible that some of these errors can be corrected at the informatics level by studying the protein domains and families. However, others eventually have to be corrected using experimental work.

INFORMATION RETRIEVAL FROM BIOLOGICAL DATABASES

As mentioned, a major goal in developing databases is to provide efficient and user-friendly access to the data stored. There are a number of retrieval systems for biological data. The most popular retrieval systems for biological databases are Entrez and Sequence Retrieval Systems (SRS) that provide access to multiple databases for retrieval of integrated search results.

To perform complex queries in a database often requires the use of Boolean operators. This is to join a series of keywords using logical terms such as AND, OR, and NOT to indicate relationships between the keywords used in a search. *AND* means that the search result must contain both words; *OR* means to search for results containing either word or both; *NOT* excludes results containing either one of the words. In addition, one can use parentheses ( ) to define a concept if multiple words and relationships are involved, so that the computer knows which part of the search to execute first. Items contained within parentheses are executed first. Quotes can be used to specify a phrase. Most search engines of public biological databases use some form of this Boolean logic.

Entrez

The NCBI developed and maintains Entrez, a biological database retrieval system. It is a gateway that allows text-based searches for a wide variety of data, including annotated genetic sequence information, structural information, as well as citations and abstracts, full papers, and taxonomic data. The key feature of Entrez is its ability to integrate information, which comes from cross-referencing between NCBI databases based on preexisting and logical relationships between individual entries. This is highly convenient: users do not have to visit multiple databases located in disparate places. For example, in a nucleotide sequence page, one may find cross-referencing links to the translated protein sequence, genome mapping data, or to the related PubMed literature information, and to protein structures if available.

Effective use of Entrez requires an understanding of the main features of the search engine. There are several options common to all NCBI databases that help to narrow the search. One option is "Limits," which helps to restrict the search to a subset of a particular database. It can also be set to restrict a search to a particular database (e.g., the field for author or publication date) or a particular type of data (e.g., chloroplast DNA/RNA). Another option is "Preview/Index," which connects different searches with the Boolean operators and uses a string of logically connected keywords to perform a new search. The search can also be limited to a particular search field (e.g., gene name or accession number). The "History" option provides a record of the previous searches so that the user can review, revise, or combine the results of earlier searches. There is also a "Clipboard" that stores search results for later viewing for a limited time. To store information in the Clipboard, the "Send to Clipboard" function should be used.

One of the databases accessible from Entrez is a biomedical literature database known as PubMed, which contains abstracts and in some cases the full text articles from nearly 4,000 journals. An important feature of PubMed is the retrieval of information based on medical subject headings (MeSH) terms. The MeSH system consists of a collection of more than 20,000 controlled and standardized vocabulary terms used for indexing articles. In other words, it is a thesaurus that helps convert search keywords into standardized terms to describe a concept. By doing so, it allows "smart" searches in which a group of accepted synonyms are employed so that the user not only gets

TABLE 2.2. Several Selected PubMed Tags and Their Brief Descriptions

Tag	Name	Description
AB	Abstract	Abstract
AD	Affiliation	Institutional affiliation and address of the first author and grant numbers
AID	Article identifier	Article ID values may include the PII (controlled publisher identifier) or doi (digital object identifier)
AU	Author	Authors
DP	Publication date	The date the article was published
JID	Journal ID	Unique journal ID in the National Library of Medicine's catalog of books, journals, and audiovisuals
LA	Language	The language in which the article was published
PL	Place of publication	Journal's country of publication
PT	Publication type	The type of material the article represents
RN	EC/RN number	Number assigned by the Enzyme Commission to designate a particular enzyme or by the Chemical Abstracts Service for Registry Numbers
SO	Source	Composite field containing bibliographic information
TA	Journal title abbreviation	Standard journal title abbreviation
TI	Title	The title of the article
VI	Volume	Journal volume

Source: www.ncbi.nlm.nih.gov/entrez/query/static/help/pmhelp.html.

exact matches, but also related matches on the same topic that otherwise might have been missed. Another way to broaden the retrieval is by using the "Related Articles" option. PubMed uses a word weight algorithm to identify related articles with similar words in the titles, abstracts, and MeSH. By using this feature, articles on the same topic that were missed in the original search can be retrieved.

For a complex search, a user can use the Boolean operators or a combination of Limits and Preview/Index features to conduct complex searches. Alternatively, field tags can be used to improve the efficiency of obtaining the search results. The tags are identifiers for each field and are placed in brackets. For example, [AU] limits the search for author name, and [JID] for journal name. PubMed uses a list of tags for literature searches. The search terms can be specified by the tags which are joined by Boolean operators. Some frequently used PubMed field tags are given in Table 2.2.

Another unique database accessible from Entrez is Online Mendelian Inheritance in Man (OMIM), which is a non-sequence-based database of human disease genes and human genetic disorders. Each entry in OMIM contains summary information about a particular disease as well as genes related to the disease. The text contains numerous hyperlinks to literature citations, primary sequence records, as well as chromosome loci of the disease genes. The database can serve as an excellent starting point to study genes related to a disease.

NCBI also maintains a taxonomy database that contains the names and taxonomic positions of over 100,000 organisms with at least one nucleotide or protein sequence represented in the GenBank database. The taxonomy database has a hierarchical classification scheme. The root level is Archaea, Eubacteria, and Eukaryota. The database allows the taxonomic tree for a particular organism to be displayed. The tree is based on molecular phylogenetic data, namely, the small ribosomal RNA data.

GenBank

GenBank is the most complete collection of annotated nucleic acid sequence data for almost every organism. The content includes genomic DNA, mRNA, cDNA, ESTs, high throughput raw sequence data, and sequence polymorphisms. There is also a GenPept database for protein sequences, the majority of which are conceptual translations from DNA sequences, although a small number of the amino acid sequences are derived using peptide sequencing techniques.

There are two ways to search for sequences in GenBank. One is using text-based keywords similar to a PubMed search. The other is using molecular sequences to search by sequence similarity using BLAST (to be described in Chapter 5).

GenBank Sequence Format

To search GenBank effectively using the text-based method requires an understanding of the GenBank sequence format. GenBank is a relational database. However, the search output for sequence files is produced as flat files for easy reading. The resulting flat files contain three sections – Header, Features, and Sequence entry (Fig. 2.3). There are many fields in the Header and Features sections. Each field has an unique identifier for easy indexing by computer software. Understanding the structure of the GenBank files helps in designing effective search strategies.

The Header section describes the origin of the sequence, identification of the organism, and unique identifiers associated with the record. The top line of the Header section is the Locus, which contains a unique database identifier for a sequence location in the database (not a chromosome locus). The identifier is followed by sequence length and molecule type (e.g., DNA or RNA). This is followed by a three-letter code for GenBank divisions. There are 17 divisions in total, which were set up simply based on convenience of data storage without necessarily having rigorous scientific basis; for example, PLN for plant, fungal, and algal sequences; PRI for primate sequences; MAM for nonprimate mammalian sequences; BCT for bacterial sequences; and EST for EST sequences. Next to the division is the date when the record was made public (which is different from the date when the data were submitted).

The following line, "DEFINITION," provides the summary information for the sequence record including the name of the sequence, the name and taxonomy of the source organism if known, and whether the sequence is complete or partial. This is followed by an accession number for the sequence, which is a unique number assigned to a piece of DNA when it was first submitted to GenBank and is permanently associated with that sequence. This is the number that should be cited in publications. It has two different formats: two letters with five digits or one letter with six digits. For a nucleotide sequence that has been translated into a protein sequence,

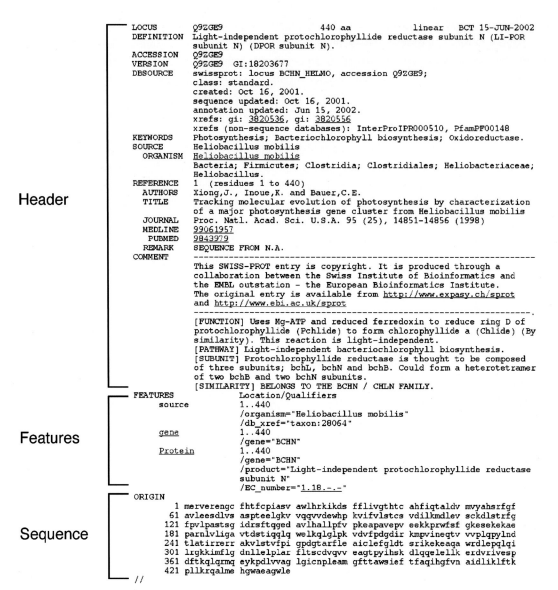

Header

Features

Sequence

```
LOCUS       Q9ZGE9                          440 aa          linear   BCT 15-JUN-2002
DEFINITION  Light-independent protochlorophyllide reductase subunit N (LI-POR
            subunit N) (DPOR subunit N).
ACCESSION   Q9ZGE9
VERSION     Q9ZGE9  GI:18203677
DBSOURCE    swissprot: locus BCHN_HELMO, accession Q9ZGE9;
            class: standard.
            created: Oct 16, 2001.
            sequence updated: Oct 16, 2001.
            annotation updated: Jun 15, 2002.
            xrefs: gi: 3820536, gi: 3820556
            xrefs (non-sequence databases): InterProIPR000510, PfamPF00148
KEYWORDS    Photosynthesis; Bacteriochlorophyll biosynthesis; Oxidoreductase.
SOURCE      Heliobacillus mobilis
  ORGANISM  Heliobacillus mobilis
            Bacteria; Firmicutes; Clostridia; Clostridiales; Heliobacteriaceae;
            Heliobacillus.
REFERENCE   1  (residues 1 to 440)
  AUTHORS   Xiong,J., Inoue,K. and Bauer,C.E.
  TITLE     Tracking molecular evolution of photosynthesis by characterization
            of a major photosynthesis gene cluster from Heliobacillus mobilis
  JOURNAL   Proc. Natl. Acad. Sci. U.S.A. 95 (25), 14851-14856 (1998)
   MEDLINE  99061957
   PUBMED   9843979
  REMARK    SEQUENCE FROM N.A.
COMMENT     -----------------------------------------------------------------
            This SWISS-PROT entry is copyright. It is produced through a
            collaboration between the Swiss Institute of Bioinformatics and
            the EMBL outstation - the European Bioinformatics Institute.
            The original entry is available from http://www.expasy.ch/sprot
            and http://www.ebi.ac.uk/sprot
            -----------------------------------------------------------------.
            [FUNCTION] Uses Mg-ATP and reduced ferredoxin to reduce ring D of
            protochlorophyllide (Pchlide) to form chlorophyllide a (Chlide) (By
            similarity). This reaction is light-independent.
            [PATHWAY] Light-independent bacteriochlorophyll biosynthesis.
            [SUBUNIT] Protochlorophyllide reductase is thought to be composed
            of three subunits; bchL, bchN and bchB. Could form a heterotetramer
            of two bchB and two bchN subunits.
            [SIMILARITY] BELONGS TO THE BCHN / CHLN FAMILY.
FEATURES             Location/Qualifiers
     source          1..440
                     /organism="Heliobacillus mobilis"
                     /db_xref="taxon:28064"
     gene            1..440
                     /gene="BCHN"
     Protein         1..440
                     /gene="BCHN"
                     /product="Light-independent protochlorophyllide reductase
                     subunit N"
                     /EC_number="1.18.-.-"
ORIGIN
        1 merverengc fhtfcpiasv awlhrkikds fflivgthtc ahfiqtaldv mvyahsrfgf
       61 avleesdlvs aspteelgkv vqqvvdewhp kvifvlstcs vdilkmdlev sckdlstrfg
      121 fpvlpastsg idrsftqged avlhallpfv pkeapavepv eekkprwfsf gkesekekae
      181 parnlvliga vtdstiqqlq welkqlglpk vdvfpdgdir kmpvineqtv vvplqpylnd
      241 tlatirrerr akvlstvfpi gpdgtarfle aiclefgldt srikekeaqa wrdlepqlqi
      301 lrgkkimflg dnllelplar fltscdvqvv eagtpyihsk dlqqelellk erdvrivesp
      361 dftkqlqrmq eykpdlvvag lgicnpleam gfttawsief tfaqihgfvn aidliklftk
      421 pllrqalme hgwaeagwle
//
```

Figure 2.3: NCBI GenBank/GenPept format showing the three major components of a sequence file.

a new accession number is given in the form of a string of alphanumeric characters. In addition to the accession number, there is also a version number and a gene index (gi) number. The purpose of these numbers is to identify the current version of the sequence. If the sequence annotation is revised at a later date, the accession number remains the same, but the version number is incremented as is the gi number. A translated protein sequence also has a different gi number from the DNA sequence it is derived from.

The next line in the Header section is the "ORGANISM" field, which includes the source of the organism with the scientific name of the species and sometimes the

tissue type. Along with the scientific name is the information of taxonomic classification of the organism. Different levels of the classification are hyperlinked to the
NCBI taxonomy database with more detailed descriptions. This is followed by the
"REFERENCE" field, which provides the publication citation related to the sequence
entry. The REFERENCE part includes author and title information of the published
work (or tentative title for unpublished work). The "JOURNAL" field includes the citation information as well as the date of sequence submission. The citation is often
hyperlinked to the PubMed record for access to the original literature information.
The last part of the Header is the contact information of the sequence submitter.

The "Features" section includes annotation information about the gene and gene
product, as well as regions of biological significance reported in the sequence, with
identifiers and qualifiers. The "Source" field provides the length of the sequence,
the scientific name of the organism, and the taxonomy identification number. Some
optional information includes the clone source, the tissue type and the cell line. The
"gene" field is the information about the nucleotide coding sequence and its name.
For DNA entries, there is a "CDS" field, which is information about the boundaries of
the sequence that can be translated into amino acids. For eukaryotic DNA, this field
also contains information of the locations of exons and translated protein sequences
is entered.

The third section of the flat file is the sequence itself starting with the label
"ORIGIN." The format of the sequence display can be changed by choosing options
at a Display pull-down menu at the upper left corner. For DNA entries, there is a BASE
COUNT report that includes the numbers of A, G, C, and T in the sequence. This
section, for both DNA or protein sequences, ends with two forward slashes (the "//"
symbol).

In retrieving DNA or protein sequences from GenBank, the search can be limited to
different fields of annotation such as "organism," "accession number," "authors," and
"publication date." One can use a combination of the "Limits" and "Preview/Index"
options as described. Alternatively, a number of search qualifiers can be used, each
defining one of the fields in a GenBank file. The qualifiers are similar to but not the
same as the field tags in PubMed. For example, in GenBank, [GENE] represents field
for gene name, [AUTH] for author name, and [ORGN] for organism name. Frequently
used GenBank qualifiers, which have to be in uppercase and in brackets, are listed in
Table 2.3.

Alternative Sequence Formats
FASTA. In addition to the GenBank format, there are many other sequence formats.
FASTA is one of the simplest and the most popular sequence formats because it contains plain sequence information that is readable by many bioinformatics analysis
programs. It has a single definition line that begins with a right angle bracket (>)
followed by a sequence name (Fig. 2.4). Sometimes, extra information such as gi
number or comments can be given, which are separated from the sequence name
by a "|" symbol. The extra information is considered optional and is ignored by

TABLE 2.3. Search Field Qualifiers for GenBank

Qualifier	Field Name	Definition
[ACCN]	Accession	Contains the unique accession number of the sequence or record, assigned to the nucleotide, protein, structure, or genome record.
[ALL]	All fields	Contains all terms from all searchable database fields in the database.
[AUTH]	Author name	Contains all authors from all references in the database records.
[ECNO]	EC/RN number	Number assigned by the Enzyme Commission or Chemical Abstract Service to designate a particular enzyme or chemical, respectively.
[FKEY]	Feature key	Contains the biological features assigned or annotated to the nucleotide sequences. Not available for the protein or structure databases.
[GENE]	Gene name	Contains the standard and common names of genes found in the database records.
[JOUR]	Journal name	Contains the name of the journal in which the data were published.
[KYWD]	Keyword	Contains special index terms from the controlled vocabularies associated with the GenBank, EMBL, DDBJ, SWISS-Prot, PIR, PRF, or PDB databases.
[MDAT]	Modification date	Contains the date that the most recent modification to that record is indexed in Entrez, in the format YYYY/MM/DD.
[MOLWT]	Molecular weight	Molecular weight of a protein, in daltons (Da), calculated by the method described in the Searching by Molecular Weight section of the Entrez help document.
[ORGN]	Organism	Contains the scientific and common names for the organisms associated with protein and nucleotide sequences.
[PROP]	Properties	Contains properties of the nucleotide or protein sequence. For example, the nucleotide database's properties index includes molecule types, publication status, molecule locations, and GenBank divisions.
[PROT]	Protein name	Contains the standard names of proteins found in database records.
[PDAT]	Publication date	Contains the date that records are released into Entrez, in the format YYYY/MM/DD.
[SQID]	SeqID	Contains the special string identifier for a given sequence.
[SLEN]	Sequence length	Contains the total length of the sequence.
[WORD]	Text word	Contains all of the "free text" associated with a record.
[TITL]	Title word	Includes only those words found in the definition line of a record.

Note: Some of these qualifiers are interchangeable with PubMed qualifiers.
Source: www.ncbi.nlm.nih.gov/entrez/query/static/help/helpdoc.html.

```
>gi|18203677|sp|Q9ZGE9|BCHN
MERVERENGCFHTFCPIASVAWLHRKIKDSFFLIVGTHTCAHFIQTALDVMVYAHSRFGFAVLEESDLVS
ASPTEELGKVVQQVVDEWHPKVIFVLSTCSVDILKMDLEVSCKDLSTRFGFPVLPASTSGIDRSFTQGED
AVLHALLPFVPKEAPAVEPVEEKKPRWFSFGKESEKEKAEPARNLVLIGAVTDSTIQQLQWELKQLGLPK
VDVFPDGDIRKMPVINEQTVVVPLQPYLNDTLATIRRERRAKVLSTVFPIGPDGTARFLEAICLEFGLDT
SRIKEKEAQAWRDLEPQLQILRGKKIMFLGDNLLELPLARFLTSCDVQVVEAGTPYIHSKDLQQELELLK
ERDVRIVESPDFTKQLQRMQEYKPDLVVAGLGICNPLEAMGFTTAWSIEFTFAQIHGFVNAIDLIKLFTK
PLLKRQALMEHGWAEACWLE
```
Figure 2.4: Example of a FASTA file.

sequence analysis programs. The plain sequence in standard one-letter symbols starts in the second line. Each line of sequence data is limited to sixty to eighty characters in width. The drawback of this format is that much annotation information is lost.

Abstract Syntax Notation One. Abstract Syntax Notation One (ASN.1) is a data mark-up language with a structure specifically designed for accessing relational databases. It describes sequences with each item of information in a sequence record separated by tags so that each subportion of the sequence record can be easily added to relational tables and later extracted (Fig. 2.5). Though more difficult for people to read, this format makes it easy for computers to filter and parse the data. This format also facilitates the transimission and integration of data between databases.

Conversion of Sequence Formats

In sequence analysis and phylogenetic analysis, there is a frequent need to convert between sequence formats. One of the most popular computer programs for sequence format conversion is *Readseq*, written by Don Gilbert at Indiana University. It recognizes sequences in almost any format and writes a new file in an alternative format. The web interface version of the program can be found at: http://iubio.bio.indiana.edu/cgi-bin/readseq.cgi/.

SRS

Sequence retrieval system (SRS; available at http://srs6.ebi.ac.uk/) is a retrieval system maintained by the EBI, which is comparable to NCBI Entrez. It is not as integrated as Entrez, but allows the user to query multiple databases simultaneously, another good example of database integration. It also offers direct access to certain sequence analysis applications such as sequence similarity searching and Clustal sequence alignment (see Chapter 5). Queries can be launched using "Quick Text Search" with only one query box in which to enter information. There are also more elaborate submission forms, the "Standard Query Form" and the "Extended Query Form." The standard form allows four criteria (fields) to be used, which are linked by Boolean operators. The extended form allows many more diversified criteria and fields to be used. The search results contain the query sequence and sequence annotation as well as links to literature, metabolic pathways, and other biological databases.

```
        name "Tracking molecular evolution of photosynthesis by
characterization of a major photosynthesis gene cluster from Heliobacillus
mobilis." } ,
        authors {
          names
            std {
              {
                name
                  name {
                    last "Xiong" ,
                    initials "J." } } ,
              {
                name
                  name {
                    last "Inoue" ,
                    initials "K." } } ,
              {
                name
                  name {
                    last "Bauer" ,
                    initials "C.E." } } } ,
          affil
            str "Department of Biology, Indiana University, Bloomington, IN
47405, USA." } ,
        from
          journal {
            title {
              iso-jta "Proc. Natl. Acad. Sci. U.S.A." ,
              ml-jta "Proc Natl Acad Sci U S A" ,
              issn "0027-8424" ,
              name "Proceedings of the National Academy of Sciences of the
United States of America." } ,
            imp {
              date
                std {
                  year 1998 ,
                  month 12 ,
                  day 8 } ,
              volume "95" ,
              issue "25" ,
              pages "14851-14856" ,
              language "eng" } } ,
        ids {
          pubmed 9843979 ,
          medline 99061957 } } ,
      pmid 9843979 } ,
    comment "SEQUENCE FROM N.A." } } ,
  inst {
    repr raw ,
    mol aa ,
    length 440 ,
    seq-data
      ncbieaa "MERVERENGCFHTFCPIASVAWLHRKIKDSFFLIVGTHTCAHFIQTALDVMVYAHSRFGFAVL
EESDLVSASPTEELGKVVQQVVDEWHPKVIFVLSTCSVDILKMDLEVSCKDLSTRFGFPVLPASTSGIDRSFTQGEDA
VLHALLPFVPKEAPAVEPVEEKKPRWFSFGKESEKEKAEPARNLVLIGAVTDSTIQQLQWELKQLGLPKVDVFPDGDI
RKMPVINEQTVVVPLQPYLNDTLATIRRERRAKVLSTVFPIGPDGTARFLEAICLEFGLDTSRIKEKEAQAWRDLEPQ
LQILRGKKIMFLGDNLLELPLARFLTSCDVQVVEAGTPYIHSKDLQQELELLKERDVRIVESPDFTKQLQRMQEYKPD
LVVAGLGICNPLEAMGFTTAWSIEFTFAQIHGFVNAIDLIKLFTKPLLKRQALMEHGWAEAGWLE" } ,
```

Figure 2.5: A portion of a sequence file in ASN.1 format.

SUMMARY

Databases are fundamental to modern biological research, especially to genomic studies. The goal of a biological database is two fold: information retrieval and knowledge discovery. Electronic databases can be constructed either as flat files, relational, or object oriented. Flat files are simple text files and lack any form of organization to facilitate information retrieval by computers. Relational databases organize data as tables and search information among tables with shared features. Object-oriented databases organize data as objects and associate the objects according to hierarchical relationships. Biological databases encompass all three types. Based on their content, biological databases are divided into primary, secondary, and specialized databases. Primary databases simply archive sequence or structure information; secondary databases include further analysis on the sequences or structures. Specialized databases cater to a particular research interest. Biological databases need to be interconnected so that entries in one database can be cross-linked to related entries in another database. NCBI databases accessible through Entrez are among the most integrated databases. Effective information retrieval involves the use of Boolean operators. Entrez has additional user-friendly features to help conduct complex searches. One such option is to use Limits, Preview/Index, and History to narrow down the search space. Alternatively, one can use NCBI-specific field qualifiers to conduct searches. To retrieve sequence information from NCBI GenBank, an understanding of the format of GenBank sequence files is necessary. It is also important to bear in mind that sequence data in these databases are less than perfect. There are sequence and annotation errors. Biological databases are also plagued by redundancy problems. There are various solutions to correct annotation and reduce redundancy, for example, merging redundant sequences into a single entry or store highly redundant sequences into a separate database.

FURTHER READING

Apweiler, R. 2000. Protein sequence databases. *Adv. Protein Chem.* 54:31–71.

Blaschke, C., Hirschman, L., and Valencia, A. 2002. Information extraction in molecular biology. *Brief. Bioinform.* 3:154–65.

Geer, R. C., and Sayers, E. W. 2003. Entrez: Making use of its power. *Brief. Bioinform.* 4:179–84.

Hughes, A. E. 2001. Sequence databases and the Internet. *Methods Mol. Biol.* 167:215–23.

Patnaik, S. K., and Blumenfeld, O. O. 2001. Use of on-line tools and databases for routine sequence analyses. *Anal. Biochem.* 289:1–9.

Stein, L. D. 2003. Integrating biological databases. *Nat. Rev. Genet.* 4:337–45.

Sequence Alignment

CHAPTER THREE

Pairwise Sequence Alignment

Sequence comparison lies at the heart of bioinformatics analysis. It is an important first step toward structural and functional analysis of newly determined sequences. As new biological sequences are being generated at exponential rates, sequence comparison is becoming increasingly important to draw functional and evolutionary inference of a new protein with proteins already existing in the database. The most fundamental process in this type of comparison is sequence alignment. This is the process by which sequences are compared by searching for common character patterns and establishing residue–residue correspondence among related sequences. Pairwise sequence alignment is the process of aligning two sequences and is the basis of database similarity searching (see Chapter 4) and multiple sequence alignment (see Chapter 5). This chapter introduces the basics of pairwise alignment.

EVOLUTIONARY BASIS

DNA and proteins are products of evolution. The building blocks of these biological macromolecules, nucleotide bases, and amino acids form linear sequences that determine the primary structure of the molecules. These molecules can be considered molecular fossils that encode the history of millions of years of evolution. During this time period, the molecular sequences undergo random changes, some of which are selected during the process of evolution. As the selected sequences gradually accumulate mutations and diverge over time, traces of evolution may still remain in certain portions of the sequences to allow identification of the common ancestry. The presence of evolutionary traces is because some of the residues that perform key functional and structural roles tend to be preserved by natural selection; other residues that may be less crucial for structure and function tend to mutate more frequently. For example, active site residues of an enzyme family tend to be conserved because they are responsible for catalytic functions. Therefore, by comparing sequences through alignment, patterns of conservation and variation can be identified. The degree of sequence conservation in the alignment reveals evolutionary relatedness of different sequences, whereas the variation between sequences reflects the changes that have occurred during evolution in the form of substitutions, insertions, and deletions.

Identifying the evolutionary relationships between sequences helps to characterize the function of unknown sequences. When a sequence alignment reveals *significant* similarity among a group of sequences, they can be considered as belonging to the same family (protein families will be further described in Chapter 7). If one member within the family has a known structure and function, then that information can be transferred to those that have not yet been experimentally characterized. Therefore, sequence alignment can be used as basis for prediction of structure and function of uncharacterized sequences.

Sequence alignment provides inference for the relatedness of two sequences under study. If the two sequences share significant similarity, it is extremely unlikely that the extensive similarity between the two sequences has been acquired randomly, meaning that the two sequences must have derived from a common evolutionary origin. When a sequence alignment is generated correctly, it reflects the evolutionary relationship of the two sequences: regions that are aligned but not identical represent residue substitutions; regions where residues from one sequence correspond to nothing in the other represent insertions or deletions that have taken place on one of the sequences during evolution. It is also possible that two sequences have derived from a common ancestor, but may have diverged to such an extent that the common ancestral relationships are not recognizable at the sequence level. In that case, the distant evolutionary relationships have to be detected using other methods (see Chapter 15).

SEQUENCE HOMOLOGY VERSUS SEQUENCE SIMILARITY

An important concept in sequence analysis is sequence homology. When two sequences are descended from a common evolutionary origin, they are said to have a *homologous relationship* or share *homology*. A related but different term is *sequence similarity*, which is the percentage of aligned residues that are similar in physiochemical properties such as size, charge, and hydrophobicity.

It is important to distinguish sequence homology from the related term sequence similarity because the two terms are often confused by some researchers who use them interchangeably in scientific literature. To be clear, *sequence homology* is an inference or a conclusion about a common ancestral relationship drawn from sequence similarity comparison when the two sequences share a high enough degree of similarity. On the other hand, *similarity* is a direct result of observation from the sequence alignment. Sequence similarity can be quantified using percentages; homology is a qualitative statement. For example, one may say that two sequences share 40% similarity. It is incorrect to say that the two sequences share 40% homology. They are either homologous or nonhomologous.

Generally, if the sequence similarity level is high enough, a common evolutionary relationship can be inferred. In dealing with real research problems, the issue of at what similarity level can one infer homologous relationships is not always clear. The answer depends on the type of sequences being examined and sequence lengths. Nucleotide sequences consist of only four characters, and therefore, unrelated sequences have

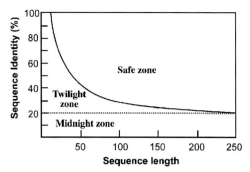

Figure 3.1: The three zones of protein sequence alignments. Two protein sequences can be regarded as homologous if the percentage sequence identity falls in the safe zone. Sequence identity values below the zone boundary, but above 20%, are considered to be in the twilight zone, where homologous relationships are less certain. The region below 20% is the midnight zone, where homologous relationships cannot be reliably determined. (*Source:* Modified from Rost 1999).

at least a 25% chance of being identical. For protein sequences, there are twenty possible amino acid residues, and so two unrelated sequences can match up 5% of the residues by random chance. If gaps are allowed, the percentage could increase to 10–20%. Sequence length is also a crucial factor. The shorter the sequence, the higher the chance that some alignment is attributable to random chance. The longer the sequence, the less likely the matching at the same level of similarity is attributable to random chance.

This suggests that shorter sequences require higher cutoffs for inferring homologous relationships than longer sequences. For determining a homology relationship of two protein sequences, for example, if both sequences are aligned at full length, which is 100 residues long, an identity of 30% or higher can be safely regarded as having close homology. They are sometimes referred to as being in the "safe zone" (Fig. 3.1). If their identity level falls between 20% and 30%, determination of homologous relationships in this range becomes less certain. This is the area often regarded as the "twilight zone," where remote homologs mix with randomly related sequences. Below 20% identity, where high proportions of nonrelated sequences are present, homologous relationships cannot be reliably determined and thus fall into the "midnight zone." It needs to be stressed that the percentage identity values only provide a tentative guidance for homology identification. This is not a precise rule for determining sequence relationships, especially for sequences in the twilight zone. A statistically more rigorous approach to determine homologous relationships is introduced in the section on the Statistical Significance of Sequence Alignment.

SEQUENCE SIMILARITY VERSUS SEQUENCE IDENTITY

Another set of related terms for sequence comparison are sequence similarity and sequence identity. Sequence similarity and sequence identity are synonymous for nucleotide sequences. For protein sequences, however, the two concepts are very different. In a protein sequence alignment, *sequence identity* refers to the percentage of matches of the same amino acid residues between two aligned sequences. *Similarity* refers to the percentage of aligned residues that have similar physicochemical characteristics and can be more readily substituted for each other.

There are two ways to calculate the sequence similarity/identity. One involves the use of the overall sequence lengths of both sequences; the other normalizes by the size of the shorter sequence. The first method uses the following formula:

$$S = [(L_s \times 2)/(L_a + L_b)] \times 100 \qquad \text{(Eq. 3.1)}$$

where S is the percentage sequence similarity, L_s is the number of aligned residues with similar characteristics, and L_a and L_b are the total lengths of each individual sequence. The sequence identity ($I\%$) can be calculated in a similar fashion:

$$I = [(L_i \times 2)/(L_a + L_b)] \times 100 \qquad \text{(Eq. 3.2)}$$

where L_i is the number of aligned identical residues.

The second method of calculation is to derive the percentage of identical/similar residues over the full length of the smaller sequence using the formula:

$$I(S)\% = L_{i(s)}/L_a \% \qquad \text{(Eq. 3.3)}$$

where L_a is the length of the shorter of the two sequences.

METHODS

The overall goal of pairwise sequence alignment is to find the best pairing of two sequences, such that there is maximum correspondence among residues. To achieve this goal, one sequence needs to be shifted relative to the other to find the position where maximum matches are found. There are two different alignment strategies that are often used: global alignment and local alignment.

Global Alignment and Local Alignment

In *global alignment*, two sequences to be aligned are assumed to be generally similar over their entire length. Alignment is carried out from beginning to end of both sequences to find the best possible alignment across the entire length between the two sequences. This method is more applicable for aligning two closely related sequences of roughly the same length. For divergent sequences and sequences of variable lengths, this method may not be able to generate optimal results because it fails to recognize highly similar local regions between the two sequences.

Local alignment, on the other hand, does not assume that the two sequences in question have similarity over the entire length. It only finds local regions with the highest level of similarity between the two sequences and aligns these regions without regard for the alignment of the rest of the sequence regions. This approach can be

```
seq1    EARDF-NQYYSSIKRSGSIQ
         . :  .::::::::.  . .
seq2    LPKLFIDQYYSSIKRTMG-H
```

global sequence alignment

```
seq1    NQYYSSIKRS
        .::::::::.
seq2    DQYYSSIKRT
```

local sequence alignment

Figure 3.2: An example of pairwise sequence comparison showing the distinction between global and local alignment. The global alignment (*top*) includes all residues of both sequences. The region with the highest similarity is highlighted in a box. The local alignment only includes portions of the two sequences that have the highest regional similarity. In the line between the two sequences, ":" indicates identical residue matches and "." indicates similar residue matches.

used for aligning more divergent sequences with the goal of searching for conserved patterns in DNA or protein sequences. The two sequences to be aligned can be of different lengths. This approach is more appropriate for aligning divergent biological sequences containing only modules that are similar, which are referred to as *domains* or *motifs*. Figure 3.2 illustrates the differences between global and local pairwise alignment.

Alignment Algorithms

Alignment algorithms, both global and local, are fundamentally similar and only differ in the optimization strategy used in aligning similar residues. Both types of algorithms can be based on one of the three methods: the dot matrix method, the dynamic programming method, and the word method. The dot matrix and dynamic programming methods are discussed herein. The word method, which is used in fast database similarity searching, is introduced in Chapter 4.

Dot Matrix Method

The most basic sequence alignment method is the dot matrix method, also known as the *dot plot method.* It is a graphical way of comparing two sequences in a two-dimensional matrix. In a dot matrix, two sequences to be compared are written in the horizontal and vertical axes of the matrix. The comparison is done by scanning each residue of one sequence for similarity with all residues in the other sequence. If a residue match is found, a dot is placed within the graph. Otherwise, the matrix positions are left blank. When the two sequences have substantial regions of similarity, many dots line up to form contiguous diagonal lines, which reveal the sequence alignment. If there are interruptions in the middle of a diagonal line, they indicate insertions or deletions. Parallel diagonal lines within the matrix represent repetitive regions of the sequences (Fig. 3.3).

A problem exists when comparing large sequences using the dot matrix method, namely, the high noise level. In most dot plots, dots are plotted all over the graph,

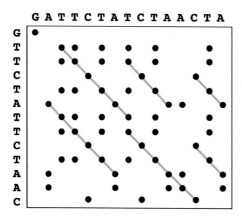

Figure 3.3: Example of comparing two sequences using dot plots. Lines linking the dots in diagonals indicate sequence alignment. Diagonal lines above or below the main diagonal represent internal repeats of either sequence.

obscuring identification of the true alignment. For DNA sequences, the problem is particularly acute because there are only four possible characters in DNA and each residue therefore has a one-in-four chance of matching a residue in another sequence. To reduce noise, instead of using a single residue to scan for similarity, a filtering technique has to be applied, which uses a "window" of fixed length covering a stretch of residue pairs. When applying filtering, windows slide across the two sequences to compare all possible stretches. Dots are only placed when a stretch of residues equal to the window size from one sequence matches completely with a stretch of another sequence. This method has been shown to be effective in reducing the noise level. The window is also called a *tuple*, the size of which can be manipulated so that a clear pattern of sequence match can be plotted. However, if the selected window size is too long, sensitivity of the alignment is lost.

There are many variations of using the dot plot method. For example, a sequence can be aligned with itself to identify internal repeat elements. In the self-comparison, there is a main diagonal for perfect matching of each residue. If repeats are present, short parallel lines are observed above and below the main diagonal. Self-complementarity of DNA sequences (also called *inverted repeats*) – for example, those that form the stems of a hairpin structure – can also be identified using a dot plot. In this case, a DNA sequence is compared with its reverse-complemented sequence. Parallel diagonals represent the inverted repeats. For comparing protein sequences, a weighting scheme has to be used to account for similarities of physicochemical properties of amino acid residues.

The dot matrix method gives a direct visual statement of the relationship between two sequences and helps easy identification of the regions of greatest similarities. One particular advantage of this method is in identification of sequence repeat regions based on the presence of parallel diagonals of the same size vertically or horizontally in the matrix. The method thus has some applications in genomics. It is useful in identifying chromosomal repeats and in comparing gene order conservation between two closely related genomes (see Chapter 17). It can also be used in identifying nucleic acid secondary structures through detecting self-complementarity of a sequence (see Chapter 16).

The dot matrix method displays all possible sequence matches. However, it is often up to the user to construct a full alignment with insertions and deletions by linking nearby diagonals. Another limitation of this visual analysis method is that it lacks statistical rigor in assessing the quality of the alignment. The method is also restricted to pairwise alignment. It is difficult for the method to scale up to multiple alignment. The following are examples of web servers that provide pairwise sequence comparison using dot plots.

Dotmatcher (bioweb.pasteur.fr/seqanal/interfaces/dotmatcher.html) and Dottup (bioweb.pasteur.fr/seqanal/interfaces/dottup.html) are two programs of the EMBOSS package, which have been made available online. Dotmatcher aligns and displays dot plots of two input sequences (DNA or proteins) in FASTA format. A window of specified length and a scoring scheme are used. Diagonal lines are only plotted over the position of the windows if the similarity is above a certain threshold. Dottup aligns sequences using the word method (to be described in Chapter 4) and is capable of handling genome-length sequences. Diagonal lines are only drawn if exact matches of words of specified length are found.

Dothelix (www.genebee.msu.su/services/dhm/advanced.html) is a dot matrix program for DNA or protein sequences. The program has a number of options for length threshold (similar to window size) and implements scoring matrices for protein sequences. In addition to drawing diagonal lines with similarity scores above a certain threshold, the program displays actual pairwise alignment.

MatrixPlot (www.cbs.dtu.dk/services/MatrixPlot/) is a more sophisticated matrix plot program for alignment of protein and nucleic acid sequences. The user has the option of adding information such as sequence logo profiles (see Chapter 7) and distance matrices from known three-dimensional structures of proteins or nucleic acids. Instead of using dots and lines, the program uses colored grids to indicate alignment or other user-defined information.

Dynamic Programming Method

Dynamic programming is a method that determines optimal alignment by matching two sequences for all possible pairs of characters between the two sequences. It is fundamentally similar to the dot matrix method in that it also creates a two-dimensional alignment grid. However, it finds alignment in a more quantitative way by converting a dot matrix into a scoring matrix to account for matches and mismatches between sequences. By searching for the set of highest scores in this matrix, the best alignment can be accurately obtained.

Dynamic programming works by first constructing a two-dimensional matrix whose axes are the two sequences to be compared. The residue matching is according to a particular scoring matrix. The scores are calculated one row at a time. This starts with the first row of one sequence, which is used to scan through the entire length of the other sequence, followed by scanning of the second row. The matching scores are calculated. The scanning of the second row takes into account the scores already obtained in the first round. The best score is put into the bottom right corner of an intermediate matrix (Fig. 3.4). This process is iterated until values for all the cells are filled. Thus, the scores are accumulated along the diagonal going from the upper left corner to the lower right corner. Once the scores have been accumulated in matrix, the next step is to find the path that represents the optimal alignment. This is done by tracing back through the matrix in reverse order from the lower right-hand corner of the matrix toward the origin of the matrix in the upper left-hand corner. The best matching path is the one that has the maximum total score (see Fig. 3.4). If two or more paths reach the same highest score, one is chosen arbitrarily to represent the best alignment. The path can also move horizontally or vertically at a certain point, which corresponds to introduction of a gap or an insertion or deletion for one of the two sequences.

Gap Penalties. Performing optimal alignment between sequences often involves applying gaps that represent insertions and deletions. Because in natural evolutionary processes insertion and deletions are relatively rare in comparison to substitutions, introducing gaps should be made more difficult computationally, reflecting the rarity of insertional and deletional events in evolution. However, assigning penalty values can be more or less arbitrary because there is no evolutionary theory to determine a precise cost for introducing insertions and deletions. If the penalty values are set too low, gaps can become too numerous to allow even nonrelated sequences to be matched up with high similarity scores. If the penalty values are set too high, gaps may become too difficult to appear, and reasonable alignment cannot be achieved, which is also unrealistic. Through empirical studies for globular proteins, a set of penalty values have been developed that appear to suit most alignment purposes. They are normally implemented as default values in most alignment programs.

Another factor to consider is the cost difference between opening a gap and extending an existing gap. It is known that it is easier to extend a gap that has already been started. Thus, gap opening should have a much higher penalty than gap extension. This is based on the rationale that if insertions and deletions ever occur, several adjacent residues are likely to have been inserted or deleted together. These differential gap penalties are also referred to as *affine gap penalties*. The normal strategy is to use preset gap penalty values for introducing and extending gaps. For example, one may use a $-12/-1$ scheme in which the gap opening penalty is -12 and the gap extension penalty -1. The total gap penalty (W) is a linear function of gap length, which is calculated using the formula:

$$W = \gamma + \delta \times (k - 1) \qquad\qquad\qquad \text{(Eq. 3.4)}$$

where γ is the gap opening penalty, δ is the gap extension penalty, and k is the length of the gap. Besides the affine gap penalty, a *constant gap penalty* is sometimes also used, which assigns the same score for each gap position regardless whether it is

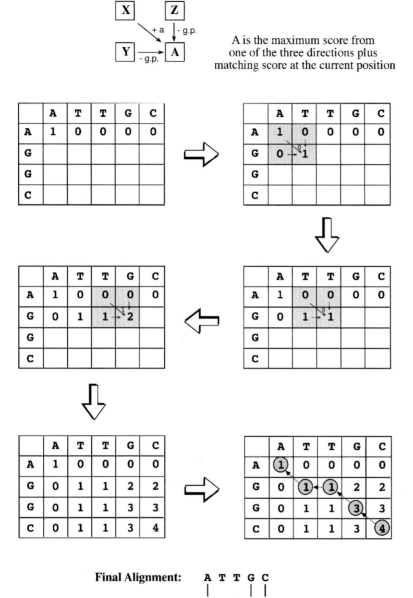

Figure 3.4: Example of pairwise alignment of two sequences using dynamic programming. The score for the lower right square **(A)** of a 2 × 2 matrix is the maximum score from the one of other three neighboring squares (X, Y, and Z) plus and minus the exact single residue match score (a) for the lower right corner and the gap penalty (g.p.), respectively. A matrix is set up for the two short sequences. A simple scoring system is applied in which an identical match is assigned a score of 1, a mismatch a score 0, and gap penalty (see below) is −1. The scores in the matrix are filled one row at a time and one cell at a time beginning from top to bottom. The best scores are filled to the lower right corner of a submatrix (grey boxes) according to this rule. When all the cells are filled with scores, a best alignment is determined through a trace-back procedure to search for the path with the best total score. When a path moves horizontally or vertically, a penalty is applied.

opening or extending. However, this penalty scheme has been found to be less realistic than the affine penalty.

Gaps at the terminal regions are often treated with no penalty because in reality many true homologous sequences are of different lengths. Consequently, end gaps can be allowed to be free to avoid getting unrealistic alignments.

Dynamic Programming for Global Alignment. The classical global pairwise alignment algorithm using dynamic programming is the Needleman–Wunsch algorithm. In this algorithm, an optimal alignment is obtained over the entire lengths of the two sequences. It must extend from the beginning to the end of both sequences to achieve the highest total score. In other words, the alignment path has to go from the bottom right corner of the matrix to the top left corner. The drawback of focusing on getting a maximum score for the full-length sequence alignment is the risk of missing the best local similarity. This strategy is only suitable for aligning two closely related sequences that are of the same length. For divergent sequences or sequences with different domain structures, the approach does not produce optimal alignment. One of the few web servers dedicated to global pairwise alignment is GAP.

GAP (http://bioinformatics.iastate.edu/aat/align/align.html) is a web-based pairwise global alignment program. It aligns two sequences without penalizing terminal gaps so similar sequences of unequal lengths can be aligned. To be able to insert long gaps in the alignment, such gaps are treated with a constant penalty. This feature is useful in aligning cDNA to exons in genomic DNA containing the same gene.

Dynamic Programming for Local Alignment. In regular sequence alignment, the divergence level between the two sequences to be aligned is not easily known. The sequence lengths of the two sequences may also be unequal. In such cases, identification of regional sequence similarity may be of greater significance than finding a match that includes all residues. The first application of dynamic programming in local alignment is the Smith–Waterman algorithm. In this algorithm, positive scores are assigned for matching residues and zeros for mismatches. No negative scores are used. A similar tracing-back procedure is used in dynamic programming. However, the alignment path may begin and end internally along the main diagonal. It starts with the highest scoring position and proceeds diagonally up to the left until reaching a cell with a zero. Gaps are inserted if necessary. In this case, affine gap penalty is often used. Occasionally, several optimally aligned segments with best scores are obtained. As in the global alignment, the final result is influenced by the choice of scoring systems (to be described next) used. The goal of local alignment is to get the highest alignment score locally, which may be at the expense of the highest possible overall score for a full-length alignment. This approach may be suitable for aligning divergent sequences or sequences with multiple domains that may be of different origins. Most commonly used pairwise alignment web servers apply the local alignment strategy, which include SIM, SSEARCH, and LALIGN.

SIM (http://bioinformatics.iastate.edu/aat/align/align.html) is a web-based program for pairwise alignment using the Smith–Waterman algorithm that finds the best scored nonoverlapping local alignments between two sequences. It is able to handle tens of kilobases of genomic sequence. The user has the option to set a scoring matrix and gap penalty scores. A specified number of best scored alignments are produced.

SSEARCH (http://pir.georgetown.edu/pirwww/search/pairwise.html) is a simple web-based programs that uses the Smith–Waterman algorithm for pairwise alignment of sequences. Only one best scored alignment is given. There is no option for scoring matrices or gap penalty scores.

LALIGN (www.ch.embnet.org/software/LALIGN_form.html) is a web-based program that uses a variant of the Smith–Waterman algorithm to align two sequences. Unlike SSEARCH, which returns the single best scored alignment, LALIGN gives a specified number of best scored alignments. The user has the option to set the scoring matrix and gap penalty scores. The same web interface also provides an option for global alignment performed by the ALIGN program.

SCORING MATRICES

In the dynamic programming algorithm presented, the alignment procedure has to make use of a scoring system, which is a set of values for quantifying the likelihood of one residue being substituted by another in an alignment. The scoring systems is called a *substitution matrix* and is derived from statistical analysis of residue substitution data from sets of reliable alignments of highly related sequences.

Scoring matrices for nucleotide sequences are relatively simple. A positive value or high score is given for a match and a negative value or low score for a mismatch. This assignment is based on the assumption that the frequencies of mutation are equal for all bases. However, this assumption may not be realistic; observations show that transitions (substitutions between purines and purines or between pyrimidines and pyrimidines) occur more frequently than transversions (substitutions between purines and pyrimidines). Therefore, a more sophisticated statistical model with different probability values to reflect the two types of mutations is needed (see the Kimura model in Chapter 10).

Scoring matrices for amino acids are more complicated because scoring has to reflect the physicochemical properties of amino acid residues, as well as the likelihood of certain residues being substituted among true homologous sequences. Certain amino acids with similar physicochemical properties can be more easily substituted than those without similar characteristics. Substitutions among similar residues are likely to preserve the essential functional and structural features. However, substitutions between residues of different physicochemical properties are more likely to cause disruptions to the structure and function. This type of disruptive substitution is less likely to be selected in evolution because it renders nonfunctional proteins.

For example, phenylalanine, tyrosine, and tryptophan all share aromatic ring structures. Because of their chemical similarities, they are easily substituted for each other without perturbing the regular function and structure of the protein. Similarly, arginine, lysine, and histidine are all large basic residues and there is a high probability of them being substituted for each other. Aspartic acid, glutamic acid, asparagine, and glutamine belong to the acid and acid amide groups and can be associated with relatively high frequencies of substitution. The hydrophobic residue group includes methionine, isoleucine, leucine, and valine. Small and polar residues include serine, threonine, and cysteine. Residues within these groups have high likelihoods of being substituted for each other. However, cysteine contains a sulfhydryl group that plays a role in metal binding, active site, and disulfide bond formation. Substitution of cysteine with other residues therefore often abolishes the enzymatic activity or destabilizes the protein structure. It is thus a very infrequently substituted residue. The small and nonpolar residues such as glycine and proline are also unique in that their presence often disrupts regular protein secondary structures (see Chapter 12). Thus, substitutions with these residues do not frequently occur. For more information on grouping amino acids based on physicochemical properties, see Table 12.1.

Amino Acid Scoring Matrices

Amino acid substitution matrices, which are 20×20 matrices, have been devised to reflect the likelihood of residue substitutions. There are essentially two types of amino acid substitution matrices. One type is based on interchangeability of the genetic code or amino acid properties, and the other is derived from empirical studies of amino acid substitutions. Although the two different approaches coincide to a certain extent, the first approach, which is based on the genetic code or the physicochemical features of amino acids, has been shown to be less accurate than the second approach, which is based on surveys of actual amino acid substitutions among related proteins. Thus, the empirical approach has gained the most popularity in sequence alignment applications and is the focus of our next discussion.

The empirical matrices, which include PAM and BLOSUM matrices, are derived from actual alignments of highly similar sequences. By analyzing the probabilities of amino acid substitutions in these alignments, a scoring system can be developed by giving a high score for a more likely substitution and a low score for a rare substitution.

For a given substitution matrix, a positive score means that the frequency of amino acid substitutions found in a data set of homologous sequences is greater than would have occurred by random chance. They represent substitutions of very similar residues or identical residues. A zero score means that the frequency of amino acid substitutions found in the homologous sequence data set is equal to that expected by chance. In this case, the relationship between the amino acids is weakly similar at best in terms of physicochemical properties. A negative score means that the frequency of amino acid substitutions found in the homologous sequence data set is less than would have occurred by random chance. This normally occurs with substitutions between dissimilar residues.

The substitution matrices apply logarithmic conversions to describe the probability of amino acid substitutions. The converted values are the so-called log-odds scores (or log-odds ratios), which are logarithmic ratios of the observed mutation frequency divided by the probability of substitution expected by random chance. The conversion can be either to the base of 10 or to the base of 2. For example, in an alignment that involves ten sequences, each having only one aligned position, nine of the sequences are F (phenylalanine) and the remaining one I (isoleucine). The observed frequency of I being substituted by F is one in ten (0.1), whereas the probability of I being substituted by F by random chance is one in twenty (0.05). Thus, the ratio of the two probabilities is 2 (0.1/0.05). After taking this ratio to the logarithm to the base of 2, this makes the log odds equal to 1. This value can then be interpreted as the likelihood of substitution between the two residues being 2^1, which is two times more frequently than by random chance.

PAM Matrices

The PAM matrices (also called Dayhoff PAM matrices) were first constructed by Margaret Dayhoff, who compiled alignments of seventy-one groups of very closely related protein sequences. PAM stands for "point accepted mutation" (although "accepted point mutation" or APM may be a more appropriate term, PAM is easier to pronounce). Because of the use of very closely related homologs, the observed mutations were not expected to significantly change the common function of the proteins. Thus, the observed amino acid mutations are considered to be accepted by natural selection.

These protein sequences were clustered based on phylogenetic reconstruction using maximum parsimony (see Chapter 11). The PAM matrices were subsequently derived based on the evolutionary divergence between sequences of the same cluster. One PAM unit is defined as 1% of the amino acid positions that have been changed. To construct a PAM1 substitution table, a group of closely related sequences with mutation frequencies corresponding to one PAM unit is chosen. Based on the collected mutational data from this group of sequences, a substitution matrix can be derived.

Construction of the PAM1 matrix involves alignment of full-length sequences and subsequent construction of phylogenetic trees using the parsimony principle. This allows computation of ancestral sequences for each internal node of the trees (see Chapter 11). Ancestral sequence information is used to count the number of substitutions along each branch of a tree. The PAM score for a particular residue pair is derived from a multistep procedure involving calculations of relative mutability (which is the number of mutational changes from a common ancestor for a particular amino acid residue divided by the total number of such residues occurring in an alignment), normalization of the expected residue substitution frequencies by random chance, and logarithmic transformation to the base of 10 of the normalized

TABLE 3.1. Correspondence of PAM Numbers with Observed Amino Acid Mutational Rates

PAM Number	Observed Mutation Rate (%)	Sequence Identity (%)
0	0	100
1	1	99
30	25	75
80	50	50
110	40	60
200	75	25
250	80	20

mutability value divided by the frequency of a particular residue. The resulting value is rounded to the nearest integer and entered into the substitution matrix, which reflects the likelihood of amino acid substitutions. This completes the log-odds score computation. After compiling all substitution probabilities of possible amino acid mutations, a 20 × 20 PAM matrix is established. Positive scores in the matrix denote substitutions occurring more frequently than expected among evolutionarily conserved replacements. Negative scores correspond to substitutions that occur less frequently than expected.

Other PAM matrices with increasing numbers for more divergent sequences are extrapolated from PAM1 through matrix multiplication. For example, PAM80 is produced by values of the PAM1 matrix multiplied by itself eighty times. The mathematical transformation accounts for multiple substitutions having occurred in an amino acid position during evolution. For example, when a mutation is observed as F replaced by I, the evolutionary changes may have actually undergone a number of intermediate steps before becoming I, such as in a scenario of F → M → L → I. For that reason, a PAM80 matrix only corresponds to 50% of observed mutational rates.

A PAM unit is defined as 1% amino acid change or one mutation per 100 residues. The increasing PAM numbers correlate with increasing PAM units and thus evolutionary distances of protein sequences (Table 3.1). For example, PAM250, which corresponds to 20% amino acid identity, represents 250 mutations per 100 residues. In theory, the number of evolutionary changes approximately corresponds to an expected evolutionary span of 2,500 million years. Thus, the PAM250 matrix is normally used for divergent sequences. Accordingly, PAM matrices with lower serial numbers are more suitable for aligning more closely related sequences. The extrapolated values of the PAM250 amino acid substitution matrix are shown in Figure 3.5.

BLOSUM Matrices

In the PAM matrix construction, the only direct observation of residue substitutions is in PAM1, based on a relatively small set of extremely closely related sequences. Sequence alignment statistics for more divergent sequences are not available. To fill

	C	S	T	P	A	G	N	D	E	Q	H	R	K	M	I	L	V	F	Y	W
C	12																			
S	0	2																		
T	-2	1	3																	
P	-3	1	0	6																
A	-2	1	1	1	2															
G	-3	1	0	-1	1	5														
N	-4	1	0	-1	0	0	2													
D	-5	0	0	-1	0	1	2	4												
E	-5	0	0	-1	0	0	1	3	4											
Q	-5	-1	-1	0	0	-1	1	2	2	4										
H	-3	-1	-1	0	-1	-2	2	1	1	3	6									
R	-4	0	-1	0	-2	-3	0	-1	-1	1	2	6								
K	-5	0	0	-1	-1	-2	1	0	0	1	0	3	5							
M	-5	-2	-1	-2	-1	-3	-2	-3	-2	-1	-2	0	0	6						
I	-2	-1	0	-2	-1	-3	-2	-2	-2	-2	-2	-2	-2	2	5					
L	-6	-3	-2	-3	-2	-4	-3	-4	-3	-2	-2	-3	-2	4	2	6				
V	-2	-1	0	-1	0	-1	-2	-2	-2	-2	-2	-2	-2	2	4	2	4			
F	-4	-3	-3	-5	-4	-5	-4	-6	-5	-5	-2	-4	-5	0	1	2	-1	9		
Y	0	-3	-3	-5	-3	-5	-2	-4	-4	-4	0	-4	-4	-2	-1	-1	-2	7	10	
W	-8	-2	-5	-6	-6	-7	-4	-7	-7	-5	-3	2	-3	-4	-5	-2	-6	0	0	17
	C	S	T	P	A	G	N	D	E	Q	H	R	K	M	I	L	V	F	Y	W

Figure 3.5: PAM250 amino acid substitution matrix. Residues are grouped according to physicochemical similarities.

in the gap, a new set of substitution matrices have been developed. This is the series of blocks amino acid substitution matrices (BLOSUM), all of which are derived based on direct observation for every possible amino acid substitution in multiple sequence alignments. These were constructed based on more than 2,000 conserved amino acid patterns representing 500 groups of protein sequences. The sequence patterns, also called *blocks*, are ungapped alignments of less than sixty amino acid residues in length. The frequencies of amino acid substitutions of the residues in these blocks are calculated to produce a numerical table, or block substitution matrix.

Instead of using the extrapolation function, the BLOSUM matrices are actual percentage identity values of sequences selected for construction of the matrices. For example, BLOSUM62 indicates that the sequences selected for constructing the matrix share an average identity value of 62%. Other BLOSUM matrices based on sequence groups of various identity levels have also been constructed. In the reversing order as the PAM numbering system, the lower the BLOSUM number, the more divergent sequences they represent.

The BLOSUM score for a particular residue pair is derived from the log ratio of observed residue substitution frequency versus the expected probability of a particular residue. The log odds is taken to the base of 2 instead of 10 as in the PAM matrices. The resulting value is rounded to the nearest integer and entered into the substitution matrix. As in the PAM matrices, positive and negative values correspond to substitutions that occur more or less frequently than expected among evolutionarily

C	9																			
S	-1	4																		
T	-1	1	5																	
P	-3	-1	-1	7																
A	0	1	0	-1	4															
G	-3	0	-2	-2	0	6														
N	-3	1	0	-2	-2	0	6													
D	-3	0	-1	-1	-2	-1	1	6												
E	-4	0	-1	-1	-1	-2	0	2	5											
Q	-3	0	-1	-1	-1	-2	0	0	2	5										
H	-3	-1	-2	-2	-2	-2	1	-1	0	0	8									
R	-3	-1	-1	-2	-1	-2	0	-2	0	1	0	5								
K	-3	0	-1	-1	-1	-2	0	-1	1	1	-1	2	5							
M	-1	-1	-1	-2	-1	-3	-2	-3	-2	0	-2	-1	-1	5						
I	-1	-2	-1	-3	-1	-4	-3	-3	-3	-3	-3	-3	-3	1	4					
L	-1	-2	-1	-3	-1	-4	-3	-4	-3	-2	-3	-2	-2	2	2	4				
V	-1	-2	0	-2	0	-3	-3	-3	-2	-2	-3	-3	-2	1	3	1	4			
F	-2	-2	-2	-4	-2	-3	-3	-3	-3	-3	-1	-3	-3	0	0	0	-1	6		
Y	-2	-2	-2	-3	-2	-3	-2	-3	-2	-1	2	-2	-2	-1	-1	-1	-1	3	7	
W	-2	-3	-2	-4	-3	-2	-4	-4	-3	-2	-2	-3	-3	-1	-3	-2	-3	1	2	11
	C	S	T	P	A	G	N	D	E	Q	H	R	K	M	I	L	V	F	Y	W

Figure 3.6: BLOSUM62 amino acid substitution matrix.

conserved replacements. The values of the BLOSUM62 matrix are shown in Figure 3.6.

Comparison between PAM and BLOSUM

There are a number of differences between PAM and BLOSUM. The principal difference is that the PAM matrices, except PAM1, are derived from an evolutionary model whereas the BLOSUM matrices consist of entirely direct observations. Thus, the BLOSUM matrices may have less evolutionary meaning than the PAM matrices. This is why the PAM matrices are used most often for reconstructing phylogenetic trees. However, because of the mathematical extrapolation procedure used, the PAM values may be less realistic for divergent sequences. The BLOSUM matrices are entirely derived from local sequence alignments of conserved sequence blocks, whereas the PAM1 matrix is based on the global alignment of full-length sequences composed of both conserved and variable regions. This is why the BLOSUM matrices may be more advantageous in searching databases and finding conserved domains in proteins.

Several empirical tests have shown that the BLOSUM matrices outperform the PAM matrices in terms of accuracy of local alignment. This could be largely because the BLOSUM matrices are derived from a much larger and more representative dataset than the one used to derive the PAM matrices. This renders the values for the BLOSUM matrices more reliable. To compensate for the deficiencies in the PAM system, newer matrices using the same approach have been devised based on much larger data sets. These include the Gonnet matrices and the Jones–Taylor–Thornton matrices, which

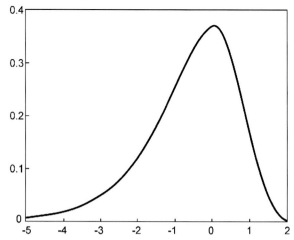

Figure 3.7: Gumble extreme value distribution for alignment scores. The distribution can be expressed as $P = 1 - e^{-Kmne-\lambda x}$, where m and n are the sequence lengths, λ is a scaling factor for the scoring matrix used, and K is a constant that depends on the scoring matrix and gap penalty combination that is used. The x-axis of the curve indicates the standard deviation of the distribution; the y-axis indicates the alignment scores in arbitrary units.

have been shown to have equivalent performance to BLOSUM in regular alignment, but are particularly robust in phylogenetic tree construction.

STATISTICAL SIGNIFICANCE OF SEQUENCE ALIGNMENT

When given a sequence alignment showing a certain degree of similarity, it is often important to ask whether the observed sequence alignment can occur by random chance or the alignment is indeed statistically sound. The truly statistically significant sequence alignment will be able to provide evidence of homology between the sequences involved.

Solving this problem requires a statistical test of the alignment scores of two unrelated sequences of the same length. By calculating alignment scores of a large number of unrelated sequence pairs, a distribution model of the randomized sequence scores can be derived. From the distribution, a statistical test can be performed based on the number of standard deviations from the average score. Many studies have demonstrated that the distribution of similarity scores assumes a peculiar shape that resembles a highly skewed normal distribution with a long tail on one side (Fig. 3.7). The distribution matches the "Gumble extreme value distribution" for which a mathematical expression is available. This means that, given a sequence similarity value, by using the mathematical formula for the extreme distribution, the statistical significance can be accurately estimated.

The statistical test for the relatedness of two sequences can be performed using the following procedure. An optimal alignment between two given sequences is first obtained. Unrelated sequences of the same length are then generated through a randomization process in which one of the two sequences is randomly shuffled. A new alignment score is then computed for the shuffled sequence pair. More such scores are similarly obtained through repeated shuffling. The pool of alignment scores from the shuffled sequences is used to generate parameters for the extreme distribution. The original alignment score is then compared against the distribution of random alignments to determine whether the score is beyond random chance. If the score is located in the extreme margin of the distribution, that means that the alignment between the two sequences is unlikely due to random chance and is thus considered significant. A P-value is given to indicate the probability that the original alignment is due to random chance.

A P-value resulting from the test provides a much more reliable indicator of possible homologous relationships than using percent identity values. It is thus important to know how to interpret the P-values. It has been shown that if a P-value is smaller than 10^{-100}, it indicates an exact match between the two sequences. If the P-value is in the range of 10^{-50} to 10^{-100}, it is considered to be a nearly identical match. A P-value in the range of 10^{-5} to 10^{-50} is interpreted as sequences having clear homology. A P-value in the range of 10^{-1} to 10^{-5} indicates possible distant homologs. If P is larger than 10^{-1}, the two sequence may be randomly related. However, the caveat is that sometimes truly related protein sequences may lack the statistical significance at the sequence level owing to fast divergence rates. Their evolutionary relationships can nonetheless be revealed at the three-dimensional structural level (see Chapter 15).

These statistics were derived from ungapped local sequence alignments. It is not known whether the Gumble distribution applies equally well to gapped alignments. However, for all practical purposes, it is reasonable to assume that scores for gapped alignments essentially fit the same distribution. A frequently used software program for assessing statistical significance of a pairwise alignment is the PRSS program.

PRSS (Probability of Random Shuffles; www.ch.embnet.org/software/PRSS_form. html) is a web-based program that can be used to evaluate the statistical significance of DNA or protein sequence alignment. It first aligns two sequences using the Smith–Waterman algorithm and calculates the score. It then holds one sequence in its original form and randomizes the order of residues in the other sequence. The shuffled sequence is realigned with the unshuffled sequence. The resulting alignment score is recorded. This process is iterated many (normally 1,000) times to help generate data for fitting the Gumble distribution. The original alignment score is then compared against the overall score distribution to derive a P-value. The major feature of the program is that it allows partial shuffling. For example, shuffling can be restricted to residues within a local window of 25–40, whereas the residues outside the window remain unchanged.

SUMMARY

Pairwise sequence alignment is the fundamental component of many bioinformatics applications. It is extremely useful in structural, functional, and evolutionary analyses of sequences. Pairwise sequence alignment provides inference for the relatedness of two sequences. Strongly similar sequences are often homologous. However, a distinction needs to be made between sequence homology and similarity. The former is inference drawn from sequence comparison, whereas the latter relates to actual observation after sequence alignment. For protein sequences, identity values from pairwise alignment are often used to infer homology, although this approach can be rather imprecise.

There are two sequence alignment strategies, local alignment and global alignment, and three types of algorithm that perform both local and global alignments. They are the dot matrix method, dynamic programming method, and word method. The dot matrix method is useful in visually identifying similar regions, but lacks the sophistication of the other two methods. Dynamic programming is an exhaustive and quantitative method to find optimal alignments. This method effectively works in three steps. It first produces a sequence versus sequence matrix. The second step is to accumulate scores in the matrix. The last step is to trace back through the matrix in reverse order to identify the highest scoring path. This scoring step involves the use of scoring matrices and gap penalties.

Scoring matrices describe the statistical probabilities of one residue being substituted by another. PAM and BLOSUM are the two most commonly used matrices for aligning protein sequences. The PAM matrices involve the use of evolutionary models and extrapolation of probability values from alignment of close homologs to more divergent ones. In contrast, the BLOSUM matrices are derived from actual alignment. The PAM and BLOSUM serial numbers also have opposite meanings. Matrices of high PAM numbers are used to align divergent sequences and low PAM numbers for aligning closely related sequences. In practice, if one is uncertain about which matrix to use, it is advisable to test several matrices and choose the one that gives the best alignment result. Statistical significance of pairwise sequence similarity can be tested using a randomization test where score distribution follows an extreme value distribution.

FURTHER READING

Batzoglou, S. 2005. The many faces of sequence alignment. *Brief. Bioinformatics* 6:6–22.

Brenner, S. E., Chothia, C., and Hubbard, T. J. 1998. Assessing sequence comparison methods with reliable structurally identified distant evolutionary relationships. *Proc. Natl. Acad. Sci. U S A* 95:6073–8.

Chao, K.-M., Pearson, W. R., and Miller, W. 1992. Aligning two sequences within a specified diagonal band. *Comput. Appl. Biosci.* 8:481–7.

Henikoff, S., and Henikoff, J. G. 1992. Amino acid substitution matrices from protein blocks. *Proc. Natl. Acad. Sci. U S A* 89:10915–19.

Huang, X. 1994. On global sequence alignment. *Comput. Appl. Biosci.* 10:227–35.

Pagni, M., and Jongeneel, V. 2001. Making sense of score statistics for sequence alignments. *Brief. Bioinformatics* 2:51–67.

Pearson, W. R. 1996. Effective protein sequence comparison. *Methods Enzymol.* 266:227–58.

Rost, B. 1999. Twilight zone of protein sequence alignments. *Protein Eng.* 12:85–94.

States, D. J., Gish, W., and Altschul, S. F. 1991. Improved sensitivity of nucleic acid database searches using application-specific scoring matrices. *Methods* 3:66–70.

Valdar, W. S. 2002. Scoring residue conservation. *Proteins.* 48:227–41.

Vingron, M., and Waterman, M. S. 1994. Sequence alignment and penalty scores. *J. Mol. Biol.* 235:1–12.

CHAPTER FOUR

Database Similarity Searching

A main application of pairwise alignment is retrieving biological sequences in databases based on similarity. This process involves submission of a query sequence and performing a pairwise comparison of the query sequence with all individual sequences in a database. Thus, database similarity searching is pairwise alignment on a large scale. This type of searching is one of the most effective ways to assign putative functions to newly determined sequences. However, the dynamic programming method described in Chapter 3 is slow and impractical to use in most cases. Special search methods are needed to speed up the computational process of sequence comparison. The theory and applications of the database searching methods are discussed in this chapter.

UNIQUE REQUIREMENTS OF DATABASE SEARCHING

There are unique requirements for implementing algorithms for sequence database searching. The first criterion is *sensitivity*, which refers to the ability to find as many correct hits as possible. It is measured by the extent of inclusion of correctly identified sequence members of the same family. These correct hits are considered "true positives" in the database searching exercise. The second criterion is *selectivity*, also called *specificity*, which refers to the ability to exclude incorrect hits. These incorrect hits are unrelated sequences mistakenly identified in database searching and are considered "false positives." The third criterion is *speed*, which is the time it takes to get results from database searches. Depending on the size of the database, speed sometimes can be a primary concern.

Ideally, one wants to have the greatest sensitivity, selectivity, and speed in database searches. However, satisfying all three requirements is difficult in reality. What generally happens is that an increase in sensitivity is associated with decrease in selectivity. A very inclusive search tends to include many false positives. Similarly, an improvement in speed often comes at the cost of lowered sensitivity and selectivity. A compromise between the three criteria often has to be made.

In database searching, as well as in many other areas in bioinformatics, are two fundamental types of algorithms. One is the *exhaustive type*, which uses a rigorous algorithm to find the best or exact solution for a particular problem by examining all mathematical combinations. Dynamic programming is an example of the exhaustive method and is computationally very intensive. Another is the *heuristic type*, which is a computational strategy to find an empirical or near optimal solution by using rules of thumb. Essentially, this type of algorithms take shortcuts by reducing the search space according to some criteria. However, the shortcut strategy is not guaranteed to find the best or most accurate solution. It is often used because of the need for obtaining results within a realistic time frame without significantly sacrificing the accuracy of the computational output.

HEURISTIC DATABASE SEARCHING

Searching a large database using the dynamic programming methods, such as the Smith–Waterman algorithm, although accurate and reliable, is too slow and impractical when computational resources are limited. An estimate conducted nearly a decade ago had shown that querying a database of 300,000 sequences using a query sequence of 100 residues took 2–3 hours to complete with a regular computer system at the time. Thus, speed of searching became an important issue. To speed up the comparison, heuristic methods have to be used. The heuristic algorithms perform faster searches because they examine only a fraction of the possible alignments examined in regular dynamic programming.

Currently, there are two major heuristic algorithms for performing database searches: BLAST and FASTA. These methods are not guaranteed to find the optimal alignment or true homologs, but are 50–100 times faster than dynamic programming. The increased computational speed comes at a moderate expense of sensitivity and specificity of the search, which is easily tolerated by working molecular biologists. Both programs can provide a reasonably good indication of sequence similarity by identifying similar sequence segments.

Both BLAST and FASTA use a heuristic *word method* for fast pairwise sequence alignment. This is the third method of pairwise sequence alignment. It works by finding short stretches of identical or nearly identical letters in two sequences. These short strings of characters are called *words*, which are similar to the windows used in the dot matrix method (see Chapter 3). The basic assumption is that two related sequences must have at least one word in common. By first identifying word matches, a longer alignment can be obtained by extending similarity regions from the words. Once regions of high sequence similarity are found, adjacent high-scoring regions can be joined into a full alignment.

BASIC LOCAL ALIGNMENT SEARCH TOOL (BLAST)

The BLAST program was developed by Stephen Altschul of NCBI in 1990 and has since become one of the most popular programs for sequence analysis. BLAST uses heuristics to align a query sequence with all sequences in a database. The objective is to find high-scoring ungapped segments among related sequences. The existence of such segments above a given threshold indicates pairwise similarity beyond random chance, which helps to discriminate related sequences from unrelated sequences in a database.

1. Query: MRD`PYN`KLIS

2. Scan every three residues to be used in searching BLAST word database.

3. Assuming one of the words finds matches in the database.

Query	**PYN**	**PYN**	**PYN**	**PYN**	. . .
Database	**PYN**	**PFN**	**PFQ**	**PFE**	. . .

4. Calculate sums of match scores based on BLOSUM62 matrix.

Query	**PYN**	**PYN**	**PYN**	**PYN**	. . .
Database	**PYN**	**PFN**	**PFQ**	**PFE**	. . .
Sum of score	**20**	**16**	**10**	**10**	. . .

5. Find the database sequence corresponding to the best word match and extend alignment in both directions.

Query	**M R D**	`PYN`	**K L I S**
Database	**M H E**	`PYN`	**D V P W**

←——————— ———————→

extension to left   extension to right

6. Determine high scored segment above threshold (22).

Query	**M R D**	`PYN`	**K L I S**
Database	**M H E**	`PYN`	**D V P W**
	5 0 2	**20**	**-1 1 -3 -3**

HSP, total score 24

Figure 4.1: Illustration of the BLAST procedure using a hypothetical query sequence matching with a hypothetical database sequence. The alignment scoring is based on the BLOSUM62 matrix (see Chapter 3). The example of the word match is highlighted in the box.

BLAST performs sequence alignment through the following steps. The first step is to create a list of words from the query sequence. Each word is typically three residues for protein sequences and eleven residues for DNA sequences. The list includes every possible word extracted from the query sequence. This step is also called *seeding*. The second step is to search a sequence database for the occurrence of these words. This step is to identify database sequences containing the matching words. The matching of the words is scored by a given substitution matrix. A word is considered a match if it is above a threshold. The fourth step involves pairwise alignment by extending from the words in both directions while counting the alignment score using the same substitution matrix. The extension continues until the score of the alignment drops below a threshold due to mismatches (the drop threshold is twenty-two for proteins and twenty for DNA). The resulting contiguous aligned segment pair without gaps is called *high-scoring segment pair* (HSP; see working example in Fig. 4.1). In the original version of BLAST, the highest scored HSPs are presented as the final report. They are also called maximum scoring pairs.

A recent improvement in the implementation of BLAST is the ability to provide gapped alignment. In gapped BLAST, the highest scored segment is chosen to be extended in both directions using dynamic programming where gaps may be introduced. The extension continues if the alignment score is above a certain threshold; otherwise it is terminated. However, the overall score is allowed to drop below the threshold only if it is temporary and rises again to attain above threshold values. Final trimming of terminal regions is needed before producing a report of the final alignment.

Variants

BLAST is a family of programs that includes BLASTN, BLASTP, BLASTX TBLASTN, and TBLASTX. BLASTN queries nucleotide sequences with a nucleotide sequence database. BLASTP uses protein sequences as queries to search against a protein sequence database. BLASTX uses nucleotide sequences as queries and translates them in all six reading frames to produce translated protein sequences, which are used to query a protein sequence database. TBLASTN queries protein sequences to a nucleotide sequence database with the sequences translated in all six reading frames. TBLASTX uses nucleotide sequences, which are translated in all six frames, to search against a nucleotide sequence database that has all the sequences translated in six frames. In addition, there is also a bl2seq program that performs local alignment of two user-provided input sequences. The graphical output includes horizontal bars and a diagonal in a two-dimensional diagram showing the overall extent of matching between the two sequences.

The BLAST web server (www.ncbi.nlm.nih.gov/BLAST/) has been designed in such a way as to simplify the task of program selection. The programs are organized based on the type of query sequences, protein sequences, nucleotide sequences, or nucleotide sequence to be translated. In addition, programs for special purposes are grouped separately; for example, bl2seq, immunoglobulin BLAST, and VecScreen, a program for removing contaminating vector sequences. The BLAST programs specially designed for searching individual genome databases are also listed in a separate category.

The choice of the type of sequences also influences the sensitivity of the search. Generally speaking, there is a clear advantage of using protein sequences in detecting homologs. This is because DNA sequences only comprise four nucleotides, whereas protein sequences contain twenty amino acids. This means that there is at least a five-fold increase in statistical complexity for protein sequences. More importantly, amino acid substitution matrices incorporate subtle differences in physicochemical properties between amino acids, meaning that protein sequences are far more informative and sensitive in detection of homologs. This is why searches using protein sequences can yield more significant matches than using DNA sequences. For that reason, if the input sequence is a protein-encoding DNA sequence, it is preferable to use BLASTX, which translates it in six open reading frames before sequence comparisons are carried out.

If one is looking for protein homologs encoded in newly sequenced genomes, one may use TBLASTN, which translates nucleotide database sequences in all six open reading frames. This may help to identify protein coding genes that have not yet been annotated. If a DNA sequence is to be used as the query, a protein-level comparison can be done with TBLASTX. However, both programs are very computationally intensive and the search process can be very slow.

Statistical Significance

The BLAST output provides a list of pairwise sequence matches ranked by statistical significance. The significance scores help to distinguish evolutionarily related sequences from unrelated ones. Generally, only hits above a certain threshold are displayed.

Deriving the statistical measure is slightly different from that for single pairwise sequence alignment; the larger the database, the more unrelated sequence alignments there are. This necessitates a new parameter that takes into account the total number of sequence alignments conducted, which is proportional to the size of the database. In BLAST searches, this statistical indicator is known as the E-value (expectation value), and it indicates the probability that the resulting alignments from a database search are caused by random chance. The E-value is related to the P-value used to assess significance of single pairwise alignment (see Chapter 3). BLAST compares a query sequence against all database sequences, and so the E-value is determined by the following formula:

$$E = m \times n \times P$$
(Eq. 4.1)

where m is the total number of residues in a database, n is the number of residues in the query sequence, and P is the probability that an HSP alignment is a result of random chance. For example, aligning a query sequence of 100 residues to a database containing a total of 10^{12} residues results in a P-value for the ungapped HSP region in one of the database matches of 1×1^{-20}. The E-value, which is the product of the three values, is $100 \times 10^{12} \times 10^{-20}$, which equals 10^{-6}. It is expressed as $1e - 6$ in BLAST output. This indicates that the probability of this database sequence match occurring due to random chance is 10^{-6}.

The E-value provides information about the likelihood that a given sequence match is purely by chance. The lower the E-value, the less likely the database match is a result of random chance and therefore the more significant the match is. Empirical interpretation of the E-value is as follows. If $E < 1e - 50$ (or 1×10^{-50}), there should be an extremely high confidence that the database match is a result of homologous relationships. If E is between 0.01 and $1e - 50$, the match can be considered a result of homology. If E is between 0.01 and 10, the match is considered not significant, but may hint at a tentative remote homology relationship. Additional evidence is needed to confirm the tentative relationship. If $E > 10$, the sequences under consideration are either unrelated or related by extremely distant relationships that fall below the limit of detection with the current method.

Because the E-value is proportionally affected by the database size, an obvious problem is that as the database grows, the E-value for a given sequence match also increases. Because the genuine evolutionary relationship between the two sequences remains constant, the decrease in credibility of the sequence match as the database grows means that one may "lose" previously detected homologs as the database enlarges. Thus, an alternative to E-value calculations is needed.

A bit score is another prominent statistical indicator used in addition to the *E*-value in a BLAST output. The *bit score* measures sequence similarity independent of query sequence length and database size and is normalized based on the raw pairwise alignment score. The bit score (S') is determined by the following formula:

$$S' = (\lambda \times S - \ln K) / \ln 2 \tag{Eq. 4.2}$$

where λ is the Gumble distribution constant, S is the raw alignment score, and K is a constant associated with the scoring matrix used. Clearly, the bit score (S') is linearly related to the raw alignment score (S). Thus, the higher the bit score, the more highly significant the match is. The bit score provides a constant statistical indicator for searching different databases of different sizes or for searching the same database at different times as the database enlarges.

Low Complexity Regions

For both protein and DNA sequences, there may be regions that contain highly repetitive residues, such as short segments of repeats, or segments that are overrepresented by a small number of residues. These sequence regions are referred to as *low complexity regions* (LCRs). LCRs are rather prevalent in database sequences; estimates indicate that LCRs account for about 15% of the total protein sequences in public databases. These elements in query sequences can cause spurious database matches and lead to artificially high alignment scores with unrelated sequences.

To avoid the problem of high similarity scores owing to matching of LCRs that obscure the real similarities, it is important to filter out the problematic regions in both the query and database sequences to improve the signal-to-noise ratio, a process known as *masking*. There are two types of masking: hard and soft. *Hard masking* involves replacing LCR sequences with an ambiguity character such as *N* for nucleotide residues or *X* for amino acid residues. The ambiguity characters are then ignored by the BLAST program, preventing the use of such regions in alignments and thus avoiding false positives. However, the drawback is that matching scores with true homologs may be lowered because of shortened alignments. *Soft masking* involves converting the problematic sequences to lower case letters, which are ignored in constructing the word dictionary, but are used in word extension and optimization of alignments.

SEG is a program that is able to detect and mask repetitive elements before executing database searches. It identifies LCRs by comparing residue frequencies of a certain region with average residue frequencies in the database. If the residue frequencies of a sequence region of the query sequence are significantly higher than the database average, the region is declared an LCR. SEG has been integrated into the BLAST web-based program. An option box for this low complexity filter needs to be selected to mask LCRs (either hard or soft masking).

RepeatMasker (http://woody.embl-heidelberg.de/repeatmask/) is an independent masking program that detects repetitive elements by comparing the query sequence with a built-in library of repetitive elements using the Smith–Waterman algorithm. If the alignment score for a sequence region is above a certain threshold, the region is declared an LCR. The corresponding residues are then masked with N's or X's.

BLAST Output Format

The BLAST output includes a graphical overview box, a matching list and a text description of the alignment (Fig. 4.2). The graphical overview box contains colored horizontal bars that allow quick identification of the number of database hits and the degrees of similarity of the hits. The color coding of the horizontal bars corresponds to the ranking of similarities of the sequence hits (red: most related; green and blue: moderately related; black: unrelated). The length of the bars represents the spans of sequence alignments relative to the query sequence. Each bar is hyperlinked to the actual pairwise alignment in the text portion of the report. Below the graphical box is a list of matching hits ranked by the E-values in ascending order. Each hit includes the accession number, title (usually partial) of the database record, bit score, and E-value.

This list is followed by the text description, which may be divided into three sections: the header, statistics, and alignment. The header section contains the gene index number or the reference number of the database hit plus a one-line description of the database sequence. This is followed by the summary of the statistics of the search output, which includes the bit score, E-value, percentages of identity, similarity ("Positives"), and gaps. In the actual alignment section, the query sequence is on the top of the pair and the database sequence is at the bottom of the pair labeled as *Subject*. In between the two sequences, matching identical residues are written out at their corresponding positions, whereas nonidentical but similar residues are labeled with "+". Any residues identified as LCRs in the query sequence are masked with Xs or Ns so that no alignment is represented in those regions.

FASTA

FASTA (FAST ALL, www.ebi.ac.uk/fasta33/) was in fact the first database similarity search tool developed, preceding the development of BLAST. FASTA uses a "hashing" strategy to find matches for a short stretch of identical residues with a length of k. The string of residues is known as *ktuples* or *ktups*, which are equivalent to words in BLAST, but are normally shorter than the words. Typically, a ktup is composed of two residues for protein sequences and six residues for DNA sequences.

The first step in FASTA alignment is to identify ktups between two sequences by using the hashing strategy. This strategy works by constructing a lookup table that shows the position of each ktup for the two sequences under consideration. The positional difference for each word between the two sequences is obtained by subtracting the position of the first sequence from that of the second sequence and is expressed as the offset. The ktups that have the same offset values are then linked to reveal a

Graphical overview

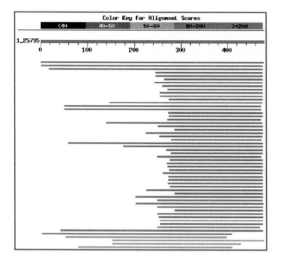

Matching list

```
                                                            Score    E
Sequences producing significant alignments:                (bits)  Value

gi|22958938|ref|ZP_00006599.1|  COG3920: Signal transduction...   896    0.0
gi|22968827|ref|ZP_00016409.1|  COG3920: Signal transduction...   390    e-107
gi|39933087|ref|NP_945363.1|    putative signal transduction h...  365    e-100
gi|17935877|ref|NP_532667.1|    two component sensor kinase [A...  175    2e-42
gi|15889280|ref|NP_354961.1|    AGR_C_3616p [Agrobacterium tum...  175    2e-42
gi|31322739|gb|AAP22926.1|      CheS3 [Rhodospirillum centenum]    158    2e-37
gi|16126793|ref|NP_421357.1|    sensor histidine kinase, putat...  157    5e-37
gi|16127400|ref|NP_421964.1|    sensor histidine kinase, putat...  155    1e-36
gi|15966187|ref|NP_386540.1|    HYPOTHETICAL PROTEIN [Sinorhiz...  155    2e-36
gi|16264804|ref|NP_437596.1|    putative two-component sensor ...  152    2e-35
gi|2808506|emb|CAA12536.1|      ExsG protein [Sinorhizobium meli... 151   2e-35
gi|13476692|ref|NP_108261.1|    two-component, sensor histidin...  149    9e-35
gi|16127278|ref|NP_421842.1|    sensor histidine kinase, putat...  149    1e-34
gi|17939110|ref|NP_535898.1|    two component sensor kinase [A...  147    4e-34
gi|13473179|ref|NP_104746.1|    hypothetical protein [Mesorhiz...  147    6e-34
gi|16119758|ref|NP_396464.1|    AGR_pAT_788p [Agrobacterium tu...  147    6e-34
gi|13488521|ref|NP_109528.1|    sensory transduction histidine...  146    1e-33
gi|16125089|ref|NP_419653.1|    sensor histidine kinase, putat...  145    1e-33
gi|22957499|ref|ZP_00005199.1|  COG3920: Signal transduction...    145    2e-33
```

Alignment output { **header** **statistics** **alignment**

```
>gi|22968827|ref|ZP_00016409.1|   COG3920: Signal transduction histidine kinase [Rhodospirillum
   rubrum]
   Length = 489

Score = 377 bits (968), Expect = e-103
Identities = 235/484 (48%), Positives = 306/484 (63%), Gaps = 14/484 (2%)

Query: 3    PAEIDELRRRLHEAEETLKAIRQGDVDALVVGASDDTDVYVIGGDPDICRSFLDMMEIGA 62
            P + ELRRRL EAEETL AIR+G+VDALV+G     +V+ IGGD +  R+F++ M+ GA
Sbjct: 4    PVVLSELRRRLAEAEETLNAIREGEVDALVIGEGGVDEVFAIGGDTESYRTFMEAMDTGA 63

Query: 63   AALDNTGRVLYANAVLADLVGRPLPELEQMRL-----SELTGDPAXXXXXXXXXXXXXXXI 117
            AA+D  GRVLYAN+ L  L+  PLP L+G L      +    +              I
Sbjct: 64   AAVDEDGRVLYANSALCRLIDHPLPTLQGKPLVSFFDARAAAEIGQMVGKTANQREKVEI 123

Query: 118  PLGVAGAER-QVMLSCGK-LRLGIVSGHAVTFTDFTEQLAAERSRQNEKAALAIIACANE 175
            L A + QV L   K +RLG V GHAVTFTD TE++ +E + + E+ A AIIA ANE
Sbjct: 124  SLKDAATKMAQVFTVSAKPVRLGLVQGHAVTFTDLTERVRSETAERAERIAAAIIASANE 183

Query: 176  PVFVCDTLGLITHXXXXXXXXXXXXXXXRPLSEVMDLSVGDGTGLL/TLGEIVAQATEGIP 235
            V VCD +G+ITH               + + L+ D  L++ G ++  A  G
Sbjct: 184  IVVVCDRVGNITHANSAAISAIYDGDLIGKMFEDAIPLTFTDAPDLMSGGALIDLALNGQA 243

Query: 236  VQGIEAVAAEGTPF--YLISAAPLQVPGEAVSGCVITMVDLSQRKAAERHQQLLLRELDH 293
            QGIEA+A       YLISAAPLQV   +SGCV+TMVDLSQRKAAE   Q LL+RELDH
Sbjct: 244  RQGIEAIATRAPKVKDYLISAAPLQVTEDQISGCVL/TMVDLSQRKAAEHQQLLLMRELDH 303

Query: 294  RVKNTLALVMSISRRTMHSEETLEGYQKAFTARIQAIAATHNLLADKSWSDISIRDVLVR 353
            RV+NTLALV+SIS RT+ +E+TL+G+ +AFT RI  LAATH+LLA KQG  W+ +S+ D++
Sbjct: 304  RVRNTLALVLSISNRTLSNEDTLQGFHQAFTQRIHGLAATHSLLAKQGWTKLSLHDIVRA 363

Query: 354  ELAPYNEGFSQRILVEVPDVEIEPRSAIALXLVIHELATNATKYGSLSTPEGQ--VRVRG 411
            ELAPY E    R+ +E +V + PR+AIALXL+ HELATNA KYG+LS     G   V VRG
Sbjct: 364  ELAPYVETDGTRLRLEGGEVALIPRAAIALXLIFHELATNAVKYGALSREGGHVLVAVRG 423

Query: 412  LPGADEPADVVCLEWLERGGPPVSEPTRSGFGQTVIRHAFYAYAEGGGAEVSFEPDGVRCR 471
            P AD  A  V +W+E GGP VS P R GFG TVI H+ AY+    GG ++SF P+GV C
Sbjct: 424  -PTADGAAMRV--DWVESGGPMVSPPQRKGFGHTVISHSLAYSSKGGTDLSFPPEGVICA 480

Query: 472  VSVP 475
            + +P
Sbjct: 481  LRIP 484
```

Figure 4.2: An example of a BLAST output showing three portions: the graphical overview box, the list of matching hits, and the text portion containing header, statistics, and the actual alignment.

1. Given two amino acid sequences for comparision:

> sequence 1   **AMPSDGL**
> sequence 2   **GPSDNAT**

2. Construct a hashing table:

amino acid	sequence position		offset
	seq 1	seq 2	
A	1	6	-5
D	5	4	1
G	6	1	5
L	7	–	–
M	2	–	–
N	–	5	–
P	3	2	1
S	4	3	1
T	–	7	–

3. Identify residues with the same offset values (highlighted in grey).

4. Find the matching word of three residues in the order of 3, 4 and 5 in one sequence and 2, 3, and 4 in the other.

5. This allows establishment of alignment between the two sequences.

> sequence 1   **AMPSDGL–**
>   **| | |**
> sequence 2   **–GPSDNAT**

Figure 4.3: The procedure of ktup identification using the hashing strategy by FASTA. Identical offset values between residues of the two sequences allow the formation of ktups.

contiguous identical sequence region that corresponds to a stretch of diagonal in a two-dimensional matrix (Fig. 4.3).

The second step is to narrow down the high similarity regions between the two sequences. Normally, many diagonals between the two sequences can be identified in the hashing step. The top ten regions with the highest density of diagonals are identified as high similarity regions. The diagonals in these regions are scored using a substitution matrix. Neighboring high-scoring segments along the same diagonal are selected and joined to form a single alignment. This step allows introducing gaps between the diagonals while applying gap penalties. The score of the gapped alignment is calculated again. In step 3, the gapped alignment is refined further using the Smith–Waterman algorithm to produce a final alignment (Fig. 4.4). The last step is to perform a statistical evaluation of the final alignment as in BLAST, which produces the E-value.

Similar to BLAST, FASTA has a number of subprograms. The web-based FASTA program offered by the European Bioinformatics Institute (www.ebi.ac.uk/) allows the use of either DNA or protein sequences as the query to search against a protein database or nucleotide database. Some available variants of the program are FASTX, which translates a DNA sequence and uses the translated protein sequence to query a protein database, and TFASTX, which compares a protein query sequence to a translated DNA database.

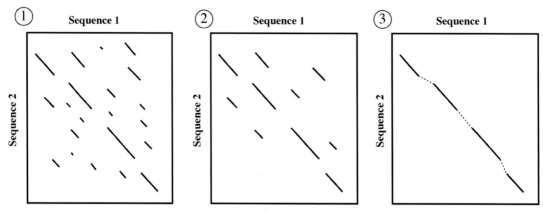

Figure 4.4: Steps of the FASTA alignment procedure. In step 1 (*left*), all possible ungapped alignments are found between two sequences with the hashing method. In step 2 (*middle*), the alignments are scored according to a particular scoring matrix. Only the ten best alignments are selected. In step 3 (*right*), the alignments in the same diagonal are selected and joined to form a single gapped alignment, which is optimized using the dynamic programming approach.

Statistical Significance

FASTA also uses *E*-values and bit scores. Estimation of the two parameters in FASTA is essentially the same as in BLAST. However, the FASTA output provides one more statistical parameter, the *Z*-score. This describes the number of standard deviations from the mean score for the database search. Because most of the alignments with the query sequence are with unrelated sequences, the higher the *Z*-score for a reported match, the further away from the mean of the score distribution, hence, the more significant the match. For a *Z*-score > 15, the match can be considered extremely significant, with certainty of a homologous relationship. If *Z* is in the range of 5 to 15, the sequence pair can be described as highly probable homologs. If *Z* < 5, their relationships is described as less certain.

COMPARISON OF FASTA AND BLAST

BLAST and FASTA have been shown to perform almost equally well in regular database searching. However, there are some notable differences between the two approaches. The major difference is in the seeding step; BLAST uses a substitution matrix to find matching words, whereas FASTA identifies identical matching words using the hashing procedure. By default, FASTA scans smaller window sizes. Thus, it gives more sensitive results than BLAST, with a better coverage rate for homologs. However, it is usually slower than BLAST. The use of low-complexity masking in the BLAST procedure means that it may have higher specificity than FASTA because potential false positives are reduced. BLAST sometimes gives multiple best-scoring alignments from the same sequence; FASTA returns only one final alignment.

DATABASE SEARCHING WITH THE SMITH–WATERMAN METHOD

As mentioned, the rigorous dynamic programming method is normally not used for database searching, because it is slow and computationally expensive. Heuristics such as BLAST and FASTA are developed for faster speed. However, the heuristic methods are limited in sensitivity and are not guaranteed to find the optimal alignment. They often fail to find alignment for distantly related sequences. It has been estimated that for some families of protein sequences, BLAST can miss 30% of truly significant hits. Recent developments in computation technologies, such as parallel processing supercomputers, have made dynamic programming a feasible approach to database searches to fill the performance gap.

For this purpose, the computer codes for the Needleman–Wunsch and Smith–Waterman algorithms have to be modified to run in a parallel processing environment so that searches can be completed within reasonable time periods. Currently, the search speed is still slower than the popular heuristic programs. Therefore, the method is not intended for routine use. Nevertheless, the availability of dynamic programming allows the maximum sensitivity for finding homologs at the sequence level. Empirical tests have indeed shown that the exhaustive method produces superior results over the heuristic methods. Below is a list of dynamic programming-based web servers for sequence database searches.

ScanPS (Scan Protein Sequence, www.ebi.ac.uk/scanps/) is a web-based program that implements a modified version of the Smith–Waterman algorithm optimized for parallel processing. The major feature is that the program allows iterative searching similar to PSI-BLAST (see Chapter 5), which builds profiles from one round of search results and uses them for the second round of database searching. Full dynamic programming is used in each cycle for added sensitivity.

ParAlign (www.paralign.org/) is a web-based server that uses parallel processors to perform exhaustive sequence comparisons using either a parallelized version of the Smith–Waterman algorithm or a heuristic program for further speed gains. The heuristic subprogram first finds exact ungapped alignments and uses them as anchors for extension into gapped alignments by combining the scores of several diagonals in the alignment matrix. The search speed of ParAlign approaches to that of BLAST, but with higher sensitivity.

SUMMARY

Database similarity searching is an essential first step in the functional characterization of novel gene or protein sequences. The major issues in database searching are sensitivity, selectivity, and speed. Speed is a particular concern in searching large databases. Thus, heuristic methods have been developed for efficient database similarity searches. The major heuristic database searching algorithms are BLAST and FASTA. They both use a word method for pairwise alignment. BLAST looks for HSPs in a database. FASTA uses a hashing scheme to identify words. The major statistical measures for significance of database matches are E-values and bit scores. A caveat for sequence database searching is to filter the LCRs using masking programs. Another caveat is to use protein sequences as the query in database searching, because they produce much more sensitive matches. In addition, it is important to keep in mind that both BLAST and FASTA are heuristic programs and are not guaranteed to find all the homologous sequences. For significant matches automatically generated by these programs, it is recommended to follow up the leads by checking the alignment using more rigorous and independent alignment programs. Advances in computational technology have also made it possible to use full dynamic programming in database searching with increased sensitivity and selectivity.

FURTHER READING

Altschul, S. F., Boguski, M. S., Gish, W., and Wootton, J. C. 1994. Issues in searching molecular sequences databases. *Nat. Genet.* 6:119–29.

Altschul, S. F., Madden, T. L., Schaffer, A. A., Zhang, J., Zhang, Z., Miller, W., and Lipman, D. J. 1997. Gapped BLAST and PSI-BLAST: A new generation of protein database search programs. *Nucleic Acids Res.* 25:3389–402.

Chen, Z. 2003. Assessing sequence comparison methods with the average precision criterion. *Bioinformatics* 19:2456–60.

Karlin, S., and Altschul, S. F. 1993. Applications and statistics for multiple high-scoring segments in molecular sequences. *Proc. Natl. Acad. Sci. U S A* 90:5873–7.

Mullan, L. J., and Williams, G. W. 2002. BLAST and go? *Brief. Bioinform.* 3:200–2.

Sansom, C. 2000. Database searching with DNA and protein sequences: An introduction. *Brief. Bioinform.* 1:22–32.

Spang, R., and Vingron, M. 1998. Statistics of large-scale sequence searching. *Bioinformatics* 14:279–84.

CHAPTER FIVE

Multiple Sequence Alignment

A natural extension of pairwise alignment is multiple sequence alignment, which is to align multiple related sequences to achieve optimal matching of the sequences. Related sequences are identified through the database similarity searching described in Chapter 4. As the process generates multiple matching sequence pairs, it is often necessary to convert the numerous pairwise alignments into a single alignment, which arranges sequences in such a way that evolutionarily equivalent positions across all sequences are matched.

There is a unique advantage of multiple sequence alignment because it reveals more biological information than many pairwise alignments can. For example, it allows the identification of conserved sequence patterns and motifs in the whole sequence family, which are not obvious to detect by comparing only two sequences. Many conserved and functionally critical amino acid residues can be identified in a protein multiple alignment. Multiple sequence alignment is also an essential pre-requisite to carrying out phylogenetic analysis of sequence families and prediction of protein secondary and tertiary structures. Multiple sequence alignment also has applications in designing degenerate polymerase chain reaction (PCR) primers based on multiple related sequences.

It is theoretically possible to use dynamic programming to align any number of sequences as for pairwise alignment. However, the amount of computing time and memory it requires increases exponentially as the number of sequences increases. As a consequence, full dynamic programming cannot be applied for datasets of more than ten sequences. In practice, heuristic approaches are most often used. In this chapter, methodologies and applications of multiple sequence alignment are discussed.

SCORING FUNCTION

Multiple sequence alignment is to arrange sequences in such a way that a maximum number of residues from each sequence are matched up according to a particular scoring function. The scoring function for multiple sequence alignment is based on the concept of sum of pairs (SP). As the name suggests, it is the sum of the scores of all possible pairs of sequences in a multiple alignment based on a particular scoring matrix. In calculating the SP scores, each column is scored by summing the scores for all possible pairwise matches, mismatches and gap costs. The score of the entire

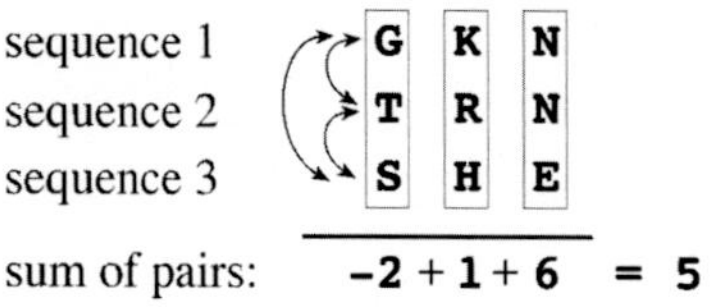

sequence 1
sequence 2
sequence 3

G	K	N
T	R	N
S	H	E

sum of pairs: **−2 + 1 + 6** **= 5**

Figure 5.1: Given a multiple alignment of three sequences, the sum of scores is calculated as the sum of the similarity scores of every pair of sequences at each position. The scoring is based on the BLOSUM62 matrix (see Chapter 3). The total score for the alignment is 5, which means that the alignment is $2^5 = 32$ times more likely to occur among homologous sequences than by random chance.

alignment is the sum of all of the column scores (Fig. 5.1). The purpose of most multiple sequence alignment algorithms is to achieve maximum SP scores.

EXHAUSTIVE ALGORITHMS

As mentioned, there are exhaustive and heuristic approaches used in multiple sequence alignment. The exhaustive alignment method involves examining all possible aligned positions simultaneously. Similar to dynamic programming in pairwise alignment, which involves the use of a two-dimensional matrix to search for an optimal alignment, to use dynamic programming for multiple sequence alignment, extra dimensions are needed to take all possible ways of sequence matching into consideration. This means to establish a multidimensional search matrix. For instance, for three sequences, a three-dimensional matrix is required to account for all possible alignment scores. Back-tracking is applied through the three-dimensional matrix to find the highest scored path that represents the optimal alignment. For aligning N sequences, an N-dimensional matrix is needed to be filled with alignment scores. As the amount of computational time and memory space required increases exponentially with the number of sequences, it makes the method computationally prohibitive to use for a large data set. For this reason, full dynamic programming is limited to small datasets of less than ten short sequences. For the same reason, few multiple alignment programs employing this "brute force" approach are publicly available. A program called DCA, which uses some exhaustive components, is described below.

DCA (Divide-and-Conquer Alignment, http://bibiserv.techfak.uni-bielefeld.de/dca/) is a web-based program that is in fact semiexhaustive because certain steps of computation are reduced to heuristics. It works by breaking each of the sequences into two smaller sections. The breaking points are determined based on regional similarity of the sequences. If the sections are not short enough, further divisions are carried out. When the lengths of the sequences reach a predefined threshold, dynamic programming is applied for aligning each set of subsequences. The resulting short alignments are joined together head to tail to yield a multiple alignment of the entire length of all sequences. This algorithm provides an option of using a more heuristic procedure (fastDCA) to choose optimal cutting points so it can more rapidly handle a greater number of sequences. It performs global alignment and requires the input sequences to be of similar lengths and domain structures. Despite the use of heuristics, the program is still extremely computationally intensive and can handle only datasets of a very limited number of sequences.

HEURISTIC ALGORITHMS

Because the use of dynamic programming is not feasible for routine multiple sequence alignment, faster and heuristic algorithms have been developed. The heuristic algorithms fall into three categories: progressive alignment type, iterative alignment type, and block-based alignment type. Each type of algorithm is described in turn.

Progressive Alignment Method

Progressive alignment depends on the stepwise assembly of multiple alignment and is heuristic in nature. It speeds up the alignment of multiple sequences through a multi-step process. It first conducts pairwise alignments for each possible pair of sequences using the Needleman–Wunsch global alignment method and records these similarity scores from the pairwise comparisons. The scores can either be percent identity or similarity scores based on a particular substitution matrix. Both scores correlate with the evolutionary distances between sequences. The scores are then converted into evolutionary distances to generate a distance matrix for all the sequences involved. A simple phylogenetic analysis is then performed based on the distance matrix to group sequences based on pairwise distance scores. As a result, a phylogenetic tree is generated using the neighbor-joining method (see Chapter 11). The tree reflects evolutionary proximity among all the sequences.

It needs to be emphasized that the resulting tree is an approximate tree and does not have the rigor of a formally constructed phylogenetic tree (see Chapter 11). Nonetheless, the tree can be used as a guide for directing realignment of the sequences. For that reason, it is often referred to as a *guide tree*. According to the guide tree, the two most closely related sequences are first re-aligned using the Needleman–Wunsch algorithm. To align additional sequences, the two already aligned sequences are converted to a consensus sequence with gap positions fixed. The consensus is then treated as a single sequence in the subsequent step.

In the next step, the next closest sequence based on the guide tree is aligned with the consensus sequence using dynamic programming. More distant sequences or sequence profiles are subsequently added one at a time in accordance with their relative positions on the guide tree. After realignment with a new sequence using dynamic programming, a new consensus is derived, which is then used for the next round of alignment. The process is repeated until all the sequences are aligned (Fig. 5.2).

Probably the most well-known progressive alignment program is Clustal. Some of its important features are introduced next.

Clustal (www.ebi.ac.uk/clustalw/) is a progressive multiple alignment program available either as a stand-alone or on-line program. The stand-alone program, which runs on UNIX and Macintosh, has two variants, ClustalW and ClustalX. The W version provides a simple text-based interface and the X version provides a more user-friendly graphical interface.

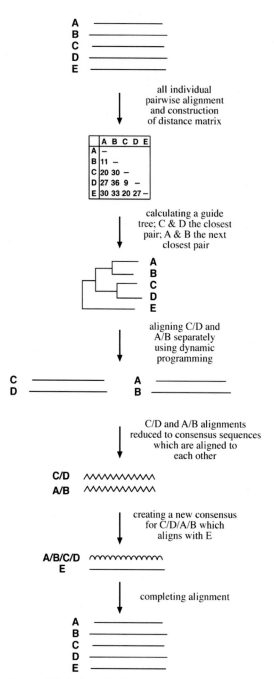

Figure 5.2: Schematic of a typical progressive alignment procedure (e.g., Clustal). Angled wavy lines represent consensus sequences for sequence pairs A/B and C/D. Curved wavy lines represent a consensus for A/B/C/D.

One of the most important features of this program is the flexibility of using substitution matrices. Clustal does not rely on a single substitution matrix. Instead, it applies different scoring matrices when aligning sequences, depending on degrees of similarity. The choice of a matrix depends on the evolutionary distances measured from the guide tree. For example, for closely related sequences that are aligned in the initial steps, Clustal automatically uses the BLOSUM62 or PAM120 matrix. When more divergent sequences are aligned in later steps of the progressive alignment, the BLOSUM45 or PAM250 matrices may be used instead.

Another feature of Clustal is the use of adjustable gap penalties that allow more insertions and deletions in regions that are outside the conserved domains, but fewer in conserved regions. For example, a gap near a series of hydrophobic residues carries more penalties than the one next to a series of hydrophilic or glycine residues, which are common in loop regions. In addition, gaps that are too close to one another can be penalized more than gaps occurring in isolated loci.

The program also applies a weighting scheme to increase the reliability of aligning divergent sequences (sequences with less than 25% identity). This is done by down-weighting redundant and closely related groups of sequences in the alignment by a certain factor. This scheme is useful in preventing similar sequences from dominating the alignment. The weight factor for each sequence is determined by its branch length on the guide tree. The branch lengths are normalized by how many times sequences share a basal branch from the root of the tree. The obtained value for each sequence is subsequently used to multiply the raw alignment scores of residues from that sequence so to achieve the goal of decreasing the matching scores of frequent characters in a multiple alignment and thereby increasing the ones of infrequent characters.

Drawbacks and Solutions

The progressive alignment method is not suitable for comparing sequences of different lengths because it is a global alignment–based method. As a result of the use of affine gap penalties (see Chapter 3), long gaps are not allowed, and, in some cases, this may limit the accuracy of the method. The final alignment result is also influenced by the order of sequence addition. Another major limitation is the "greedy" nature of the algorithm: it depends on initial pairwise alignment. Once gaps introduced in the early steps of alignment, they are fixed. Any errors made in these steps cannot be corrected. This problem of "once an error, always an error" can propagate throughout the entire alignment. In other words, the final alignment could be far from optimal. The problem can be more glaring when dealing with divergent sequences. To alleviate some of the limitations, a new generation of algorithms have been developed, which specifically target some of the problems of the Clustal program.

T-Coffee (Tree-based Consistency Objective Function for alignment Evaluation; www.ch.embnet.org/software/TCoffee.html) performs progressive sequence alignments as in Clustal. The main difference is that, in processing a query, T-Coffee performs both global and local pairwise alignment for all possible pairs involved. The global pairwise alignment is performed using the Clustal program. The local pairwise alignment is generated by the Lalign program, from which the top ten scored alignments are selected. The collection of local and global sequence alignments are pooled to form a library. The consistency of the alignments is evaluated. For every pair of residues in a pair of sequences, a consistency score is calculated for both global and local alignments. Each pairwise alignment is further aligned with a possible third sequence. The result is used to refine the original pairwise alignment based on a consistency criterion in a process known as *library extension.* Based on the refined pairwise alignments, a distance matrix is built to derive a guide tree, which is then used to direct a full multiple alignment using the progressive approach.

Because an optimal initial alignment is chosen from many alternative alignments, T-Coffee avoids the problem of getting stuck in the suboptimal alignment regions, which minimizes errors in the early stages of alignment assembly. Benchmark assessment has shown that T-Coffee indeed outperforms Clustal when aligning moderately divergent sequences. However, it is also slower than Clustal because of the extra computing time necessary for the calculation of consistency scores. T-Coffee provides a graphical output of the alignment results, with colored boxes to display degree of agreement in the alignment library for various sequence regions.

DbClustal (http://igbmc.u-strasbg.fr:8080/DbClustal/dbclustal.html) is a Clustal-based database search algorithm for protein sequences that combines local and global alignment features. It first performs a BLASTP search for a query sequence. The resulting sequence alignment pairs above a certain threshold are analyzed to obtain *anchor points*, which are common conserved regions, by using a program called Ballast. A global alignment is subsequently generated by Clustal, which is weighted toward the anchor points. Since the anchor points are derived from local alignments, this strategy minimizes errors caused by the global alignment. The resulting multiple alignment is further evaluated by NorMD, which removes unrelated or badly aligned sequences from the multiple alignment. Thus, the final alignment should be more accurate than using Clustal alone. It also allows the incorporation of very long gaps for insertions and terminal extensions.

Poa (Partial order alignments, www.bioinformatics.ucla.edu/poa/) is a progressive alignment program that does not rely on guide trees. Instead, the multiple alignment is assembled by adding sequences in the order they are given. Instead of using regular sequence consensus, a partial order graph is used to represent a growing multiple alignment, in which identical residues in a column are condensed to a node resembling a knot on a rope and divergent residues are allowed to remain as such, allowing the rope to "bubble" (Fig. 5.3). The graph profile preserves all the information from the original alignment. Each time a new sequence is added, it is aligned with every sequence within the partial order graph individually using the Smith–Waterman algorithm. This allows the formation of a modified graph model, which is then used for the next cycle of pairwise alignment. By building such a graph profile, the algorithm maintains the information of the original sequences and eliminates the problem of error fixation as in the Clustal alignment. Poa is local alignment-based and has been shown to produce more accurate alignments than Clustal. Another advantage of this

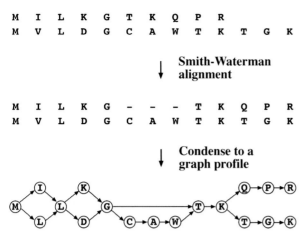

```
M   I   L   K   G   T   K   Q   P   R
M   V   L   D   G   C   A   W   T   K   T   G   K
```

↓ **Smith-Waterman alignment**

```
M   I   L   K   G   –   –   –   T   K   Q   P   R
M   V   L   D   G   C   A   W   T   K   T   G   K
```

↓ **Condense to a graph profile**

Figure 5.3: Conversion of a sequence alignment into a graphical profile in the Poa algorithm. Identical residues in the alignment are condensed as nodes in the partial order graph.

algorithm is its speed. It is reported to be able to align 5,000 sequences in 4 hours using a regular PC workstation. It is available both as an online program and as a stand-alone UNIX program.

PRALINE (http://ibivu.cs.vu.nl/programs/pralinewww/) is a web-based progressive alignment program. It first performs preprocessing of the input sequences by building profiles for each sequence. Profiles (see Chapter 6) can be interpreted as probability description of a multiple alignment. By default, the profiles are automatically generated using PSI-BLAST database searching (see Chapter 6). Each preprocessed profile is then used for multiple alignment using the progressive approach. However, this method does not use a guide tree in the successive enlargement of the alignment, but rather considers the closest neighbor to be joined to a larger alignment by comparing the profile scores. Because the profiles already incorporate information of distant relatives of each input sequence, this approach allows more accurate alignment of distantly related sequences in the original dataset. In addition, the program also has the feature to incorporate protein secondary structure information which is derived from state-of-the-art secondary structure prediction programs, such as PROF or SSPRO (see Chapter 14). The secondary structure information is used to modify the profile scores to help constrain sequence matching to the structured regions. PRALINE is perhaps the most sophisticated and accurate alignment program available. Because of the high complexity of the algorithm, its obvious drawback is the extremely slow computation.

Iterative Alignment

The iterative approach is based on the idea that an optimal solution can be found by repeatedly modifying existing suboptimal solutions. The procedure starts by producing a low-quality alignment and gradually improves it by iterative realignment through well-defined procedures until no more improvements in the alignment scores

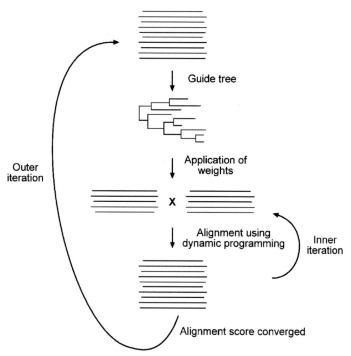

Figure 5.4: Schematic of iterative alignment procedure for PRRN, which involves two sets of iterations.

can be achieved. Because the order of the sequences used for alignment is different in each iteration, this method may alleviate the "greedy" problem of the progressive strategy. However, this method is also heuristic in nature and does not have guarantees for finding the optimal alignment. An example of iterative alignment is given below.

PRRN (http://prrn.ims.u-tokyo.ac.jp/) is a web-based program that uses a double-nested iterative strategy for multiple alignment. It performs multiple alignment through two sets of iterations: inner iteration and outer iteration. In the *outer iteration*, an initial random alignment is generated that is used to derive a UPGMA tree (see Chapter 11). Weights are subsequently applied to optimize the alignment. In the *inner iteration*, the sequences are randomly divided into two groups. Randomized alignment is used for each group in the initial cycle, after which the alignment positions in each group are fixed. The two groups, each treated as a single sequence, are then aligned to each other using global dynamic programming. The process is repeated through many cycles until the total SP score no longer increases. At this point, the resulting alignment is used to construct a new UPGMA tree. New weights are applied to optimize alignment scores. The newly optimized alignment is subject to further realignment in the inner iteration. This process is repeated over many cycles until there is no further improvement in the overall alignment scores (Fig. 5.4).

Block-Based Alignment

The progressive and iterative alignment strategies are largely global alignment based and may therefore fail to recognize conserved domains and motifs (see Chapter 7) among highly divergent sequences of varying lengths. For such divergent sequences that share only regional similarities, a local alignment based approach has to be used. The strategy identifies a block of ungapped alignment shared by all the sequences, hence, the block-based local alignment strategy. Two block-based alignment programs are introduced below.

DIALIGN2 (http://bioweb.pasteur.fr/seqanal/interfaces/dialign2.html) is a web-based program designed to detect local similarities. It does not apply gap penalties and thus is not sensitive to long gaps. The method breaks each of the sequences down to smaller segments and performs all possible pairwise alignments between the segments. High-scoring segments, called *blocks*, among different sequences are then compiled in a progressive manner to assemble a full multiple alignment. It places emphasis on block-to-block comparison rather than residue-to-residue comparison. The sequence regions between the blocks are left unaligned. The program has been shown to be especially suitable for aligning divergent sequences with only local similarity.

Match-Box (www.sciences.fundp.ac.be/biologie/bms/matchbox_submit.shtml) is a web-based server that also aims to identify conserved blocks (or boxes) among sequences. The program compares segments of every nine residues of all possible pairwise alignments. If the similarity of particular segments is above a certain threshold across all sequences, they are used as an anchor to assemble multiple alignments; residues between blocks are unaligned. The server requires the user to submit a set of sequences in the FASTA format and the results are returned by e-mail.

PRACTICAL ISSUES

Protein-Coding DNA Sequences

As mentioned in the Chapter 4, alignment at the protein level is more sensitive than at the DNA level. Sequence alignment directly at the DNA level can often result in frameshift errors because in DNA alignment gaps are introduced irrespective of codon boundaries. Therefore, in the process of achieving maximum sequence similarity at the DNA level, mismatches of genetic codons occur that violate the accepted evolutionary scenario that insertions or deletions occur in units of codons. The resulting alignment can thus be biologically unrealistic. The example in Figure 5.5 shows how such errors can occur when two sequences are being compared at the protein and DNA levels.

For that reason, sequence alignment at the protein level is much more informative for functional and evolutionary analysis. However, there are occasions when sequence

Protein alignment

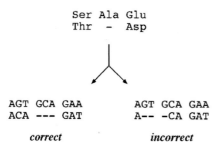

```
Ser Ala Glu
Thr  -  Asp
```

Figure 5.5: Comparison of alignment at the protein level and DNA level. The DNA alignment on the left is the correct one and consistent with amino acid sequence alignment, whereas the DNA alignment on the right, albeit more optimal in matching similar residues, is incorrect because it disregards the codon boundaries.

```
AGT GCA GAA        AGT GCA GAA
ACA --- GAT        A-- -CA GAT

  correct            incorrect
```

DNA alignment

alignment at the DNA level is often necessary, for example, in designing PCR primers and in constructing DNA-based molecular phylogenetic trees.

Because conducting alignment directly at the DNA level often leads to errors, DNA can be translated into an amino acid sequence before carrying out alignment to avoid the errors of inserting gaps within codon boundaries. After alignment of the protein sequences, the alignment can be converted back to DNA alignment while ensuring that codons of the DNA sequences line up based on corresponding amino acids. The following are two web-based programs that allow easy conversion from protein alignment to DNA alignment.

RevTrans (www.cbs.dtu.dk/services/RevTrans/) takes a set of DNA sequences, translates them, aligns the resulting protein sequences, and uses the protein alignment as a scaffold for constructing the corresponding DNA multiple alignment. It also allows the user to provide multiple protein alignment for greater control of the alignment process. PROTA2DNA (http://bioweb.pasteur.fr/seqanal/interfaces/protal2dna.html) aligns DNA sequences corresponding to a protein multiple alignment.

Editing

No matter how good an alignment program seems, the automated alignment often contains misaligned regions. It is imperative that the user check the alignment carefully for biological relevance and edit the alignment if necessary. This involves introducing or removing gaps to maximize biologically meaningful matches. Sometimes, portions that are ambiguously aligned and deemed to be incorrect have to deleted. In manual editing, empirical evidence or mere experience is needed to make corrections on an alignment. One can simply use a word processor to edit the text-based alignment. There are also dedicated software programs that assist in the process.

BioEdit (www.mbio.ncsu.edu/BioEdit/bioedit.html) is a multifunctional sequence alignment editor for Windows. It has a coloring scheme for nucleotide or amino acid residues that facilitates manual editing. In addition, it is able to do BLAST searches, plasmid drawing, and restriction mapping.

Rascal (http://igbmc.u-strasbg.fr/PipeAlign/Rascal/rascal.html) is a web-based program that automatically refines a multiple sequence alignment. It is part of the PipeAlign package. It is able to identify misaligned regions and realign them to improve the quality of the alignment. It works by dividing the input alignment into several regions horizontally and vertically to identify well-aligned and poorly aligned regions using an internal scoring scheme. Regions below certain thresholds are considered misaligned and are subsequently realigned using the progressive approach. The overall quality of the alignment is then reassessed. If necessary, certain regions are further realigned. The program also works in conjunction with NorMD, which validates the refined alignment and identifies potentially unrelated sequences for removal.

Format Conversion

In many bioinformatics analyses, in particular, phylogenetic analysis, it is often necessary to convert various formats of sequence alignments to the one acceptable by an application program. The task of format conversion requires a program to be able to read a multiple alignment in one format and rewrite it into another while maintaining the original alignment. This is a different task from simply converting the format of individual unaligned sequences. The BioEdit program mentioned is able to save an alignment in a variety of different formats. In addition, the Readseq program mentioned in Chapter 2 is able to perform format conversion of multiple alignment.

Readseq (http://iubio.bio.indiana.edu/cgi-bin/readseq.cgi/) is a web-based program that is able to do both simple sequence format conversion as well as alignment format conversions. The program can handle formats such as MSF, Phylip, Clustal, PAUP, and Pretty.

SUMMARY

Multiple sequence alignment is an essential technique in many bioinformatics applications. Many algorithms have been developed to achieve optimal alignment. Some programs are exhaustive in nature; some are heuristic. Because exhaustive programs are not feasible in most cases, heuristic programs are commonly used. These include progressive, iterative, and block-based approaches. The progressive method is a stepwise assembly of multiple alignment according to pairwise similarity. A prominent example is Clustal, which is characterized by adjustable scoring matrices and gap penalties as well as by the application of weighting schemes. The major shortcoming of the program is its "greediness," which relates to error fixation in the early steps of computation. To remedy the problem, T-Coffee and DbClustal have been developed that combine both global and local alignment to generate more sensitive alignment. Another improvement on the traditional progressive approach is to use graphic profiles, as in Poa, which eliminate the problem of error fixation. Praline is profile based and has the capacity to restrict alignment based on protein structure information and is thus much more accurate than Clustal. The iterative approach works by repetitive refinement of suboptimal alignments. The block-based method focuses on identifying regional similarities. It is important to keep in mind that no alignment program is absolutely guaranteed to find correct alignment, especially when the number of sequences is large and the divergence level is high. The alignment resulting from automated alignment programs often contains errors. The best approach is to perform alignment using a combination of multiple alignment programs. The alignment result can be further refined manually or using Rascal. Protein-encoding DNA sequences should preferably be aligned at the protein level first, after which the alignment can be converted back to DNA alignment.

FURTHER READING

Apostolico, A., and Giancarlo, R. 1998. Sequence alignment in molecular biology. *J. Comput. Biol.* 5:173–96.

Gaskell, G. J. 2000. Multiple sequence alignment tools on the Web. *Biotechniques* 29:60–2.

Gotoh, O. 1999. Multiple sequence alignment: Algorithms and applications. *Adv. Biophys.* 36:159–206.

Lecompte, O., Thompson, J. D., Plewniak, F., Thierry, J., and Poch, O. 2001. Multiple alignment of complete sequences (MACS) in the post-genomic era. *Gene* 270:17–30.

Morgenstern, B. 1999. DIALIGN 2: improvement of the segment-to-segment approach to multiple sequence alignment. *Bioinformatics* 15:211–8.

Morgenstern, B., Dress, A., and Werner T. 1996. Multiple DNA and protein sequence alignment based on segment-to-segment comparison. *Proc. Natl. Acad. Sci. U S A* 93:12098–103.

Mullan, L. J. 2002. Multiple sequence alignment – The gateway to further analysis. *Brief. Bioinform.* 3:303–5.

Nicholas, H. B. Jr., Ropelewski, A. J., and Deerfield, D. W. II. 2002. Strategies for multiple sequence alignment. *Biotechniques* 32:572–91.

Notredame, C. 2002. Recent progress in multiple sequence alignment: A survey. *Pharmacogenomics* 3:131–44.

Notredame, C., Higgins, D. G., and Heringa, J. 2000. T-Coffee: A novel method for fast and accurate multiple sequence alignment. *J. Mol. Biol.* 302:205–17.

Thompson, J. D., Higgins, D. G., and Gibson, T. J. 1994. CLUSTAL W: Improving the sensitivity of progressive multiple sequence alignment through sequence weighting, position-specific gap penalties and weight matrix choice. *Nucleic Acids Res.* 22:4673–80.

Thompson, J. D., Plewniak, F., and Poch, O. 1999. A comprehensive comparison of multiple sequence alignment programs. *Nucleic Acids Res.* 27:2682–90.

CHAPTER SIX

Profiles and Hidden Markov Models

One of the applications of multiple sequence alignments in identifying related sequences in databases is by construction of position-specific scoring matrices (PSSMs), profiles, and hidden Markov models (HMMs). These are statistical models that reflect the frequency information of amino acid or nucleotide residues in a multiple alignment. Thus, they can be treated as consensus for a given sequence family. However, the "consensus" is not exactly a single sequence, but rather a model that captures not only the observed frequencies but also predicted frequencies of unobserved characters. The purpose of establishing the mathematical models is to allow partial matches with a query sequence so they can be used to detect more distant members of the same sequence family, resulting in an increased sensitivity of database searches. This chapter covers the basics of these statistical models followed by discussion of their applications.

POSITION-SPECIFIC SCORING MATRICES

A PSSM is defined as a table that contains probability information of amino acids or nucleotides at each position of an ungapped multiple sequence alignment. The matrix resembles the substitution matrices discussed in Chapter 3, but is more complex in that it contains positional information of the alignment. In such a table, the rows represent residue positions of a particular multiple alignment and the columns represent the names of residues or vice versa (Fig. 6.1). The values in the table represent log odds scores of the residues calculated from the multiple alignment.

To construct a matrix, raw frequencies of each residue at each column position from a multiple alignment are first counted. The frequencies are normalized by dividing positional frequencies of each residue by overall frequencies so that the scores are length and composition independent. The values are converted to the probability values by taking to the logarithm (normally to the base of 2). In this way, the matrix values become log odds scores of residues occurring at each alignment position. In this matrix, a positive score represents identical residue or similar residue match; a negative score represents a nonconserved sequence match.

This constructed matrix can be considered a distilled representation for the entire group of related sequences, providing a quantitative description of the degree of sequence conservation at each position of a multiple alignment. The probabilistic model can then be used like a single sequence for database searching and alignment

Position	1 2 3 4 5 6
Sequence 1	**ATGTCG**
Sequence 2	**AAGACT**
Sequence 3	**TACTCA**
Sequence 4	**CGGAGG**
Sequence 5	**AACCTG**

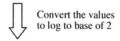 Convert multiple alignment to a raw frequency table

Pos.	1	2	3	4	5	6	Overall freq.
A	0.6	0.6	—	0.4	—	0.2	0.30
T	0.2	0.2	—	0.4	0.2	0.2	0.20
G	—	0.2	0.6	—	0.2	0.6	0.27
C	0.2	—	0.4	0.2	0.6	—	0.23

Normalize the values by dividing them by overall freq.

Pos.	1	2	3	4	5	6	Overall freq.
A	2.0	2.0	—	1.33	—	0.67	0.30
T	1.0	1.0	—	2.0	1.0	1.0	0.20
G	—	0.74	2.22	—	0.74	2.22	0.27
C	0.87	—	1.74	0.87	2.61	—	0.23

Convert the values to log to base of 2

Pos.	1	2	3	4	5	6
A	1.0	1.0	—	0.41	—	-0.58
T	0.0	0.0	—	1.0	0.0	0.0
G	—	-0.43	1.15	—	-0.43	1.15
C	-0.2	—	0.8	-0.2	1.38	—

Figure 6.1: Example of construction of a PSSM from a multiple alignment of nucleotide sequences. The process involves counting raw frequencies of each nucleotide at each column position, normalization of the frequencies by dividing positional frequencies of each nucleotide by overall frequencies and converting the values to log odds scores.

or can be used to test how well a particular target sequence fits into the sequence group.

For example, given the matrix shown in Figure 6.1, which is derived from a DNA multiple alignment, one can ask the question, how well does the new sequence AACTCG fit into the matrix? To answer the question, the probability values of the sequence at

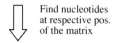

Match **AACTCG** in the matrix

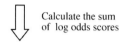

Pos.	1	2	3	4	5	6
A	1.0	1.0	—	0.41	—	-0.58
T	0.0	0.0	—	1.0	0.0	0.0
G	—	-0.43	1.15	—	-0.43	1.15
C	-0.2	—	0.8	-0.2	1.38	—

Calculate the sum
of log odds scores

1.0 + 1.0 + 0.8 + 1.0 + 1.38 + 1.15 = 6.33

Figure 6.2: Example of calculation of how well a new sequence fits into the PSSM produced in Figure 6.1. The matching positions for the new sequence AACTCG are circled in the matrix.

respective positions of the matrix can be added up to produce the sum of the scores (Fig. 6.2). In this case, the total match score for the sequence is 6.33. Because the matrix values have been taken to the logarithm to the base of 2, the score can be interpreted as the probability of the sequence fitting the matrix as $2^{6.33}$, or 80 times more likely than by random chance. Consequently, the new sequence can be confidently classified as a member of the sequence family.

The probability values in a PSSM depend on the number of sequences used to compile the matrix. Because the matrix is often constructed from the alignment of an insufficient number of closely related sequences, to increase the predictive power of the model, a weighting scheme similar to the one used in the Clustal algorithm (see Chapter 5) is used that downweights overrepresented, closely related sequences and upweights underrepresented and divergent ones, so that more divergent sequences can be included. Application of such a weighting scheme makes the matrix less biased and able to detect more distantly related sequences.

PROFILES

Actual multiple sequence alignments often contain gaps of varying lengths. When gap penalty information is included in the matrix construction, a profile is created. In other words, a profile is a PSSM with penalty information regarding insertions and deletions for a sequence family. However, in the literature, *profile* is often used interchangeably with PSSM, even though the two terms in fact have subtle but significant differences.

As in sequence alignment, gap penalty scores in a profile matrix are often arbitrarily set. Thus, to achieve an optimal alignment between a query sequence and a profile, a series of gap parameters have to be tested.

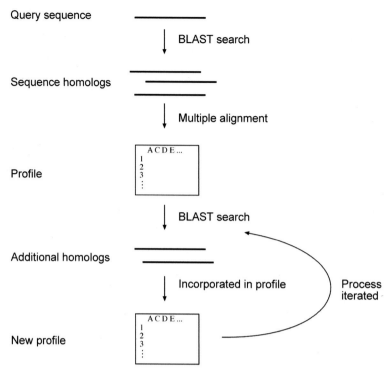

Query sequence

BLAST search

Sequence homologs

Multiple alignment

Profile

ACDE...
1
2
3
:

BLAST search

Additional homologs

Incorporated in profile   Process
                                           iterated

New profile

ACDE...
1
2
3
:

Figure 6.3: Schematic diagram of PSI-BLAST, an iterative process used to identify distant homologs.

PSI-BLAST

Profiles can be used in database searching to find remote sequence homologs. However, to manually construct a profile from a multiple alignment and calculate scores for matching sequences from a large database is tedious and involves significant expertise. It is desirable to have a program to establish profiles and use them to search against sequence databases in an automated way. Such a program is fortunately available as PSI-BLAST, a variant of BLAST, provided by the National Center for Biotechnology Information.

Position-specific iterated BLAST (PSI-BLAST) builds profiles and performs database searches in an iterative fashion. It first uses a single query protein sequence to perform a normal BLASTP search to generate initial similarity hits. The high-scoring hits are used to build a multiple sequence alignment, from which a profile is created. The profile is then used in the second round of searching to identify more members of the same family that may match with the profile. When new sequence hits are identified, they are combined with the previous multiple alignment to generate a new profile, which is in turn used in subsequent cycles of database searching. The process is repeated until no new sequence hits are found (Fig. 6.3).

The main feature of PSI-BLAST is that profiles are constructed automatically and are fine-tuned in each successive cycle. The program also employs a weighting scheme in the profile construction in each iteration to increase sensitivity. Another measure

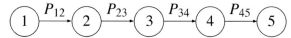

Figure 6.4: A simple representation of a Markov chain, which consists of a linear chain of events or states (numbered) linked by transition probability values between events (states).

of PSI-BLAST to increase sensitivity is the use of pseudocounts (to be discussed next) to provide extra weight to unobserved residues to make the profile more inclusive.

The optimization of profile parameters makes PSI-BLAST a very sensitive search strategy to detect weak but biologically significant similarities between sequences. It has been estimated that the profile-based approach is able to identify three times more homologs than regular BLAST, which mainly fall within the range of less than 30% sequence identity.

However, the high sensitivity of PSI-BLAST is also its pitfall; it is associated with low selectivity caused by the false-positives generated in the automated profile construction process. If unrelated sequences are erroneously included, profiles become biased. This allows further errors to be incorporated in subsequent cycles. This problem is known as *profile drift*. A partial solution to this problem is to let the user visually inspect results in each iteration and reject certain sequences that are known to be unrelated based on external knowledge. In addition, it is also prudent to conduct only a limited number of cycles instead of reaching full convergence. Typically, three to five iterations of PSI-BLAST are sufficient to find most distant homologs at the sequence level.

MARKOV MODEL AND HIDDEN MARKOV MODEL

Markov Model

A more efficient way of computing matching scores between a sequence and a sequence profile is through the use of HMMs, which are statistical models originally developed for use in speech recognition. This statistical tool was subsequently found to be ideal for describing sequence alignments. To understand HMMs, it is important to have some general knowledge of Markov models.

A Markov model, also known as *Markov chain,* describes a sequence of events that occur one after another in a chain. Each event determines the probability of the next event (Fig. 6.4). A Markov chain can be considered as a process that moves in one direction from one state to the next with a certain probability, which is known as *transition probability.* A good example of a Markov model is the signal change of traffic lights in which the state of the current signal depends on the state of the previous signal (e.g., green light switches on after red light, which switches on after yellow light).

Biological sequences written as strings of letters can be described by Markov chains as well; each letter representing a state is linked together with transitional probability values. The description of biological sequences using Markov chains allows the calculation of probability values for a given residue according to the unique distribution frequencies of nucleotides or amino acids.

There are several different types of Markov models used to describe datasets of different complexities. In each type of Markov model, different mathematical solutions are derived. A *zero-order Markov model* describes the probability of the current state independent of the previous state. This is typical for a random sequence, in which every residue occurs with an equal frequency. A *first-order Markov model* describes the probability of the current state being determined by the previous state. This corresponds to the unique frequencies of two linked residues (dimer) occurring simultaneously. Similarly, a *second-order Markov model* describes the situation in which the probability of the current state is determined by the previous two states. This corresponds to the unique trimer frequencies (three linked residues occurring simultaneously as in the case of a codon) in biological sequences. For example, in a protein-coding sequence, the frequency of unique trimers should be different from that in a noncoding or random sequence. This discrepancy can be described by the second-order Markov model. In addition, even higher orders of Markov models are available for biological sequence analysis (see Chapter 8).

Hidden Markov Model

In a Markov model, all states in a linear sequence are directly observable. In some situations, some nonobserved factors influence state transition calculations. To include such factors in calculations requires the use of more sophisticated models: HMMs. An HMM combines two or more Markov chains with only one chain consisting of observed states and the other chains made up of unobserved (or "hidden") states that influence the outcome of the observed states (Fig. 6.5). For example, in a gapped alignment, gaps do not correspond to any residues and are considered as unobservable states. However, gaps indirectly influence the transition probability of the observed states.

In an HMM, as in a Markov chain, the probability going from one state to another state is the *transition probability.* Each state may be composed of a number elements or symbols. For nucleotide sequences, there are four possible symbols – A, T, G, and C – in each state. For amino acid sequences, there twenty symbols. The probability value associated with each symbol in each state is called *emission probability.* To calculate the total probability of a particular path of the model, both transition and emission probabilities linking all the "hidden" as well as observed states need to be taken into account. Figure 6.6 provides a simple example of how to use two states of a partial HMM to represent (or generate) a sequence.

To develop a functional HMM that can be used to best represent a sequence alignment, the statistical model has to be "trained," which is a process to obtain the optimal statistical parameters in the HMM. The training process involves calculation of the frequencies of residues in each column in the multiple alignment built from a set of related sequences. The frequency values are used to fill the emission and transition probability values in the model. Similar to the construction of a PSSM, once an HMM is established based on the training sequences, it can be used to determine how well an unknown sequence matches the model.

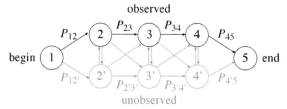

Figure 6.5: A simplified HMM involving two interconnected Markov chains with observed states and common "begin" and "end" states. The observed states are colored in black and unobserved states in grey. The transition probability values between observed states or between unobserved states are labeled. The probability values between the observed and hidden states are unlabeled.

To use an HMM to describe gapped multiple sequence alignment, a character in the alignment can be in one of three states, match (a mismatch can be quantitatively expressed as low probability of a match), insertion, and deletion. "Match" states are observed states, whereas the insertions and deletions, designated as "insert" and "delete" states, are "hidden" as far as transitions between match states are concerned.

To represent the three states in an HMM, a special graphical representation has been traditionally used. In this representation, transitions from state to state proceed from left to right via various paths through the model representing all possible combinations of matches, mismatches, and gaps to generate an alignment. Each path is associated with a unique probability value (Fig. 6.7).

The circles on top of the insert state indicate self-looping, which allows insertions of any number of residues to fit into the model. In addition, there is a beginning state and an end state. There are many possible combinations of states or paths to travel through the model, from the beginning state to the end state. Each path generates a unique sequence, which includes insertions or deletions, with a probability value. For a given HMM, there may be only one optimal path that generates the most probable sequence representing an optimal sequence family alignment.

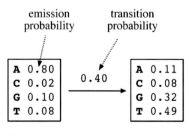

STATE 1   STATE 2

Figure 6.6: Graphic illustration of a simplified partial HMM for DNA sequences with emission and transition probability values. Both probability values are used to calculate the total probability of a particular path of the model. For example, to generate the sequence AG, the model has to progress from A from STATE 1 to G in STATE 2, the probability of this path is $0.80 \times 0.40 \times 0.32 = 0.102$. Obviously, there are $4 \times 4 = 16$ different sequences this simple model can generate. The one that has the highest probability is AT.

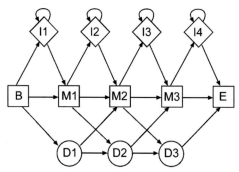

Figure 6.7: A typical architecture of a hidden Markov model representing a multiple sequence alignment. Squares indicate match states (M), diamonds insert states (I), and circles delete states (D). The beginning and end of the match states are indicated by B and E, respectively. The states are connected by arrowed lines with transition probability values.

Score Computation

To find an optimal path within an HMM that matches a query sequence with the highest probability, a matrix of probability values for every state at every residue position needs to be constructed (Fig. 6.8). Several algorithms are available to determine the most probable path for this matrix. One such algorithm is the Viterbi algorithm, which works in a similar fashion as in dynamic programming for sequence alignment (see Chapter 3). It constructs a matrix with the maximum emission probability values of all the symbols in a state multiplied by the transition probability for that state. It then uses a trace-back procedure going from the lower right corner to the upper left corner to find the path with the highest values in the matrix. Another frequently used algorithm is the forward algorithm, which constructs a matrix using the sum of multiple emission states instead of the maximum, and calculates the most likely path from the upper left corner of the matrix to the lower right corner. In other words, it proceeds in an opposite direction to the Viterbi algorithm. In practice, both methods have equal performance in finding an optimal alignment.

	M_0	I_1	D_1	M_1	I_2	D_2	M_2	I_3	D_3	M_3
S_0	0.26	0.00	0.00	0.00	0.00	0.00	0.00	0.00	0.00	0.00
S_1	0.00	0.00	0.00	0.19	0.03	0.00	0.00	0.00	0.00	0.00
S_2	0.00	0.00	0.00	0.00	0.00	0.00	0.37	0.01	0.00	0.00
S_3	0.00	0.00	0.00	0.00	0.00	0.00	0.00	0.00	0.00	0.42

Figure 6.8: Score matrix constructed from a simple HMM with the optimal score path chosen by the Vertibi algorithm. M, I, and D represent match, insert, and delete states. State 0 (S_0) is the beginning state; S_3 is the end state. The probability value for each state (S) is the maximum emission probability of each state multiplied by the transition probability to that state. The Viterbi algorithm works in a trace-back procedure by traveling from the lower right corner to the upper left corner to find the highest scored path.

In HMM construction, as in profile construction, there is always an issue of limited sampling size, which causes overrepresentation of observed characters while ignoring the unobserved characters. This problem is known as *overfitting*. To make sure that the HMM model generated from the training set is representative of not only the training set sequences, but also of other members of the family not yet sampled, some level of "smoothing" is needed, but not to the extent that it distorts the observed sequence patterns in the training set. This smoothing method is called *regularization*.

One of the regularization methods involves adding an extra amino acid called a *pseudocount,* which is an artificial value for an amino acid that is not observed in the training set. When probabilities are computed, pseudocounts are treated like real counts. The modified profile enhances the predictive power of the profile and HMM.

To automate the process of regularization, various mathematical models have been developed to simulate the amino acid distribution in a sequence alignment. These preconstructed models aim to correct the observed amino acid distribution derived from a limited sequence alignment. A well-known statistical model for this purpose is the *Dirichlet mixture*, derived from prior distributions of amino acids found in a large number of conserved protein domains. This is essentially a weighting scheme that gives pseudocounts to amino acids and makes the distribution more reasonable.

Applications

An advantage of HMMs over profiles is that the probability modeling in HMMs has more predictive power. This is because an HMM is able to differentiate between insertion and deletion states, whereas in profile calculation, a single gap penalty score that is often subjectively determined represents either an insertion or deletion. Because the handling of insertions and deletions is a major problem in recognizing highly divergent sequences, HMMs are therefore more robust in describing subtle patterns of a sequence family than standard profile analysis.

HMMs are very useful in many aspects of bioinformatics. Although an HMM has to be trained based on multiple sequence alignment, once it is trained, it can in turn be used for the construction of multiple alignment of related sequences. HMMs can be used for database searching to detect distant sequence homologs. As to be discussed in Chapter 7, HMMs are also used in protein family classification through motif and pattern identification. Advanced gene and promoter prediction (see Chapters 8 and 9), transmembrane protein prediction (see Chapter 14), as well as protein fold recognition (see Chapter 15), also employ HMMs.

HMMer (http://hmmer.wustl.edu/) is an HMM package for sequence analysis available in the public domain. It is a suite of UNIX programs that work in conjunction to perform an HMM analysis. It creates profile HMMs from a sequence alignment using the subprogram *hmmbuild*. Another subprogram, *hmmcalibrate*, calibrates search statistics for the newly generated HMMs by fitting the scores to the Gumble extreme value distribution (see Chapter 3). The subprogram *hmmemit* generates probability distribution based on profile HMMs. The program *hmmsearch* searches a sequence database for matching sequences with a profile HMM.

SUMMARY

PSSMs, profiles, and HMMs are statistical models that represent the consensus of a sequence family. Because they allow partial matches, they are more sensitive in detecting remote homologs than regular sequence alignment methods. A PSSM by definition is a scoring table derived from ungapped multiple sequence alignment. A profile is similar to PSSM, but also includes probability information for gaps derived from gapped multiple alignment. An HMM is similar to profiles but differentiates insertions from deletions in handling gaps.

The probability calculation in HMMs is more complex than in profiles. It involves traveling through a special architecture of various observed and hidden states to describe a gapped multiple sequence alignment. As a result of flexible handling of gaps, HMM is more sensitive than profiles in detecting remote sequence homologs. All three types of models require training because the statistical parameters have to be determined according to alignment of sequence families. PSI-BLAST is an example of the practical application of profiles in database searches to detect remote homologs in a database. The automated nature of PSI-BLAST has stimulated a widespread use of profile-based homolog detection.

FURTHER READING

Altschul, S. F., and Koonin, E. V. 1998. Iterated profile searches with PSI-BLAST – A tool for discovery in protein databases. *Trends Biochem. Sci.* 23:444–7.

Baldi, P., Chauvin, Y., Hunkapiller, T., and McClure M. A. 1994. Hidden Markov models of biological primary sequence information. *Proc. Natl. Acad. Sci. U S A* 91:1059–63.

Eddy, S. R. 1996. Hidden Markov models. *Curr. Opin. Struct. Biol.* 6:361–5.

Eddy, S. R. 1998. Profile hidden Markov models. *Bioinformatics* 14:755–63.

Jones, D. T., and Swindells, M. B. 2002. Getting the most from PSI-BLAST. *Trends Biochem. Sci.* 27:161–4.

Panchenko, A. R., and Bryant, S. H. 2002. A comparison of position-specific score matrices based on sequence and structure alignments. *Protein Sci.* 11:361–70.

CHAPTER SEVEN

Protein Motif and Domain Prediction

An important aspect of biological sequence characterization is identification of motifs and domains. It is an important way to characterize unknown protein functions because a newly obtained protein sequence often lacks significant similarity with database sequences of known functions over their entire length, which makes functional assignment difficult. In this case, biologists can gain insight of the protein function based on identification of short consensus sequences related to known functions. These consensus sequence patterns are termed *motifs* and *domains*.

A *motif* is a short conserved sequence pattern associated with distinct functions of a protein or DNA. It is often associated with a distinct structural site performing a particular function. A typical motif, such as a Zn-finger motif, is ten to twenty amino acids long. A *domain* is also a conserved sequence pattern, defined as an independent functional and structural unit. Domains are normally longer than motifs. A domain consists of more than 40 residues and up to 700 residues, with an average length of 100 residues. A domain may or may not include motifs within its boundaries. Examples of domains include transmembrane domains and ligand-binding domains.

Motifs and domains are evolutionarily more conserved than other regions of a protein and tend to evolve as units, which are gained, lost, or shuffled as one module. The identification of motifs and domains in proteins is an important aspect of the classification of protein sequences and functional annotation. Because of evolutionary divergence, functional relationships between proteins often cannot be distinguished through simple BLAST or FASTA database searches. In addition, proteins or enzymes often perform multiple functions that cannot be fully described using a single annotation through sequence database searching. To resolve these issues, identification of the motifs and domains becomes very useful.

Identification of motifs and domains heavily relies on multiple sequence alignment as well as profile and hidden Markov model (HMM) construction (see Chapters 5 and 6). This chapter focuses on some fundamental issues relating to protein motif and domain databases as well as classification of protein sequences using full length sequences. In addition, computational tools for discovering subtle motifs from divergent sequences are also introduced.

IDENTIFICATION OF MOTIFS AND DOMAINS IN MULTIPLE SEQUENCE ALIGNMENT

Motifs and domains are first constructed from multiple alignment of related sequences. Based on the multiple sequence alignment, commonly conserved regions can be identified. The regions considered motifs and domains then serve as diagnostic features for a protein family. The consensus sequence information of motifs and domains can be stored in a database for later searches of the presence of similar sequence patterns from unknown sequences. By scanning the presence of known motifs or domains in a query sequence, associated functional features in a query sequence can be revealed rapidly, which is often not possible by simply matching full-length sequences in the primary databases.

There are generally two approaches to representing the consensus information of motifs and domains. The first is to reduce the multiple sequence alignment from which motifs or domains are derived to a consensus sequence pattern, known as a *regular expression*. For example, the protein phosphorylation motif can be expressed as [ST]-X-[RK]. The second approach is to use a statistical model such as a profile or HMM to include probability information derived from the multiple sequence alignment.

MOTIF AND DOMAIN DATABASES USING REGULAR EXPRESSIONS

A regular expression is a concise way of representing a sequence family by a string of characters. When domains and motifs are written as regular expressions, the following basic rules to describe a sequence pattern are used: When a position is restricted to a single conserved amino acid residue, it is indicated as such using the standard, one-letter code. When a position represents multiple alternative conserved residues, the residues to be included are placed within brackets. If the position excludes certain residues, residues to be excluded are placed in curly braces; nonspecific residues present in a given position in the pattern are indicated by an X; if a sequence element within the pattern is repetitive, the number of pattern repetitions is indicated within parentheses; and each position is linked by a hyphen. For example, a motif written as E-X(2)-[FHM]-X(4)-{P}-L can be interpreted as an E followed by two unspecific residues which are followed by an F, or H or M residue which is followed by another four unspecific residues followed by a non-P residue and a final L.

There are two mechanisms of matching regular expressions with a query sequence. One is exact matching and the other is fuzzy matching. In exact matching, there must be a strict match of sequence patterns. Any variations in the query sequence from the predefined patterns are not allowed. Searching a motif database using this approach results in either a match or nonmatch. This way of searching has a good chance of missing truly relevant motifs that have slight variations, thus generating false-negative results. Another limitation with using exact matching is that, as new sequences of a motif are being accumulated, the rigid regular expression tends to become obsolete if not updated regularly to reflect the changes.

Fuzzy matches, also called *approximate matches,* provide more permissive matching by allowing more flexible matching of residues of similar biochemical properties. For example, if an original alignment only contains phenylalanine at a particular position, fuzzy matching allows other aromatic residues (including unobserved tyrosine and tryptophan) in a sequence to match with the expression. This method is able to include more variant forms of a motif with a conserved function. However, associated with the more relaxed matching is the inevitable increase of the noise level and false positives. This is especially the case for short motifs. This is partly because the rule of matching is based on assumptions not actual observations.

Motif databases have commonly been used to classify proteins, provide functional assignment, and identify structural and evolutionary relationships. Two databases that mainly employ regular expressions for the purpose of searching sequence patterns are described next.

PROSITE (www.expasy.ch/prosite/) is the first established sequence pattern database and is still widely used. It primarily uses a single consensus pattern or "sequence signature" to characterize a protein function and a sequence family. The consensus sequence patterns are derived from conserved regions of protein sequence alignments and are represented with regular expressions. The functional information of these patterns is primarily based on published literature. To search the database with a query sequence, PROSITE uses exact matches to the sequence patterns. In addition to regular expressions, the database also constructs profiles to complement some of the sequence patterns. The major pitfall with the PROSITE patterns is that some of the sequence patterns are too short to be specific. The problem with these short sequence patterns is that the resulting match is very likely to be a result of random events. Another problem is that the database is relatively small and motif searches often yield no results when there are in fact true motif matches present (false negatives). Overall, PROSITE has a greater than 20% error rate. Thus, either a match or nonmatch in PROSITE should be treated with caution.

Emotif (http://motif.stanford.edu/emotif/emotif-search.html) is a motif database that uses multiple sequence alignments from both the BLOCKS and PRINTS databases with an alignment collection much larger than PROSITE. It identifies patterns by allowing fuzzy matching of regular expressions. Therefore, it produces fewer false negatives than PROSITE.

MOTIF AND DOMAIN DATABASES USING STATISTICAL MODELS

The major limitation of regular expressions is that this method does not take into account sequence probability information about the multiple alignment from which it is modeled. If a regular expression is derived from an incomplete sequence set, it has less predictive power because many more sequences with the same type of motifs are not represented.

Unlike regular expressions, position-specific scoring matrices (PSSMs), profiles, and HMMs (see Chapter 6) preserve the sequence information from a multiple sequence alignment and express it with probabilistic models. In addition, these statistical models allow partial matches and compensate for unobserved sequence patterns using pseudocounts. Thus, these statistical models have stronger predictive power than the regular expression based approach, even when they are derived from a limited set of sequences. Using such a powerful scoring system can enhance the sensitivity of motif discovery and detect more divergent but truly related sequences.

The following programs mainly use the profile/HMM method extensively for sequence pattern construction.

PRINTS (http://bioinf.man.ac.uk/dbbrowser/PRINTS/) is a protein fingerprint database containing ungapped, manually curated alignments corresponding to the most conserved regions among related sequences. This program breaks down a motif into even smaller nonoverlapping units called *fingerprints,* which are represented by unweighted PSSMs. To define a motif, at least a majority of fingerprints are required to match with a query sequence. A query that has simultaneous high-scoring matches to a majority of fingerprints belonging to a motif is a good indication of containing the functional motif. The drawbacks of PRINTS are 1) the difficulty to recognize short motifs when they reach the size of single fingerprints and 2) a relatively small database, which restricts detection of many motifs.

BLOCKS (http://blocks.fhcrc.org/blocks) is a database that uses multiple alignments derived from the most conserved, ungapped regions of homologous protein sequences. The alignments are automatically generated using the same data sets used for deriving the BLOSUM matrices (see Chapter 3). The derived ungapped alignments are called *blocks.* The blocks, which are usually longer than motifs, are subsequently converted to PSSMs. A weighting scheme and pseudocounts are subsequently applied to the PSSMs to account for underrepresented and unobserved residues in alignments. Because blocks often encompass motifs, the functional annotation of blocks is thus consistent with that for the motifs. A query sequence can be used to align with precomputed profiles in the database to select the highest scored matches. Because of the use of the weighting scheme, the signal-to-noise ratio is improved relative to PRINTS.

ProDom (http://prodes.toulouse.inra.fr/prodom/2002.1/html/form.php) is a domain database generated from sequences in the SWISSPROT and TrEMBL databases (see Chapter 2). The domains are built using recursive iterations of PSI-BLAST. The automatically generated sequence pattern database is designed to be an exhaustive collection of domains without their functions necessarily being known.

Pfam (http://pfam.wustl.edu/hmmsearch.shtml) is a database with protein domain alignments derived from sequences in SWISSPROT and TrEMBL. Each motif or domain is represented by an HMM profile generated from the seed alignment of a number of conserved homologous proteins. Since the probability scoring mechanism is more complex in HMM than in a profile-based approach (see Chapter 6), the use of HMM yields further increases in sensitivity of the database matches. The Pfam database is composed of two parts, Pfam-A and Pfam-B. Pfam-A involves manual alignments and Pfam-B, automatic alignment in a way similar to ProDom. The functional annotation of motifs in Pfam-A is often related to that in PROSITE. Pfam-B only contains sequence families not covered in Pfam-A. Because of the automatic nature, Pfam-B has a much larger coverage but is also more error prone because some HMMs are generated from unrelated sequences.

SMART (Simple Modular Architecture Research Tool; http://smart.embl-heidel berg.de/) contains HMM profiles constructed from manually refined protein domain alignments. Alignments in the database are built based on tertiary structures whenever available or based on PSI-BLAST profiles. Alignments are further checked and refined by human annotators before HMM profile construction. Protein functions are also manually curated. Thus, the database may be of better quality than Pfam with more extensive functional annotations. Compared to Pfam, the SMART database contains an independent collection of HMMs, with emphasis on signaling, extracellular, and chromatin-associated motifs and domains. Sequence searching in this database produces a graphical output of domains with well-annotated information with respect to cellular localization, functional sites, superfamily, and tertiary structure.

InterPro (www.ebi.ac.uk/interpro/) is an integrated pattern database designed to unify multiple databases for protein domains and functional sites. The database integrates information from PROSITE, Pfam, PRINTS, ProDom, and SMART databases. The sequence patterns from the five databases are further processed. Only overlapping motifs and domains in a protein sequence derived by all five databases are included. The InterPro entries use a combination of regular expressions, fingerprints, profiles, and HMMs in pattern matching. However, an InterPro search does not obviate the need to search other databases because of its unique criteria of motif inclusion and thus may have lower sensitivity than exhaustive searches in individual databases. A popular feature of this database is a graphical output that summarizes motif matches and has links to more detailed information.

Reverse PSI-BLAST (RPS-BLAST; www.ncbi.nlm.nih.gov/BLAST/) is a web-based server that uses a query sequence to search against a pre-computed profile database generated by PSI-BLAST. This is opposite of PSI-BLAST that builds profiles from matched database sequences, hence a "reverse" process. It performs only one iteration of regular BLAST searching against a database of PSI-BLAST profiles to find the high-scoring gapped matches.

CDART (Conserved Domain Architecture; www.ncbi.nlm.nih.gov/BLAST/) is a domain search program that combines the results from RPS-BLAST, SMART, and Pfam. The resulting domain architecture of a query sequence can be graphically presented along with related sequences. The program is now an integral part of the regular BLAST search function. As with InterPro, CDART is not a substitute for individual database searches because it often misses certain features that can be found in SMART and Pfam.

Caveats

Because of underlying differences in database construction and patten matching methods, each pattern database has its strengths and weaknesses. The coverage of these databases overlaps only to a certain extent. If a particular motif search returns nothing from a particular database search, it does not mean that the sequence contains no patterns. It may be a result of the limited coverage of a particular database or an error in the database. Also keep in mind that there are many misannotated sequences in databases, which hinder the detection of true motifs. Alternatively, the nonmatch may be a result of insensitive sequence matching methods. Therefore, it is advisable to use a combination of multiple databases in motif searching to get the greatest coverage and consensus functional information. In cases of inconsistency of results when using several different databases, a majority rule can be a good way to discriminate between the matches.

PROTEIN FAMILY DATABASES

The databases mentioned classify proteins based on the presence of motifs and domains. Another way of classifying proteins is based on near full-length sequence comparison. The latter classification scheme requires clustering of proteins based on overall sequence similarities. The clustering criteria include statistical scores in sequence alignments or orthologous relationships. Protein family databases derived from this approach do not depend on the presence of particular sequence signatures and thus can be more comprehensive. However, the disadvantage is that there are more ambiguity and artifacts in protein classification. Two examples of protein family databases based on clustering and phylogenetic classification are presented.

COG (Cluster of Orthologous Groups; www.ncbi.nlm.nih.gov/COG/) is a protein family database based on phylogenetic classification. It is constructed by comparing protein sequences encoded in forty-three completely sequenced genomes, which are mainly from prokaryotes, representing thirty major phylogenetic lineages. Through all-against-all sequence comparisons among the genomes, orthologous proteins shared by three or more lineages are identified and clustered together as orthologous groups. Each group should have at least one representative from Archea, Bacteria, and Eukarya. Orthologs are included in a cluster as long as they satisfy the criterion of being the mutual best hits in BLAST searches among the genomes.

Because orthologous proteins shared by three or more lineages are considered to have descended through a vertical evolutionary scenario, if the function of one of the members is known, functionality of other members can be assigned. Similarly, a query sequence can be assigned function if it has significant similarity matches with any member of the cluster. Currently, there are 4,873 clusters in the COG databases derived from unicellular organisms. The interface for sequence searching in the COG database is the COGnitor program, which is based on gapped

BLAST. An eukaryotic version of the program is now available, known as KOG (www.ncbi.nlm.nih.gov/COG/new/kognitor.html).

ProtoNet (www.protonet.cs.huji.ac.il/) is a database of clusters of homologous proteins similar to COG. Orthologous protein sequences in the SWISSPROT database are clustered based on pairwise sequence comparisons between all possible protein pairs using BLAST. Protein relatedness is defined by the *E*-values from the BLAST alignments. This produces different levels of protein similarity, yielding a hierarchical organization of protein groups. The most closely related sequences are grouped into the lowest level clusters. More distant protein groups are merged into higher levels of clusters. The outcome of this cluster merging is a tree-like structure of functional categories. A query protein sequence can be submitted to the server for cluster identification and functional annotation. The database further provides gene ontology information (see Chapter 16) for protein cluster at each level as well as keywords from InterPro domains for functional prediction.

MOTIF DISCOVERY IN UNALIGNED SEQUENCES

For a set of closely related sequences, commonly shared motifs can be discovered by using the multiple sequence alignment–based methods. Often, however, distantly related sequences that share common motifs cannot be readily aligned. For example, the sequences for the helix-turn-helix motif in transcription factors can be subtly different enough that traditional multiple sequence alignment approaches fail to generate a satisfactory answer. For detecting such subtle motifs, more sophisticated algorithms such as expectation maximization (EM) and Gibbs sampling are used.

Expectation Maximization

The EM procedure can be used to find hidden motifs using a method that is somewhat different from profiles and PSSMs. The method works by first making a random or guessed alignment of the sequences to generate a trial PSSM. The trial PSSM is then used to compare with each sequence individually. The log odds scores of the PSSM are modified in each iteration to maximize the alignment of the matrix to each sequence. During the iterations, the sequence pattern for the conserved motifs is gradually "recruited" to the PSSM (Fig. 7.1). The drawback of the EM method is that the procedure stops prematurely if the scores reach convergence, a problem known as a *local optimum*. In addition, the final result is sensitive to the initial alignment.

Gibbs Motif Sampling

Another way to find conserved patterns from unaligned sequences is to use the Gibbs sampling method. Similar to the EM method, the Gibbs sampling algorithm makes an initial guessed alignment of all but one sequence. A trial PSSM is built to represent the alignment. The matrix is then aligned to the left-out sequence. The matrix scores are subsequently adjusted to achieve the best alignment with the left-out sequence. This process is repeated many times until there is no further improvement on the matrix

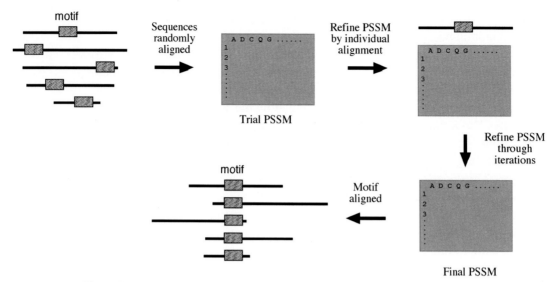

Figure 7.1: Schematic diagram of the EM algorithm.

scores. The end result is that after a number of iterations, the most probable patterns can be incorporated into a final PSSM. This procedure is less susceptible to premature termination due to the local optimum problem.

MEME (Multiple EM for motif elicitation, http://meme.sdsc.edu/meme/website/meme-intro.html) is a web-based program that uses the EM algorithm to find motifs either for DNA or protein sequences. It uses a modified EM algorithm to avoid the local minimum problem. In constructing a probability matrix, it allows multiple starting alignments and does not assume that there are motifs in every sequence. The computation is a two-step procedure. In the first step, the user provides approximately 20 unaligned sequences. The program applies EM to generate a sequence motif, which is an ungapped local sequence alignment. In the second step, segments from the query sequences with the same length as the motif are reapplied with the EM procedure to optimize the alignment between the subsequences and the motif. A segment with the highest score from the second iteration is selected as the optimum motif.

Gibbs sampler (http://bayesweb.wadsworth.org/gibbs/gibbs.html) is a web-based program that uses the Gibbs sampling approach to look for short, partially conserved gap-free segments for either DNA or protein sequences. To ensure accuracy, more than twenty sequences of the exact same length should be used.

SEQUENCE LOGOS

A multiple sequence alignment or a motif is often represented by a graphic representation called a *logo*. In a logo, each position consists of stacked letters representing the residues appearing in a particular column of a multiple alignment (Fig. 7.2). The overall height of a logo position reflects how conserved the position is, and the height of

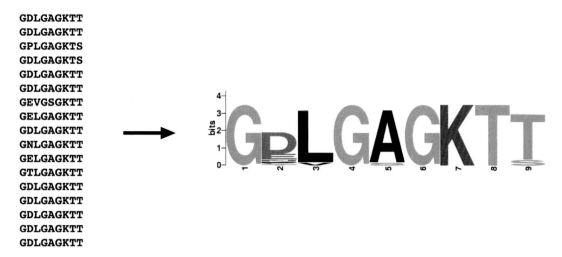

```
GDLGAGKTT
GDLGAGKTT
GPLGAGKTS
GDLGAGKTS
GDLGAGKTT
GDLGAGKTT
GEVGSGKTT
GELGAGKTT
GDLGAGKTT
GNLGAGKTT
GELGAGKTT
GTLGAGKTT
GDLGAGKTT
GDLGAGKTT
GDLGAGKTT
GDLGAGKTT
GDLGAGKTT
```

Figure 7.2: Example of multiple alignment representation using a logo (produced using the WebLogo program).

each letter in a position reflects the relative frequency of the residue in the alignment. Conserved positions have fewer residues and bigger symbols, whereas less conserved positions have a more heterogeneous mixture of smaller symbols stacked together. In general, a sequence logo provides a clearer description of a consensus sequence.

WebLogo (http://weblogo.berkeley.edu/) is an interactive program for generating sequence logos. A user needs to enter the sequence alignment in FASTA format to allow the program to compute the logos. A graphic file is returned to the user as a result.

SUMMARY

Sequence motifs and domains represent conserved, functionally important portions of proteins. Identifying domains and motifs is a crucial step in protein functional assignment. Domains correspond to contiguous regions in protein three-dimensional structures and serve as units of evolution. Motifs are highly conserved segments in multiple protein alignments that may be associated with particular biological functions. Databases for motifs and domains can be constructed based on multiple sequence alignment of related sequences. The derived motifs can be represented as regular expressions or profiles or HMMs. The mechanism of matching regular expressions with query sequences can be either exact matches or fuzzy matches. There are many databases constructed based on profiles or HMMs. Examples include Pfam, ProDom, and SMART. However, differences between databases render different sensitivities in detecting sequence motifs from unknown sequences. Thus, searching using multiple database tools is recommended.

In addition to motifs and domains, proteins can be classified based on overall sequence similarities. This type of classification makes use of either clustering or phylogenetic algorithms. Some examples are COG and ProtoNet. They are powerful tools in functional annotation of new protein sequences. Subtle motifs from divergent sequences can be discovered using the EM and Gibbs sampling approaches. Sequence logos are an effective way to represent motifs.

FURTHER READING

Attwood, T. K. 2000. The quest to deduce protein function from sequence: The role of pattern databases. *Int. J. Biochem. Cell. Biol.* 32:139–55.

Attwood, T. K. 2002. The PRINTS database: A resource for identification of protein families. *Brief. Bioinform.* 3:252–63.

Biswas, M., O'Rourke, J. F., Camon, E., Fraser, G., Kanapin, A., Karavidopoulou, Y., Kersey, P., et al. Applications of InterPro in protein annotation and genome analysis. *Brief. Bioinform.* 3:285–95.

Copley, R. R., Ponting, C. P., Schultz, J., and Bork, P. 2002. Sequence analysis of multidomain proteins: Past perspectives and future directions. *Adv. Protein Chem.* 61:75–98.

Kanehisa, M., and Bork, P. 2003. Bioinformatics in the post-sequence era. *Nat. Genet.* 33 (Suppl): 305–10.

Kong, L., and Ranganathan, S. 2004. Delineation of modular proteins: Domain boundary prediction from sequence information. *Brief. Bioinform.* 5:179–92.

Kriventseva, E. V., Biswas, M., and Apweiler R. 2001. Clustering and analysis of protein families. *Curr. Opin. Struct. Biol.* 11:334–9.

Liu, J., and Rost, B. 2003. Domains, motifs and clusters in the protein universe. *Curr. Opin. Chem. Biol.* 7:5–11.

Peri, S., Ibarrola, N., Blagoev, B., Mann, M., and Pandey, A. 2001. Common pitfalls in bioinformatics-based analyses: Look before you leap. *Trends Genet.* 17:541–5.

Servant, F., Bru, C., Carrere, S., Courcelle, E., Gouzy, J., Peyruc, D., and Kahn, D. 2002. ProDom: Automated clustering of homologous domains. *Brief. Bioinform.* 3:246–51.

Sigrist, C. J., Cerutti, L., Hulo, N., Gattiker, A., Falquet, L., Pagni, M., Bairoch, A., and Bucher, P. 2002. PROSITE: A documented database using patterns and profiles as motif descriptors. *Brief. Bioinform.* 3:265–74.

Wu, C. H., Huang, H., Yeh, L. S., and Barker, W. C. 2003. Protein family classification and functional annotation. *Comput. Biol. Chem.* 27:37–47.

Gene and Promoter Prediction

CHAPTER EIGHT

Gene Prediction

With the rapid accumulation of genomic sequence information, there is a pressing need to use computational approaches to accurately predict gene structure. Computational gene prediction is a prerequisite for detailed functional annotation of genes and genomes. The process includes detection of the location of open reading frames (ORFs) and delineation of the structures of introns as well as exons if the genes of interest are of eukaryotic origin. The ultimate goal is to describe all the genes computationally with near 100% accuracy. The ability to accurately predict genes can significantly reduce the amount of experimental verification work required.

However, this may still be a distant goal, particularly for eukaryotes, because many problems in computational gene prediction are still largely unsolved. Gene prediction, in fact, represents one of the most difficult problems in the field of pattern recognition. This is because coding regions normally do not have conserved motifs. Detecting coding potential of a genomic region has to rely on subtle features associated with genes that may be very difficult to detect.

Through decades of research and development, much progress has been made in prediction of prokaryotic genes. A number of gene prediction algorithms for prokaryotic genomes have been developed with varying degrees of success. Algorithms for eukarytotic gene prediction, however, are still yet to reach satisfactory results. This chapter describes a number of commonly used prediction algorithms, their theoretical basis, and limitations. Because of the significant differences in gene structures of prokaryotes and eukaryotes, gene prediction for each group of organisms is discussed separately. In addition, because of the predominance of protein coding genes in a genome (as opposed to rRNA and tRNA genes), the discussion focuses on the prediction of protein coding sequences.

CATEGORIES OF GENE PREDICTION PROGRAMS

The current gene prediction methods can be classified into two major categories, ab initio–based and homology-based approaches. The ab initio–based approach predicts genes based on the given sequence alone. It does so by relying on two major features associated with genes. The first is the existence of gene signals, which include start and stop codons, intron splice signals, transcription factor binding sites, ribosomal binding sites, and polyadenylation (poly-A) sites. In addition, the triplet codon structure limits the coding frame length to multiples of three, which can be used as a condition for gene prediction. The second feature used by ab initio algorithms is gene content, which is statistical description of coding regions. It has been observed that nucleotide composition and statistical patterns of the coding regions tend to vary significantly from those of the noncoding regions. The unique features can be detected by employing probabilistic models such as Markov models or hidden Markov models (HMMs; see Chapter 6) to help distinguish coding from noncoding regions.

The homology-based method makes predictions based on significant matches of the query sequence with sequences of known genes. For instance, if a translated DNA sequence is found to be similar to a known protein or protein family from a database search, this can be strong evidence that the region codes for a protein. Alternatively, when possible exons of a genomic DNA region match a sequenced cDNA, this also provides experimental evidence for the existence of a coding region.

Some algorithms make use of both gene-finding strategies. There are also a number of programs that actually combine prediction results from multiple individual programs to derive a consensus prediction. This type of algorithms can therefore be considered as consensus based.

GENE PREDICTION IN PROKARYOTES

Prokaryotes, which include bacteria and Archaea, have relatively small genomes with sizes ranging from 0.5 to 10 Mbp (1 Mbp = 10^6 bp). The gene density in the genomes is high, with more than 90% of a genome sequence containing coding sequence. There are very few repetitive sequences. Each prokaryotic gene is composed of a single contiguous stretch of ORF coding for a single protein or RNA with no interruptions within a gene.

More detailed knowledge of the bacterial gene structure can be very useful in gene prediction. In bacteria, the majority of genes have a start codon ATG (or AUG in mRNA; because prediction is done at the DNA level, T is used in place of U), which codes for methionine. Occasionally, GTG and TTG are used as alternative start codons, but methionine is still the actual amino acid inserted at the first position. Because there may be multiple ATG, GTG, or TGT codons in a frame, the presence of these codons at the beginning of the frame does not necessarily give a clear indication of the translation initiation site. Instead, to help identify this initiation codon, other features associated with translation are used. One such feature is the ribosomal binding site, also called the *Shine-Delgarno sequence*, which is a stretch of purine-rich sequence complementary to 16S rRNA in the ribosome (Fig. 8.1). It is located immediately downstream of the transcription initiation site and slightly upstream of the translation start codon. In many bacteria, it has a consensus motif of AGGAGGT. Identification of the ribosome binding site can help locate the start codon.

At the end of the protein coding region is a stop codon that causes translation to stop. There are three possible stop codons, identification of which is straightforward. Many prokaryotic genes are transcribed together as one operon. The end of the operon is characterized by a transcription termination signal called *ρ-independent terminator*. The terminator sequence has a distinct stem-loop secondary structure

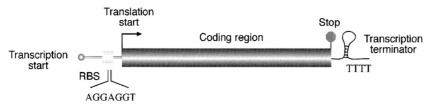

Figure 8.1: Structure of a typical prokaryotic gene structure. *Abbreviation:* RBS, ribosome binding site.

followed by a string of Ts. Identification of the terminator site, in conjunction with promoter site identification (see Chapter 9), can sometimes help in gene prediction.

Conventional Determination of Open Reading Frames

Without the use of specialized programs, prokaryotic gene identification can rely on manual determination of ORFs and major signals related to prokaryotic genes. Prokaryotic DNA is first subject to conceptual translation in all six possible frames, three frames forward and three frames reverse. Because a stop codon occurs in about every twenty codons by chance in a noncoding region, a frame longer than thirty codons without interruption by stop codons is suggestive of a gene coding region, although the threshold for an ORF is normally set even higher at fifty or sixty codons. The putative frame is further manually confirmed by the presence of other signals such as a start codon and Shine–Delgarno sequence. Furthermore, the putative ORF can be translated into a protein sequence, which is then used to search against a protein database. Detection of homologs from this search is probably the strongest indicator of a protein-coding frame.

In the early stages of development of gene prediction algorithms, genes were predicted by examining the nonrandomness of nucleotide distribution. One method is based on the nucleotide composition of the third position of a codon. In a coding sequence, it has been observed that this position has a preference to use G or C over A or T. By plotting the GC composition at this position, regions with values significantly above the random level can be identified, which are indicative of the presence of ORFs (Fig. 8.2). In practice, because genes can be in any of the six frames, the statistical patterns are computed for all possible frames. In addition to codon bias, there is a similar method called TESTCODE (implemented in the commercial GCG package) that exploits the fact that the third codon nucleotides in a coding region tend to repeat themselves. By plotting the repeating patterns of the nucleotides at this position, coding and noncoding regions can be differentiated (see Fig. 8.2). The results of the two methods are often consistent. The two methods are often used in conjunction to confirm the results of each other.

These statistical methods, which are based on empirical rules, examine the statistics of a single nucleotide (either G or C). They identify only typical genes and tend to miss atypical genes in which the rule of codon bias is not strictly followed. To improve the prediction accuracies, the new generation of prediction algorithms use more sophisticated statistical models.

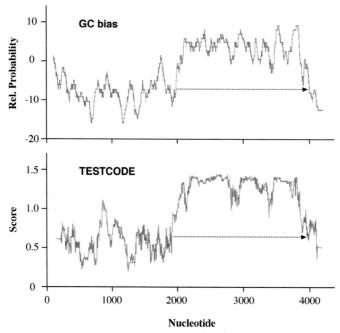

Figure 8.2: Coding frame detection of a bacterial gene using either the GC bias or the TESTCODE method. Both result in similar identification of a reading frame (*dashed arrows*).

Gene Prediction Using Markov Models and Hidden Markov Models

Markov models and HMMs can be very helpful in providing finer statistical description of a gene (see Chapter 6). A Markov model describes the probability of the distribution of nucleotides in a DNA sequence, in which the conditional probability of a particular sequence position depends on k previous positions. In this case, k is the order of a Markov model. A zero-order Markov model assumes each base occurs independently with a given probability. This is often the case for noncoding sequences. A first-order Markov model assumes that the occurrence of a base depends on the base preceding it. A second-order model looks at the preceding two bases to determine which base follows, which is more characteristic of codons in a coding sequence.

The use of Markov models in gene finding exploits the fact that oligonucleotide distributions in the coding regions are different from those for the noncoding regions. These can be represented with various orders of Markov models. Since a fixed-order Markov chain describes the probability of a particular nucleotide that depends on previous k nucleotides, the longer the oligomer unit, the more nonrandomness can be described for the coding region. Therefore, the higher the order of a Markov model, the more accurately it can predict a gene.

Because a protein-encoding gene is composed of nucleotides in triplets as codons, more effective Markov models are built in sets of three nucleotides, describing non-random distributions of trimers or hexamers, and so on. The parameters of a Markov model have to be trained using a set of sequences with known gene locations. Once the parameters of the model are established, it can be used to compute the nonrandom

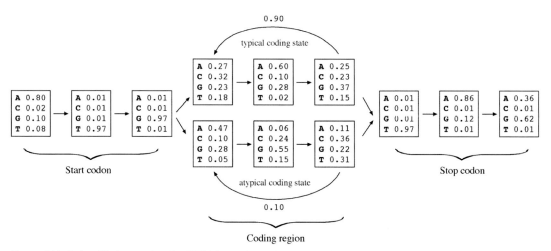

Figure 8.3: A simplified second-order HMM for prokaryotic gene prediction that includes a statistical model for start codons, stop codons, and the rest of the codons in a gene sequence represented by a typical model and an atypical model.

distributions of trimers or hexamers in a new sequence to find regions that are compatible with the statistical profiles in the learning set.

Statistical analyses have shown that pairs of codons (or amino acids at the protein level) tend to correlate. The frequency of six unique nucleotides appearing together in a coding region is much higher than by random chance. Therefore, a fifth-order Markov model, which calculates the probability of hexamer bases, can detect nucleotide correlations found in coding regions more accurately and is in fact most often used.

A potential problem of using a fifth-order Markov chain is that if there are not enough hexamers, which happens in short gene sequences, the method's efficacy may be limited. To cope with this limitation, a variable-length Markov model, called an *interpolated Markov model* (IMM), has been developed. The IMM method samples the largest number of sequence patterns with k ranging from 1 to 8 (dimers to ninemers) and uses a weighting scheme, placing less weight on rare k-mers and more weight on more frequent k-mers. The probability of the final model is the sum of probabilities of all weighted k-mers. In other words, this method has more flexibility in using Markov models depending on the amount of data available. Higher-order models are used when there is a sufficient amount of data and lower-order models are used when the amount of data is smaller.

It has been shown that the gene content and length distribution of prokaryotic genes can be either typical or atypical. Typical genes are in the range of 100 to 500 amino acids with a nucleotide distribution typical of the organism. Atypical genes are shorter or longer with different nucleotide statistics. These genes tend to escape detection using the typical gene model. This means that, to make the algorithm capable of fully describing all genes in a genome, more than one Markov model is needed. To combine different Markov models that represent typical and atypical nucleotide distributions creates an HMM prediction algorithm. A simplified HMM for gene finding is shown in Fig. 8.3.

The following describes a number of HMM/IMM-based gene finding programs for prokaryotic organisms.

GeneMark (http://opal.biology.gatech.edu/GeneMark/) is a suite of gene prediction programs based on the fifth-order HMMs. The main program – GeneMark.hmm – is trained on a number of complete microbial genomes. If the sequence to be predicted is from a nonlisted organism, the most closely related organism can be chosen as the basis for computation. Another option for predicting genes from a new organism is to use a self-trained program GeneMarkS as long as the user can provide at least 100 kbp of sequence on which to train the model. If the query sequence is shorter than 100 kbp, a GeneMark heuristic program can be used with some loss of accuracy. In addition to predicting prokaryotic genes, GeneMark also has a variant for eukaryotic gene prediction using HMM.

Glimmer (Gene Locator and Interpolated Markov Modeler, www.tigr.org/softlab/ glimmer/glimmer.html) is a UNIX program from TIGR that uses the IMM algorithm to predict potential coding regions. The computation consists of two steps, namely model building and gene prediction. The model building involves training by the input sequence, which optimizes the parameters of the model. In an actual gene prediction, the overlapping frames are "flagged" to alert the user for further inspection. Glimmer also has a variant, GlimmerM, for eukaryotic gene prediction.

FGENESB (www.softberry.com/berry.phtml?topic=gfindb) is a web-based program that is also based on fifth-order HMMs for detecting coding regions. The program is specifically trained for bacterial sequences. It uses the Vertibi algorithm (see Chapter 6) to find an optimal match for the query sequence with the intrinsic model. A linear discriminant analysis (LDA) is used to further distinguish coding signals from noncoding signals.

These programs have been shown to be reasonably successful in finding genes in a genome. The common problem is imprecise prediction of translation initiation sites because of inefficient identification of ribosomal binding sites. This problem can be remedied by identifying the ribosomal binding site associated with a start codon. A number of algorithms have been developed solely for this purpose. RBSfinder is one such algorithm.

RBSfinder (ftp://ftp.tigr.org/pub/software/RBSfinder/) is a UNIX program that uses the prediction output from Glimmer and searches for the Shine–Delgarno sequences in the vicinity of predicted start sites. If a high-scoring site is found by the intrinsic probabilistic model, a start codon is confirmed; otherwise the program moves to other putative translation start sites and repeats the process.

Performance Evaluation

The accuracy of a prediction program can be evaluated using parameters such as sensitivity and specificity. To describe the concept of sensitivity and specificity accurately, four features are used: true positive (TP), which is a correctly predicted feature; false positive (FP), which is an incorrectly predicted feature; false negative (FN), which is a missed feature; and true negative (TN), which is the correctly predicted absence of

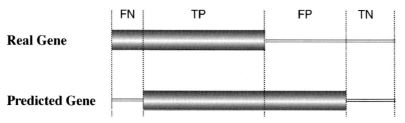

Figure 8.4: Definition of four basic measures of gene prediction accuracy at the nucleotide level. *Abbreviations:* FN, false negative; TP, true positive; FP, false positive; TN, true negative.

a feature (Fig. 8.4). Using these four terms, sensitivity (Sn) and specificity (Sp) can be described by the following formulas:

$$Sn = TP/(TP + FN) \qquad \text{(Eq. 8.1)}$$

$$Sp = TP/(TP + FP) \qquad \text{(Eq. 8.2)}$$

According to these formulas, *sensitivity* is the proportion of true signals predicted among all possible true signals. It can be considered as the ability to include correct predictions. In contrast, *specificity* is the proportion of true signals among all signals that are predicted. It represents the ability to exclude incorrect predictions. A program is considered accurate if both sensitivity and specificity are simultaneously high and approach a value of 1. In a case in which sensitivity is high but specificity is low, the program is said to have a tendency to overpredict. On the other hand, if the sensitivity is low but specificity high, the program is too conservative and lacks predictive power.

Because neither sensitivity nor specificity alone can fully describe accuracy, it is desirable to use a single value to summarize both of them. In the field of gene finding, a single parameter known as the correlation coefficient (CC) is often used, which is defined by the following formula:

$$CC = \frac{TP \bullet TN - FP \bullet FN}{\sqrt{(TP + FP)(TN + FN)(FP + TN)}} \qquad \text{(Eq. 8.3)}$$

The value of the CC provides an overall measure of accuracy, which ranges from -1 to $+1$, with $+1$ meaning always correct prediction and -1 meaning always incorrect prediction. Table 8.1 shows a performance analysis using the Glimmer program as an example.

GENE PREDICTION IN EUKARYOTES

Eukaryotic nuclear genomes are much larger than prokaryotic ones, with sizes ranging from 10 Mbp to 670 Gbp (1 Gbp $= 10^9$ bp). They tend to have a very low gene density. In humans, for instance, only 3% of the genome codes for genes, with about 1 gene per 100 kbp on average. The space between genes is often very large and rich in repetitive sequences and transposable elements.

Most importantly, eukaryotic genomes are characterized by a mosaic organization in which a gene is split into pieces (called *exons*) by intervening noncoding sequences

TABLE 8.1. Performance Analysis of the Glimmer Program for Gene Prediction of Three Genomes

Species	GC (%)	FN	FP	Sensitivity	Specificity
Campylobacter jejuni	30.5	10	19	99.3	98.7
Haemophilus influenzae	38.2	3	54	99.8	96.1
Helicobacter pylori	38.9	6	39	99.5	97.2

Note: The data sets were from three bacterial genomes (Aggarwal and Ramaswamy, 2002). *Abbreviations:* FN, false negative; FP, false positive.

(called *introns*) (Fig. 8.5). The nascent transcript from a eukaryotic gene is modified in three different ways before becoming a mature mRNA for protein translation. The first is capping at the 5′ end of the transcript, which involves methylation at the initial residue of the RNA. The second event is splicing, which is the process of removing introns and joining exons. The molecular basis of splicing is still not completely understood. What is known currently is that the splicing process involves a large RNA-protein complex called spliceosome. The reaction requires intermolecular interactions between a pair of nucleotides at each end of an intron and the RNA component of the spliceosome. To make the matter even more complex, some eukaryotic genes can have their transcripts spliced and joined in different ways to generate more than one transcript per gene. This is the phenomenon of alternative splicing. As to be discussed in more detail in Chapter 16, alternative splicing is a major mechanism for generating functional diversity in eukaryotic cells. The third modification is polyadenylation, which is the addition of a stretch of As (∼250) at the 3′ end of the RNA.

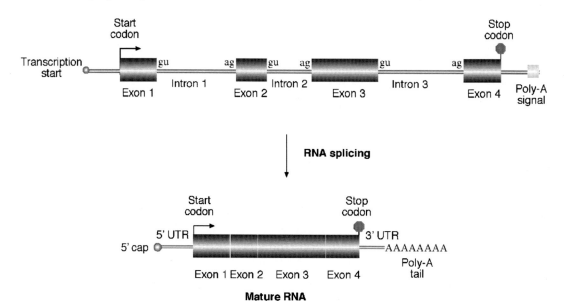

Figure 8.5: Structure of a typical eukaryotic RNA as primary transcript from genomic DNA and as mature RNA after posttranscriptional processing. *Abbreviations:* UTR, untranslated region; poly-A, polyadenylation.

This process is controlled by a poly-A signal, a conserved motif slightly downstream of a coding region with a consensus CAATAAA(T/C).

The main issue in prediction of eukaryotic genes is the identification of exons, introns, and splicing sites. From a computational point of view, it is a very complex and challenging problem. Because of the presence of split gene structures, alternative splicing, and very low gene densities, the difficulty of finding genes in such an environment is likened to finding a needle in a haystack. The needle to be found actually is broken into pieces and scattered in many different places. The job is to gather the pieces in the haystack and reproduce the needle in the correct order.

The good news is that there are still some conserved sequence features in eukaryotic genes that allow computational prediction. For example, the splice junctions of introns and exons follow the GT–AG rule in which an intron at the 5′ splice junction has a consensus motif of GTAAGT; and at the 3′ splice junction is a consensus motif of $(Py)_{12}$NCAG (see Fig. 8.5). Some statistical patterns useful for prokaryotic gene finding can be applied to eukaryotic systems as well. For example, nucleotide compositions and codon bias in coding regions of eukaryotes are different from those of the noncoding regions. Hexamer frequencies in coding regions are also higher than in the noncoding regions. Most vertebrate genes use ATG as the translation start codon and have a uniquely conserved flanking sequence call a *Kozak sequence* (CCGCCATGG). In addition, most of these genes have a high density of CG dinucleotides near the transcription start site. This region is referred to as a CpG island (*p* refers to the phosphodiester bond connecting the two nucleotides), which helps to identify the transcription initiation site of a eukaryotic gene. The poly-A signal can also help locate the final coding sequence.

Gene Prediction Programs

To date, numerous computer programs have been developed for identifying eukaryotic genes. They fall into all three categories of algorithms: ab initio based, homology based, and consensus based. Most of these programs are organism specific because training data sets for obtaining statistical parameters have to be derived from individual organisms. Some of the algorithms are able to predict the most probable exons as well as suboptimal exons providing information for possible alternative spliced transcription products.

Ab Initio–Based Programs

The goal of the ab initio gene prediction programs is to discriminate exons from noncoding sequences and subsequently join the exons together in the correct order. The main difficulty is correct identification of exons. To predict exons, the algorithms rely on two features, gene signals and gene content. Signals include gene start and stop sites and putative splice sites, recognizable consensus sequences such as poly-A sites. *Gene content* refers to coding statistics, which includes nonrandom nucleotide distribution, amino acid distribution, synonymous codon usage, and hexamer frequencies. Among these features, the hexamer frequencies appear to be most discriminative for

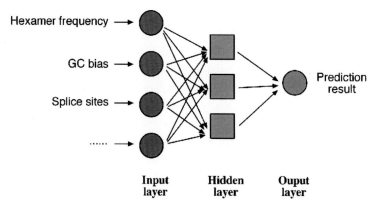

Figure 8.6: Architecture of a neural network for eukaryotic gene prediction.

coding potentials. To derive an assessment for this feature, HMMs can be used, which require proper training. In addition to HMMs, neural network-based algorithms are also common in the gene prediction field. This begs the question of what is a neural network algorithm. A brief introduction is given next.

Prediction Using Neural Networks. A *neural network* (or *artificial neural network*) is a statistical model with a special architecture for pattern recognition and classification. It is composed of a network of mathematical variables that resemble the biological nervous system, with variables or nodes connected by weighted functions that are analogous to synapses (Fig. 8.6). Another aspect of the model that makes it look like a biological neural network is its ability to "learn" and then make predictions after being trained. The network is able to process information and modify parameters of the weight functions between variables during the training stage. Once it is trained, it is able to make automatic predictions about the unknown.

In gene prediction, a neural network is constructed with multiple layers; the input, output, and hidden layers. The input is the gene sequence with intron and exon signals. The output is the probability of an exon structure. Between input and output, there may be one or several hidden layers where the machine learning takes place. The machine learning process starts by feeding the model with a sequence of known gene structure. The gene structure information is separated into several classes of features such as hexamer frequencies, splice sites, and GC composition during training. The weight functions in the hidden layers are adjusted during this process to recognize the nucleotide patterns and their relationship with known structures. When the algorithm predicts an unknown sequence after training, it applies the same rules learned in training to look for patterns associated with the gene structures.

The frequently used ab initio programs make use of neural networks, HMMs, and discriminant analysis, which are described next.

GRAIL (Gene Recognition and Assembly Internet Link; http://compbio.ornl.gov/public/tools/) is a web-based program that is based on a neural network algorithm. The program is trained on several statistical features such as splice junctions, start

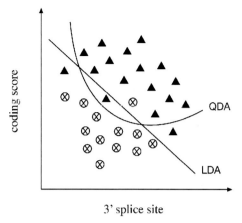

Figure 8.7: Comparison of two discriminant analysis, LDA and QDA. ▲ coding features; ⊗ noncoding features.

and stop codons, poly-A sites, promoters, and CpG islands. The program scans the query sequence with windows of variable lengths and scores for coding potentials and finally produces an output that is the result of exon candidates. The program is currently trained for human, mouse, *Arabidopsis, Drosophila,* and *Escherichia coli* sequences.

Prediction Using Discriminant Analysis. Some gene prediction algorithms rely on discriminant analysis, either LDA or quadratic discriminant analysis (QDA), to improve accuracy. LDA works by plotting a two-dimensional graph of coding signals versus all potential 3′ splice site positions and drawing a diagonal line that best separates coding signals from noncoding signals based on knowledge learned from training data sets of known gene structures (Fig. 8.7). QDA draws a curved line based on a quadratic function instead of drawing a straight line to separate coding and noncoding features. This strategy is designed to be more flexible and provide a more optimal separation between the data points.

FGENES (Find Genes; www.softberry.com/) is a web-based program that uses LDA to determine whether a signal is an exon. In addition to FGENES, there are many variants of the program. Some programs, such as FGENESH, make use of HMMs. There are others, such as FGENESH_C, that are similarity based. Some programs, such as FGENESH+, combine both ab initio and similarity-based approaches.

MZEF (Michael Zhang's Exon Finder; http://argon.cshl.org/genefinder/) is a web-based program that uses QDA for exon prediction. Despite the more complex mathematical functions, the expected increase in performance has not been obvious in actual gene prediction.

Prediction Using HMMs. GENSCAN (http://genes.mit.edu/GENSCAN.html) is a web-based program that makes predictions based on fifth-order HMMs. It combines hexamer frequencies with coding signals (initiation codons, TATA box, cap site, poly-A, etc.) in prediction. Putative exons are assigned a probability score (P) of being a true exon. Only predictions with $P > 0.5$ are deemed reliable. This program is trained for sequences from vertebrates, *Arabidopsis,* and maize. It has been used extensively in annotating the human genome (see Chapter 17).

HMMgene (www.cbs.dtu.dk/services/HMMgene) is also an HMM-based web program. The unique feature of the program is that it uses a criterion called the *conditional maximum likelihood* to discriminate coding from noncoding features. If a sequence already has a subregion identified as coding region, which may be based on similarity with cDNAs or proteins in a database, these regions are locked as coding regions. An HMM prediction is subsequently made with a bias toward the locked region and is extended from the locked region to predict the rest of the gene coding regions and even neighboring genes. The program is in a way a hybrid algorithm that uses both ab initio-based and homology-based criteria.

Homology-Based Programs

Homology-based programs are based on the fact that exon structures and exon sequences of related species are highly conserved. When potential coding frames in a query sequence are translated and used to align with closest protein homologs found in databases, near perfectly matched regions can be used to reveal the exon boundaries in the query. This approach assumes that the database sequences are correct. It is a reasonable assumption in light of the fact that many homologous sequences to be compared with are derived from cDNA or expressed sequence tags (ESTs) of the same species. With the support of experimental evidence, this method becomes rather efficient in finding genes in an unknown genomic DNA.

The drawback of this approach is its reliance on the presence of homologs in databases. If the homologs are not available in the database, the method cannot be used. Novel genes in a new species cannot be discovered without matches in the database. A number of publicly available programs that use this approach are discussed next.

GenomeScan (http://genes.mit.edu/genomescan.html) is a web-based server that combines GENSCAN prediction results with BLASTX similarity searches. The user provides genomic DNA and protein sequences from related species. The genomic DNA is translated in all six frames to cover all possible exons. The translated exons are then used to compare with the user-supplied protein sequences. Translated genomic regions having high similarity at the protein level receive higher scores. The same sequence is also predicted with a GENSCAN algorithm, which gives exons probability scores. Final exons are assigned based on combined score information from both analyses.

EST2Genome (http://bioweb.pasteur.fr/seqanal/interfaces/est2genome.html) is a web-based program purely based on the sequence alignment approach to define intron–exon boundaries. The program compares an EST (or cDNA) sequence with a genomic DNA sequence containing the corresponding gene. The alignment is done using a dynamic programming–based algorithm. One advantage of the approach is the ability to find very small exons and alternatively spliced exons that are very difficult to predict by any ab initio–type algorithms. Another advantage is that there is no need for model training, which provides much more flexibility for gene prediction. The limitation is that EST or cDNA sequences often contain errors or even introns if the transcripts are not completely spliced before reverse transcription.

SGP-1 (Syntenic Gene Prediction; http://195.37.47.237/sgp-1/) is a similarity-based web program that aligns two genomic DNA sequences from closely related organisms. The program translates all potential exons in each sequence and does pairwise alignment for the translated protein sequences using a dynamic programming approach. The near-perfect matches at the protein level define coding regions. Similar to EST2Genome, there is no training needed. The limitation is the need for two homologous sequences having similar genes with similar exon structures; if this condition is not met, a gene escapes detection from one sequence when there is no counterpart in another sequence.

TwinScan (http://genes.cs.wustl.edu/) is also a similarity-based gene-finding server. It is similar to GenomeScan in that it uses GenScan to predict all possible exons from the genomic sequence. The putative exons are used for BLAST searching to find closest homologs. The putative exons and homologs from BLAST searching are aligned to identify the best match. Only the closest match from a genome database is used as a template for refining the previous exon selection and exon boundaries.

Consensus-Based Programs

Because different prediction programs have different levels of sensitivity and specificity, it makes sense to combine results of multiple programs based on consensus. This idea has prompted development of consensus-based algorithms. These programs work by retaining common predictions agreed by most programs and removing inconsistent predictions. Such an integrated approach may improve the specificity by correcting the false positives and the problem of overprediction. However, since this procedure punishes novel predictions, it may lead to lowered sensitivity and missed predictions. Two examples of consensus-based programs are given next.

GeneComber (www.bioinformatics.ubc.ca/genecomber/index.php) is a web server that combines HMMgene and GenScan prediction results. The consistency of both prediction methods is calculated. If the two predictions match, the exon score is reinforced. If not, exons are proposed based on separate threshold scores.

DIGIT (http://digit.gsc.riken.go.jp/cgi-bin/index.cgi) is another consensus-based web server. It uses prediction from three ab initio programs – FGENESH, GENSCAN, and HMMgene. It first compiles all putative exons from the three gene-finders and assigns ORFs with associated scores. It then searches a set of exons with the highest additive score under the reading frame constraints. During this process, a Bayesian procedure and HMMs are used to infer scores and search the optimal exon set which gives the final designation of gene structure.

Performance Evaluation

Because of extra layers of complexity for eukaryotic gene prediction, the sensitivity and specificity have to be defined on the levels of nucleotides, exons, and entire genes.

TABLE 8.2. Accuracy Comparisons for a Number of Ab Initio Gene Prediction Programs at Nucleotide and Exon Levels

	Nucleotide level			Exon level				
	Sn	Sp	CC	Sn	Sp	(Sn + Sp)/2	ME	WE
FGENES	0.86	0.88	0.83	0.67	0.67	0.67	0.12	0.09
GeneMark	0.87	0.89	0.83	0.53	0.54	0.54	0.13	0.11
Genie	0.91	0.90	0.88	0.71	0.70	0.71	0.19	0.11
GenScan	0.95	0.90	0.91	0.70	0.70	0.70	0.08	0.09
HMMgene	0.93	0.93	0.91	0.76	0.77	0.76	0.12	0.07
Morgan	0.75	0.74	0.74	0.46	0.41	0.43	0.20	0.28
MZEF	0.70	0.73	0.66	0.58	0.59	0.59	0.32	0.23

Note: The data sets used were single mammalian gene sequences (performed by Sanja Rogic, from www.cs.ubc.ca/~rogic/evaluation/tablesgen.html.
Abbreviations: Sn, sensitivity; Sp, specificity; CC, correlation coefficient; ME, missed exons; WE, wrongly predicted exons.

The sensitivity at the exon and gene level is the proportion of correctly predicted exons or genes among actual exons or genes. The specificity at the two levels is the proportion of correctly predicted exons or genes among all predictions made. For exons, instead of using CC, an average of sensitivity and specificity at the exon level is used instead. In addition, the proportion of missed exons and missed genes as well as wrongly predicted exons and wrong genes, which have no overlaps with true exons or genes, often have to be indicated.

By introducing these measures, the criteria for prediction accuracy evaluation become more stringent (Table 8.2). For example, a correct exon requires all nucleotides belonging to the exon to be predicted correctly. For a correctly predicted gene, all nucleotides and all exons have to be predicted correctly. One single error at the nucleotide level can negate the entire gene prediction. Consequently, the accuracy values reported on the levels of exons and genes are much lower than those for nucleotides.

When a new gene prediction program is published, the accuracy level is usually reported. However, the reported performance should be treated with caution because the accuracy is usually estimated based on particular datasets, which may have been optimized for the program. The datasets used are also mainly composed of short genomic sequences with simple gene structures. When the programs are used in gene prediction for truly unknown eukaryotic genomic sequences, the accuracy can become much lower. Because of the lack of unbiased and realistic datasets and objective comparison for eukaryotic gene prediction, it is difficult to know the true accuracy of the current prediction tools.

At present, no single software program is able to produce consistent superior results. Some programs may perform well on certain types of exons (e.g., internal or single exons) but not others (e.g., initial and terminal exons). Some are sensitive to the G-C content of the input sequences or to the lengths of introns and exons. Most programs make overpredictions when genes contain long introns. In sum, they all suffer from the problem of generating a high number of false positives and false negatives. This is especially true for ab initio–based algorithms. For complex genomes such as the human genome, most popular programs can predict no more than 40% of the genes exactly right. Drawing consensus from results by multiple prediction programs may enhance performance to some extent.

SUMMARY

Computational prediction of genes is one of the most important steps of genome sequence analysis. For prokaryotic genomes, which are characterized by high gene density and noninterrupted genes, prediction of genes is easier than for eukaryotic genomes. Current prokaryotic gene prediction algorithms, which are based on HMMs, have achieved reasonably good accuracy. Many difficulties still persist for eukaryotic gene prediction. The difficulty mainly results from the low gene density and split gene structure of eukaryotic genomes. Current algorithms are either ab initio based, homology based, or a combination of both. For ab initio–based eukaryotic gene prediction, the HMM type of algorithm has overall better performance in differentiating intron–exon boundaries. The major limitation is the dependency on training of the statistical models, which renders the method to be organism specific. The homology-based algorithms in combination with HMMs may yield improved accuracy. The method is limited by the availability of identifiable sequence homologs in databases. The combined approach that integrates statistical and homology information may generate further improved performance by detecting more genes and more exons correctly. With rapid advances in computational techniques and understanding of the splicing mechanism, it is hoped that reliable eukaryotic gene prediction can become more feasible in the near future.

FURTHER READING

Aggarwal, G., and Ramaswamy, R. 2002. Ab initio gene identification: Prokaryote genome annotation with GeneScan and GLIMMER. *J. Biosci.* 27:7–14.

Ashurst, J. L., and Collins, J. E. 2003. Gene annotation: Prediction and testing. *Annu. Rev. Genomics Hum. Genet.* 4:69–88.

Azad, R. K., and Borodovsky, M. 2004. Probabilistic methods of identifying genes in prokaryotic genomes: Connections to the HMM theory. *Brief. Bioinform.* 5:118–30.

Cruveiller, S., Jabbari, K., Clay, O., and Bemardi, G. 2003. Compositional features of eukaryotic genomes for checking predicted genes. *Brief. Bioinform.* 4:43–52.

Davuluri, R. V., and Zhang, M. Q. 2003. "Computer software to find genes in plant genomic DNA." In *Plant Functional Genomics*, edited by E. Grotewold, 87–108. Totowa, NJ: Human Press.

Guigo, R., and Wiehe, T. 2003. "Gene prediction accuracy in large DNA sequences." In *Frontiers in Computational Genomics*, edited by M. Y. Galperin and E. V. Koonin, 1–33. Norfolk, UK: Caister Academic Press.

Guigo, R., Dermitzakis, E. T., Agarwal, P., Ponting, C. P., Parra, G., Reymond, A., Abril, J. F., et al R. 2003. Comparison of mouse and human genomes followed by experimental verification yields an estimated 1,019 additional genes. *Proc. Natl. Acad. Sci. USA* 100:1140–5.

Li, W., and Godzik, A. 2002. Discovering new genes with advanced homology detection. *Trends Biotechnol.* 20:315–16.

Makarov, V. 2002. Computer programs for eukaryotic gene prediction. *Brief. Bioinform.* 3:195–9.

Mathe, C., Sagot, M. F., Schiex, T., and Rouze, P. 2002. Current methods of gene prediction, their strengths and weaknesses. *Nucleic Acids Res.* 30:4103–17.

Parra, G., Agarwal, P., Abril, J. F., Wiehe, T., Fickett, J. W., and Guigo, R. 2003. Comparative gene prediction in human and mouse. *Genome Res.* 13:108–17.

Wang, J., Li, S., Zhang, Y., Zheng, H., Xu, Z., Ye, J., Yu, J., and Wong, G. K. 2003. Vertebrate gene predictions and the problem of large genes. *Nat. Rev. Genet.* 4:741–9.

Wang, Z., Chen, Y., and Li, Y. 2004. A brief review of computational gene prediction methods. *Geno. Prot. Bioinfo.* 4:216–21.

Zhang, M. Q. 2002. Computational prediction of eukaryotic protein coding genes. *Nat. Rev. Genetics.* 3:698–709.

CHAPTER NINE

Promoter and Regulatory Element Prediction

An issue related to gene prediction is promoter prediction. Promoters are DNA elements located in the vicinity of gene start sites (which should not be confused with the translation start sites) and serve as binding sites for the gene transcription machinery, consisting of RNA polymerases and transcription factors. Therefore, these DNA elements directly regulate gene expression. Promoters and regulatory elements are traditionally determined by experimental analysis. The process is extremely time consuming and laborious. Computational prediction of promoters and regulatory elements is especially promising because it has the potential to replace a great deal of extensive experimental analysis.

However, computational identification of promoters and regulatory elements is also a very difficult task, for several reasons. First, promoters and regulatory elements are not clearly defined and are highly diverse. Each gene seems to have a unique combination of sets of regulatory motifs that determine its unique temporal and spatial expression. There is currently a lack of sufficient understanding of all the necessary regulatory elements for transcription. Second, the promoters and regulatory elements cannot be translated into protein sequences to increase the sensitivity for their detection. Third, promoter and regulatory sites to be predicted are normally short (six to eight nucleotides) and can be found in essentially any sequence by random chance, thus resulting in high rates of false positives associated with theoretical predictions.

Current solutions for providing preliminary identification of these elements are to combine a multitude of features and use sophisticated algorithms that give either ab initio–based predictions or predictions based on evolutionary information or experimental data. These computational approaches are described in detail in this chapter following a brief introduction to the structures of promoters and regulatory elements in both prokaryotes and eukaryotes.

PROMOTER AND REGULATORY ELEMENTS IN PROKARYOTES

In bacteria, transcription is initiated by RNA polymerase, which is a multi-subunit enzyme. The σ subunit (e.g., σ^{70}) of the RNA polymerase is the protein that recognizes specific sequences upstream of a gene and allows the rest of the enzyme complex to bind. The upstream sequence where the σ protein binds constitutes the promoter sequence. This includes the sequence segments located 35 and 10 base pairs (bp) upstream from the transcription start site. They are also referred to as the -35 and -10 boxes. For the σ^{70} subunit in *Escherichia coli*, for example, the -35 box

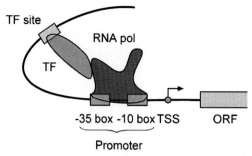

Promoter

Figure 9.1: Schematic representation of elements involved in bacterial transcription initiation. RNA polymerase binds to the promoter region, which initiates transcription through interaction with transcription factors binding at different sites. *Abbreviations:* TSS, transcription start site; ORF, reading frame; pol, polymerase; TF, transcription factor (see color plate section).

has a consensus sequence of TTGACA. The –10 box has a consensus of TATAAT. The promoter sequence may determine the expression of one gene or a number of linked genes downstream. In the latter case, the linked genes form an operon, which is controlled by the promoter.

In addition to the RNA polymerase, there are also a number of DNA-binding proteins that facilitate the process of transcription. These proteins are called *transcription factors*. They bind to specific DNA sequences to either enhance or inhibit the function of the RNA polymerase. The specific DNA sequences to which the transcription factors bind are referred to as *regulatory elements*. The regulatory elements may bind in the vicinity of the promoter or bind to a site several hundred bases away from the promoter. The reason that the regulatory proteins binding at long distance can still exert their effect is because of the flexible structure of DNA, which is able to bend and and exert its effect by bringing the transcription factors in close contact with the RNA polymerase complex (Fig. 9.1).

PROMOTER AND REGULATORY ELEMENTS IN EUKARYOTES

In eukaryotes, gene expression is also regulated by a protein complex formed between transcription factors and RNA polymerase. However, eukaryotic transcription has an added layer of complexity in that there are three different types of RNA polymerase complexes, namely RNA polymerases I, II, and III. Each polymerase transcribes different sets of genes. RNA polymerases I and III are responsible for the transcription of ribosomal RNAs and tRNAs, respectively. RNA polymerase II is exclusively responsible for transcribing protein-encoding genes (or synthesis of mRNAs).

Unlike in prokaryotes, where genes often form an operon with a shared promoter, each eukaryotic gene has its own promoter. The eukaryotic transcription machinery also requires many more transcription factors than its prokaryotic counterpart to help initiate transcription. Furthermore, eukaryotic RNA polymerase II does not directly bind to the promoter, but relies on a dozen or more transcription factors to recognize and bind to the promoter in a specific order before its own binding around the promoter.

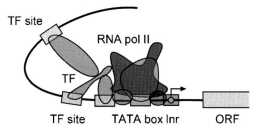

Figure 9.2: Schematic diagram of an eukaryotic promoter with transcription factors and RNA polymerase bound to the promoter. *Abbreviations:* Inr, initiator sequence; ORF, reading frame; pol, polymerase; TF, transcription factor (see color plate section).

The core of many eukaryotic promoters is a so-called TATA box, located 30 bps upstream from the transcription start site, having a consensus motif TATA(A/T)A (A/T) (Fig. 9.2.). However, not all eukaryotic promoters contain the TATA box. Many genes such as housekeeping genes do not have the TATA box in their promoters. Still, the TATA box is often used as an indicator of the presence of a promoter. In addition, many genes have a unique initiator sequence (Inr), which is a pyrimidine-rich sequence with a consensus (C/T)(C/T)CA(C/T)(C/T). This site coincides with the transcription start site. Most of the transcription factor binding sites are located within 500 bp upstream of the transcription start site. Some regulatory sites can be found tens of thousands base pairs away from the gene start site. Occasionally, regulatory elements are located downstream instead of upstream of the transcription start site. Often, a cluster of transcription factor binding sites spread within a wide range to work synergistically to enhance transcription initiation.

PREDICTION ALGORITHMS

Current algorithms for predicting promoters and regulatory elements can be categorized as either ab initio based, which make de novo predictions by scanning individual sequences; or similarity based, which make predictions based on alignment of homologous sequences; or expression profile based using profiles constructed from a number of coexpressed gene sequences from the same organism. The similarity type of prediction is also called phylogenetic footprinting. As mentioned, because RNA polymerase II transcribes the eukaryotic mRNA genes, most algorithms are thus focused on prediction of the RNA polymerase II promoter and associated regulatory elements. Each of the categories is discussed in detail next.

Ab Initio–Based Algorithms

This type of algorithm predicts prokaryotic and eukaryotic promoters and regulatory elements based on characteristic sequences patterns for promoters and regulatory elements. Some ab initio programs are signal based, relying on characteristic promoter sequences such as the TATA box, whereas others rely on content information such as hexamer frequencies. The advantage of the ab initio method is that the sequence can be applied as such without having to obtain experimental information. The limitation is the need for training, which makes the prediction programs species specific. In addition, this type of method has a difficulty in discovering new, unknown motifs.

The conventional approach to detecting a promoter or regulatory site is through matching a consensus sequence pattern represented by regular expressions (see Chapter 7) or matching a position-specific scoring matrix (PSSM; see Chapter 6) constructed from well-characterized binding sites. In either case, the consensus sequences or the matrices are relatively short, covering 6 to 10 bases. As described in Chapter 7, to determine whether a query sequence matches a weight matrix, the sequence is scanned through the matrix. Scores of matches and mismatches at all matrix positions are summed up to give a log odds score, which is then evaluated for statistical significance. This simple approach, however, often has difficulty differentiating true promoters from random sequence matches and generates high rates of false positives as a result.

To better discriminate true motifs from background noise, a new generation of algorithms has been developed that take into account the higher order correlation of multiple subtle features by using discriminant functions, neural networks, or hidden Markov models (HMMs) that are capable of incorporating more neighboring sequence information. To further improve the specificity of prediction, some algorithms selectively exclude coding regions and focus on the upstream regions (0.5 to 2.0 kb) only, which are most likely to contain promoters. In that sense, promoter prediction and gene prediction are coupled.

Prediction for Prokaryotes

One of the unique aspects in prokaryotic promoter prediction is the determination of operon structures, because genes within an operon share a common promoter located upstream of the first gene of the operon. Thus, operon prediction is the key in prokaryotic promoter prediction. Once an operon structure is known, only the first gene is predicted for the presence of a promoter and regulatory elements, whereas other genes in the operon do not possess such DNA elements.

There are a number of methods available for prokaryotic operon prediction. The most accurate is a set of simple rules developed by Wang et al. (2004). This method relies on two kinds of information: gene orientation and intergenic distances of a pair of genes of interest and conserved linkage of the genes based on comparative genomic analysis. More about gene linkage patterns across genomes is introduced in Chapters 16 and 18. A scoring scheme is developed to assign operons with different levels of confidence (Fig. 9.3). This method is claimed to produce accurate identification of an operon structure, which in turn facilitates the promoter prediction.

This newly developed scoring approach is, however, not yet available as a computer program. The prediction can be done manually using the rules, however. The few dedicated programs for prokaryotic promoter prediction do not apply the Wang et al. rule for historical reasons. The most frequently used program is BPROM.

Scoring criteria for operon prediction

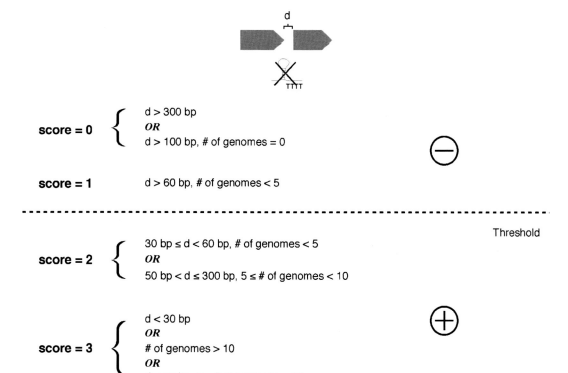

Figure 9.3: Prediction of operons in prokaryotes based on a scoring scheme developed by Wang et al. (2004). This method states that, for two adjacent genes transcribed in the same orientation and without a ρ-independent transcription termination signal in between, the score is assigned 0 if the intergenic distance is larger than 300 bp regardless of the gene linkage pattern or if the distance is larger than 100 bp with the linkage not observed in other genomes. The score is assigned 1 if the intergenic distance is larger than 60 bp with the linkage shared in less than five genomes. The score is assigned 2 if the distance of the two genes is between 30 and 60 bp with the linkage shared in less than five genomes or if the distance is between 50 and 300 bp with the linkage shared in between five to ten genomes. The score is assigned 3 if the intergenic distance is less than 30 bp regardless of the conserved linkage pattern or if the linkage is conserved in more than ten genomes regardless of the intergenic distance or if the distance is less than 50 bp with the linkage shared in between five to ten genomes. A minimum score of 2 is considered the threshold for assigning the two genes in one operon.

BPROM (www.softberry.com/berry.phtml?topic=bprom&group=programs &subgroup=gfindb) is a web-based program for prediction of bacterial promoters. It uses a linear discriminant function (see Chapter 8) combined with signal and content information such as consensus promoter sequence and oligonucleotide composition of the promoter sites. This program first predicts a given sequence for bacterial operon structures by using an intergenic distance of 100 bp as basis for distinguishing genes to be in an operon. This rule is more arbitrary than the Wang et al. rule, leading to high rates of false positives. Once the operons are assigned, the program is able to predict putative promoter sequences. Because most bacterial promoters are located within 200 bp of the protein coding region, the program is most effectively used when about

200 bp of upstream sequence of the first gene of an operon is supplied as input to increase specificity.

FindTerm   (http://sun1.softberry.com/berry.phtml?topic=findterm&group=programs&subgroup=gfindb) is a program for searching bacterial ρ-independent termination signals located at the end of operons. It is available from the same site as FGENES and BPROM. The predictions are made based on matching of known profiles of the termination signals combined with energy calculations for the derived RNA secondary structures for the putative hairpin-loop structure (see Chapter 16). The sequence region that scores best in features and energy terms is chosen as the prediction. The information can sometimes be useful in defining an operon.

Prediction for Eukaryotes
The ab initio method for predicting eukaryotic promoters and regulatory elements also relies on searching the input sequences for matching of consensus patterns of known promoters and regulatory elements. The consensus patterns are derived from experimentally determined DNA binding sites which are compiled into profiles and stored in a database for scanning an unknown sequence to find similar conserved patterns. However, this approach tends to generate very high rate of false positives owing to nonspecific matches with the short sequence patterns. Furthermore, because of the high variability of transcription factor binding sites, the simple sequence matching often misses true promoter sites, creating false negatives.

To increase the specificity of prediction, a unique feature of eukaryotic promoter is employed, which is the presence of CpG islands. It is known that many vertebrate genes are characterized by a high density of CG dinucleotides near the promoter region overlapping the transcription start site (see Chapter 8). By identifying the CpG islands, promoters can be traced on the immediate upstream region from the islands. By combining CpG islands and other promoter signals, the accuracy of prediction can be improved. Several programs have been developed based on the combined features to predict the transcription start sites in particular.

As discussed, the eukaryotic transcription initiation requires cooperation of a large number of transcription factors. *Cooperativity* means that the promoter regions tend to contain a high density of protein-binding sites. Thus, finding a cluster of transcription factor binding sites often enhances the probability of individual binding site prediction.

A number of representatives of ab initio promoter prediction algorithms that incorporate the unique properties of eukaryotic promoters are introduced next.

CpGProD (http://pbil.univ-lyon1.fr/software/cpgprod.html) is a web-based program that predicts promoters containing a high density of CpG islands in mammalian genomic sequences. It calculates moving averages of GC% and CpG ratios (observed/expected) over a window of a certain size (usually 200 bp). When the values are above a certain threshold, the region is identified as a CpG island.

Eponine (http://servlet.sanger.ac.uk:8080/eponine/) is a web-based program that predicts transcription start sites based on a series of preconstructed PSSMs of several regulatory sites, such as the TATA box, the CCAAT box, and CpG islands. The query sequence from a mammalian source is scanned through the PSSMs. The sequence stretches with high-score matching to all the PSSMs, as well as matching of the spacing between the elements, are declared transcription start sites. A Bayesian method is also used in decision making.

Cluster-Buster (http://zlab.bu.edu/cluster-buster/cbust.html) is an HMM-based, web-based program designed to find clusters of regulatory binding sites. It works by detecting a region of high concentration of known transcription factor binding sites and regulatory motifs. A query sequence is scanned with a window size of 1 kb for putative regulatory motifs using motif HMMs. If multiple motifs are detected within a window, a positive score is assigned to each motif found. The total score of the window is the sum of each motif score subtracting a gap penalty, which is proportional to the distances between motifs. If the score of a certain region is above a certain threshold, it is predicted to contain a regulatory cluster.

FirstEF (First Exon Finder; http://rulai.cshl.org/tools/FirstEF/) is a web-based program that predicts promoters for human DNA. It integrates gene prediction with promoter prediction. It uses quadratic discriminant functions (see Chapter 8) to calculate the probabilities of the first exon of a gene and its boundary sites. A segment of DNA (15 kb) upstream of the first exon is subsequently extracted for promoter prediction on the basis of scores for CpG islands.

McPromoter (http://genes.mit.edu/McPromoter.html) is a web-based program that uses a neural network to make promoter predictions. It has a unique promoter model containing six scoring segments. The program scans a window of 300 bases for the likelihoods of being in each of the coding, noncoding, and promoter regions. The input for the neural network includes parameters for sequence physical properties, such as DNA bendability, plus signals such as the TATA box, initiator box, and CpG islands. The hidden layer combines all the features to derive an overall likelihood for a site being a promoter. Another unique feature is that McPromoter does not require that certain patterns must be present, but instead the combination of all features is important. For instance, even if the TATA box score is very low, a promoter prediction can still be made if the other features score highly. The program is currently trained for *Drosophila* and human sequences.

TSSW (www.softberry.com/berry.phtml?topic=promoter) is a web program that distinguishes promoter sequences from non-promoter sequences based on a combination of unique content information such as hexamer/trimer frequencies and signal information such the TATA box in the promoter region. The values are fed to a linear discriminant function (see Chapter 8) to separate true motifs from background noise.

CONPRO (http://stl.bioinformatics.med.umich.edu/conpro) is a web-based program that uses a consensus method to identify promoter elements for human DNA.

To use the program, a user supplies the transcript sequence of a gene (cDNA). The program uses the information to search the human genome database for the position of the gene. It then uses the GENSCAN program to predict 5′ untranslated exons in the upstream region. Once the 5′-most exon is located, a further upstream region (1.5 kb) is used for promoter prediction, which relies on a combination of five promoter prediction programs, TSSG, TSSW, NNPP, PROSCAN, and PromFD. For each program, the highest score prediction is taken as the promoter in the region. If three predictions fall within a 100-bp region, this is considered a consensus prediction. If no three-way consensus is achieved, TSSG and PromFD predictions are taken. Because no coding sequence is used in prediction, specificity is improved relative to each individual program.

Phylogenetic Footprinting–Based Method

It has been observed that promoter and regulatory elements from closely related organisms such as human and mouse are highly conserved. The conservation is both at the sequence level and at the level of organization of the elements. Therefore, it is possible to obtain such promoter sequences for a particular gene through comparative analysis. The identification of conserved noncoding DNA elements that serve crucial functional roles is referred to as *phylogenetic footprinting;* the elements are called *phylogenetic footprints.* This type of method can apply to both prokaryotic and eukaryotic sequences.

The selection of organisms for comparison is an important consideration in this type of analysis. If the pair of organisms selected are too closely related, such as human and chimpanzee, the sequence difference between them may not be sufficient to filter out functional elements. On the other hand, if the organisms' evolutionary distances are too long, such as between human and fish, long evolutionary divergence may render promoter and other elements undetectable. One example of appropriate selection of species is the use of human and mouse sequences, which often yields informative results.

Another caveat of phylogenetic footprinting is to extract noncoding sequences upstream of corresponding genes and focus the comparison to this region only, which helps to prevent false positives. The predictive value of this method also depends on the quality of the subsequent sequence alignments. The advanced alignment programs introduced in Chapter 5 can be used. Even more sophisticated expectation maximization (EM) and Gibbs sampling algorithms can be used in detecting weakly conserved motifs.

There are software programs specifically designed to take advantage of the presence of phylogenetic footprints to make comparisons among a number of related species to identify putative transcription factor binding sites. The advantage in implementing the algorithms is that no training of the probabilistic models is required; hence, it is more broadly applicable. There is also a potential to discover new regulatory motifs shared among organisms. The obvious limitation is the constraint on the evolutionary distances among the orthologous sequences.

ConSite (http://mordor.cgb.ki.se/cgi-bin/CONSITE/consite) is a web server that finds putative promoter elements by comparing two orthologous sequences. The user provides two individual sequences which are aligned by ConSite using a global alignment algorithm. Alternatively, the program accepts precomputed alignment. Conserved regions are identified by calculating identity scores, which are then used to compare against a motif database of regulatory sites (TRANSFAC). High-scoring sequence segments upstream of genes are returned as putative regulatory elements.

rVISTA (http://rvista.dcode.org/) is a similar cross-species comparison tool for promoter recognition. The program uses two orthologous sequences as input and first identifies all putative regulatory motifs based on TRANSFAC matches. It then aligns the two sequences using a local alignment strategy. The motifs that have the highest percent identity in the pairwise comparison are presented graphically as regulatory elements.

PromH(W) (www.softberry.com/berry.phtml?topic=promhw&group=programs &subgroup=promoter) is a web-based program that predicts regulatory sites by pairwise sequence comparison. The user supplies two orthologous sequences, which are aligned by the program to identify conserved regions. These regions are subsequently predicted for RNA polymerase II promoter motifs in both sequences using the TSSW program. Only the conserved regions having high scored promoter motifs are returned as results.

Bayes aligner (www.bioinfo.rpi.edu/applications/bayesian/bayes/bayes_align12. pl) is a web-based footprinting program. It aligns two sequences using a Bayesian algorithm which is a unique sequence alignment method. Instead of returning a single best alignment, the method generates a distribution of a large number of alignments using a full range of scoring matrices and gap penalties. Posterior probability values, which are considered estimates of the true alignment, are calculated for each alignment. By studying the distribution, the alignment that has the highest likelihood score, which is in the extreme margin of the distribution, is chosen. Based on this unique alignment searching algorithm, weakly conserved motifs can be identified with high probability scores.

FootPrinter (http://abstract.cs.washington.edu/~blanchem/FootPrinterWeb/Foot PrinterInput2.pl) is a web-based program for phylogenetic footprinting using multiple input sequences. The user also needs to provide a phylogenetic tree that defines the evolutionary relationship of the input sequences. (One may obtain the tree information from the "Tree of Life" web site [http://tolweb.org/tree/phylogeny.html], which archives known phylogenetic trees using ribosomal RNAs as gene markers.) The program performs multiple alignment of the input sequences to identify conserved motifs. The motifs from organisms spanning over the widest evolutionary distances are identified as promoter or regulatory motifs. In other words, it identifies unusually well-conserved motifs across a set of orthologous sequences.

Expression Profiling–Based Method

Recent advances in high throughput transcription profiling analysis, such as DNA microarray analysis (see Chapter 18) have allowed simultaneous monitoring of expression of hundreds or thousands of genes. Genes with similar expression profiles are considered coexpressed, which can be identified through a clustering approach (see Chapter 18). The basis for coexpression is thought to be due to common promoters and regulatory elements. If this assumption is valid, the upstream sequences of the coexpressed genes can be aligned together to reveal the common regulatory elements recognizable by specific transcription factors.

This approach is essentially experimentally based and appears to be robust for finding transcription factor binding sites. The problem is that the regulatory elements of coexpressed genes are usually short and weak. Their patterns are difficult to discern using simple multiple sequence alignment approaches. Therefore, an advanced alignment-independent profile construction method such as EM and Gibbs motif sampling (see Chapter 7) is often used in finding the subtle sequence motifs. As a reminder, EM is a motif extraction algorithm that finds motifs by repeatedly optimizing a PSSM through comparison with single sequences. Gibbs sampling uses a similar matrix optimization approach but samples motifs with a more flexible strategy and may have a higher likelihood of finding the optimal pattern. Through matrix optimization, subtly conserved motifs can be detected from the background noise.

One of the drawbacks of this approach is that determination of the set of coexpressed genes depends on the clustering approaches, which are known to be error prone. That means that the quality of the input data may be questionable when functionally unrelated genes are often clustered together. In addition, the assumption that coexpressed genes have common regulatory elements is not always valid. Many coexpressed genes have been found to belong to parallel signaling pathways that are under the control of distinct regulatory mechanisms. Therefore, caution should always be exercised when using this method.

The following lists a small selection of motif finders using the EM or Gibbs sampling approach.

MEME (http://meme.sdsc.edu/meme/website/meme-intro.html) is the EM-based program introduced in Chapter 7 for protein motif discovery but can also be used in DNA motif finding. The use is similar to that for protein sequences.

AlignACE (http://atlas.med.harvard.edu/cgi-bin/alignace.pl) is a web-based program using the Gibbs sampling algorithm to find common motifs. The program is optimized for DNA sequence motif extraction. It automatically determines the optimal number and lengths of motifs from the input sequences.

Melina (Motif Elucidator In Nucleotide sequence Assembly; http://melina.hgc.jp/) is a web-based program that runs four individual motif-finding algorithms – MEME, GIBBS sampling, CONSENSUS, and Coresearch – simultaneously. The user compares the results to determine the consensus of motifs predicted by all four prediction methods.

INCLUSive (www.esat.kuleuven.ac.be/~dna/BioI/Software.html) is a suite of web-based tools designed to streamline the process of microarray data collection and sequence motif detection. The pipeline processes microarray data, automatically clusters genes according expression patterns, retrieves upstream sequences of coregulated genes and detects motifs using a Gibbs sampling approach (Motif Sampler). To further avoid the problem of getting stuck in a local optimum (see Chapter 7), each sequence dataset is submitted to Motif Sampler ten times. The results may vary in each run. The results from the ten runs are compiled to derive consensus motifs.

PhyloCon (Phylogenetic Consensus; http://ural.wustl.edu/~twang/PhyloCon/) is a UNIX program that combines phylogenetic footprinting with gene expression profiling analysis to identify regulatory motifs. This approach takes advantage of conservation among orthologous genes as well as conservation among coregulated genes. For each individual gene in a set of coregulated genes, multiple sequence homologs are aligned to derive profiles. Based on the gene expression data, profiles between coregulated genes are further compared to identify functionally conserved motifs among evolutionary conserved motifs. In other words, regulatory motifs are defined from both sets of analysis. This approach integrates the "single gene–multiple species" and "single species–multiple genes" methods and has been found to reduce false positives compared to either phylogenetic footprinting or simple motif extraction approaches alone.

SUMMARY

Identification of promoter and regulatory elements remains a great bioinformatic challenge. The existing algorithms can be classified as ab initio based, phylogenetic footprinting based, and expression profiling based. The true accuracy of the ab initio programs is still difficult to assess because of the lack of common benchmarks. The reported overall sensitivity and specificity levels are currently below 0.5 for most programs. For a prediction method to be acceptable, both accuracy indicators have to be consistently above 0.9 to be reliable enough for routine prediction purposes. That means that the algorithmic development in this field still has a long road ahead. To achieve better results, combining multiple prediction programs seems to be helpful in some circumstances. The comparative approach using phylogenetic footprinting is able to take a completely different approach in identifying promoter elements. The resulting prediction can be used to check against the ab initio prediction. Finally, the experimental based approach using gene expression data offers another route to finding regulatory motifs. Because the DNA motifs are often subtle, EM and Gibbs motif sampling algorithms are necessary for this purpose. Alternatively, the EM and Gibbs sampling programs can be used for phylogenetic footprinting if the input sequences are from different organisms. In essence, all three approaches are interrelated. The results from all three types of methods can be combined to further increase the reliability of the predictions.

FURTHER READING

Dubchak, I., and Pachter, L. 2002. The computational challenges of applying comparative-based computational methods to whole genomes. *Brief. Bioinform.* 3:18–22.

Hannenhalli, S., and Levy, S. 2001. Promoter prediction in the human genome. *Bioinformatics* 17(Suppl):S90–6.

Hehl, R., and Wingender, E. 2001. Database-assisted promoter analysis. *Trends Plant Sci.* 6:251–5.

Ohler, U., and Niemann, H. 2001. Identification and analysis of eukaryotic promoters: Recent computational approaches. *Trends Genet.* 17:56–60.

Ovcharenko, I., and Loots, G. G. 2003. Finding the needle in the haystack: Computational strategies for discovering regulatory sequences in genomes. *Curr. Genomics* 4:557–68.

Qiu, P. 2003. Recent advances in computational promoter analysis in understanding the transcriptional regulatory network. *Biochem. Biophys. Res. Commun.* 309:495–501.

Rombauts S., Florquin K., Lescot M., Marchal K., Rouze P., and van de Peer Y. 2003. Computational approaches to identify promoters and *cis*-regulatory elements in plant genomes. *Plant Physiol.* 132:1162–76.

Salgado, H., Moreno-Hagelsieb, G., Smith, T. F., and Collado-Vides, J. 2000. Operons in *Escherichia coli*: Genomic analyses and predictions. *Proc. Natl. Acad. Sci. U S A* 97:6652–7.

Wang, L., Trawick, J. D., Yamamoto, R., and Zamudio, C. 2004. Genome-wide operon prediction in *Staphylococcus aureus*. *Nucleic Acids Res.* 32:3689–702.

Werner, T. 2003. The state of the art of mammalian promoter recognition. *Brief. Bioinform.* 4:22–30.

Molecular Phylogenetics

CHAPTER TEN

Phylogenetics Basics

Biological sequence analysis is founded on solid evolutionary principles (see Chapter 2). Similarities and divergence among related biological sequences revealed by sequence alignment often have to be rationalized and visualized in the context of phylogenetic trees. Thus, molecular phylogenetics is a fundamental aspect of bioinformatics. In this chapter, we focus on phylogenetic tree construction. Before discussing the methods of phylogenetic tree construction, some fundamental concepts and background terminology used in molecular phylogenetics need to be described. This is followed by discussion of the initial steps involved in phylogenetic tree construction.

MOLECULAR EVOLUTION AND MOLECULAR PHYLOGENETICS

To begin the phylogenetics discussion, we need to understand the basic question, "What is evolution?" Evolution can be defined in various ways under different contexts. In the biological context, evolution can be defined as the development of a biological form from other preexisting forms or its origin to the current existing form through natural selections and modifications. The driving force behind evolution is natural selection in which "unfit" forms are eliminated through changes of environmental conditions or sexual selection so that only the fittest are selected. The underlying mechanism of evolution is genetic mutations that occur spontaneously. The mutations on the genetic material provide the biological diversity within a population; hence, the variability of individuals within the population to survive successfully in a given environment. Genetic diversity thus provides the source of raw material for the natural selection to act on.

Phylogenetics is the study of the evolutionary history of living organisms using tree-like diagrams to represent pedigrees of these organisms. The tree branching patterns representing the evolutionary divergence are referred to as *phylogeny*. Phylogenetics can be studied in various ways. It is often studied using fossil records, which contain morphological information about ancestors of current species and the timeline of divergence. However, fossil records have many limitations; they may be available only for certain species. Existing fossil data can be fragmentary and their collection is often limited by abundance, habitat, geographic range, and other factors. The descriptions of morphological traits are often ambiguous, which are due to multiple genetic factors. Thus, using fossil records to determine phylogenetic relationships can often be biased. For microorganisms, fossils are essentially nonexistent, which makes it impossible to study phylogeny with this approach.

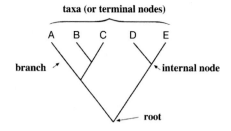

Figure 10.1: A typical bifurcating phylogenetic tree showing root, internal nodes, terminal nodes and branches.

Fortunately, molecular data that are in the form of DNA or protein sequences can also provide very useful evolutionary perspectives of existing organisms because, as organisms evolve, the genetic materials accumulate mutations over time causing phenotypic changes. Because genes are the medium for recording the accumulated mutations, they can serve as *molecular fossils.* Through comparative analysis of the molecular fossils from a number of related organisms, the evolutionary history of the genes and even the organisms can be revealed.

The advantage of using molecular data is obvious. Molecular data are more numerous than fossil records and easier to obtain. There is no sampling bias involved, which helps to mend the gaps in real fossil records. More clear-cut and robust phylogenetic trees can be constructed with the molecular data. Therefore, they have become favorite and sometimes the only information available for researchers to reconstruct evolutionary history. The advent of the genomic era with tremendous amounts of molecular sequence data has led to the rapid development of molecular phylogenetics.

The field of molecular phylogenetics can be defined as the study of evolutionary relationships of genes and other biological macromolecules by analyzing mutations at various positions in their sequences and developing hypotheses about the evolutionary relatedness of the biomolecules. Based on the sequence similarity of the molecules, evolutionary relationships between the organisms can often be inferred.

Major Assumptions

To use molecular data to reconstruct evolutionary history requires making a number of reasonable assumptions. The first is that the molecular sequences used in phylogenetic construction are homologous, meaning that they share a common origin and subsequently diverged through time. Phylogenetic divergence is assumed to be bifurcating, meaning that a parent branch splits into two daughter branches at any given point. Another assumption in phylogenetics is that each position in a sequence evolved independently. The variability among sequences is sufficiently informative for constructing unambiguous phylogenetic trees.

TERMINOLOGY

Before discussing methods for reconstruction of phylogenies, it is useful to define some frequently used terminology that characterizes a phylogenetic tree. A typical bifurcating phylogenetic tree is a graph shown in Figure 10.1. The lines in the tree are

Figure 10.2: A phylogenetic tree showing an example of bifurcation and multifurcation. Multifurcation is normally a result of insufficient evidence to fully resolve the tree or a result of an evolutionary process known as *radiation*.

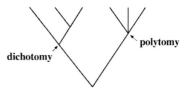

called *branches*. At the tips of the branches are present-day species or sequences known as *taxa* (the singular form is *taxon*) or operational taxonomic units. The connecting point where two adjacent branches join is called a *node*, which represents an inferred ancestor of extant taxa. The bifurcating point at the very bottom of the tree is the *root node*, which represents the common ancestor of all members of the tree.

A group of taxa descended from a single common ancestor is defined as a *clade* or *monophyletic group*. In a monophyletic group, two taxa share a unique common ancestor not shared by any other taxa. They are also referred to as *sister taxa* to each other (e.g., taxa B and C). The branch path depicting an ancestor–descendant relationship on a tree is called a *lineage*, which is often synonymous with a tree branch leading to a defined monophyletic group. When a number of taxa share more than one closest common ancestors, they do not fit the definition of a clade. In this case, they are referred to as *paraphyletic* (e.g., taxa B, C, and D).

The branching pattern in a tree is called *tree topology*. When all branches bifurcate on a phylogenetic tree, it is referred to as *dichotomy*. In this case, each ancestor divides and gives rise to two descendants. Sometimes, a branch point on a phylogenetic tree may have more than two descendents, resulting in a *multifurcating node*. The phylogeny with multifurcating branches is called *polytomy* (Fig. 10.2). A polytomy can be a result of either an ancestral taxon giving rise to more than two immediate descendants simultaneously during evolution, a process known as *radiation*, or an unresolved phylogeny in which the exact order of bifurcations cannot be determined precisely.

A phylogenetic tree can be either rooted or unrooted (Fig. 10.3). An *unrooted phylogenetic tree* does not assume knowledge of a common ancestor, but only positions the taxa to show their relative relationships. Because there is no indication of which node represents an ancestor, there is no direction of an evolutionary path in an

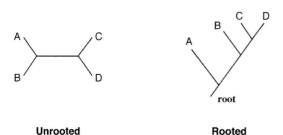

Unrooted   **Rooted**

Figure 10.3: An illustration of rooted versus unrooted trees. A phylogenetic tree without definition of a root is unrooted (*left*). The tree with a root is rooted (*right*).

unrooted tree. To define the direction of an evolution path, a tree must be rooted. In a *rooted tree*, all the sequences under study have a common ancestor or root node from which a unique evolutionary path leads to all other nodes. Obviously, a rooted tree is more informative than an unrooted one. To convert an unrooted tree to a rooted tree, one needs to first determine where the root is.

Strictly speaking, the root of the tree is not known; the common ancestor is already extinct. In practice, however, it is often desirable to define the root of a tree. There are two ways to define the root of a tree. One is to use an *outgroup*, which is a sequence that is homologous to the sequences under consideration, but separated from those sequences at an early evolutionary time. Outgroups are generally determined from independent sources of information. For example, a bird sequence can be used as a root for the phylogenetic analysis of mammals based on multiple lines of evidence that indicate that birds branched off prior to all mammalian taxa in the ingroup. Outgroups are required to be distinct from the ingroup sequences, but not too distant from the ingroup. Using too divergent sequences as an outgroup can lead to errors in tree construction. In the absence of a good outgroup, a tree can be rooted using the *midpoint rooting approach*, in which the midpoint of the two most divergent groups judged by overall branch lengths is assigned as the root. This type of rooting assumes that divergence from root to tips for both branches is equal and follows the "molecular clock" hypothesis.

Molecular clock is an assumption by which molecular sequences evolve at constant rates so that the amount of accumulated mutations is proportional to evolutionary time. Based on this hypothesis, branch lengths on a tree can be used to estimate divergence time. This assumption of uniformity of evolutionary rates, however, rarely holds true in reality.

GENE PHYLOGENY VERSUS SPECIES PHYLOGENY

One of the objectives of building phylogenetic trees based on molecular sequences is to reconstruct the evolutionary history of the species involved. However, strictly speaking, a gene phylogeny (phylogeny inferred from a gene or protein sequence) only describes the evolution of that particular gene or encoded protein. This sequence may evolve more or less rapidly than other genes in the genome or may have a different evolutionary history from the rest of the genome owing to horizontal gene transfer events (see Chapter 17). Thus, the evolution of a particular sequence does not necessarily correlate with the evolutionary path of the species. The species evolution is the combined result of evolution by multiple genes in a genome. In a species tree, the branching point at an internal node represents the speciation event whereas, in a gene tree, the internal node indicates a gene duplication event. The two events may or may not coincide. Thus, to obtain a species phylogeny, phylogenetic trees from a variety of gene families need to be constructed to give an overall assessment of the species evolution.

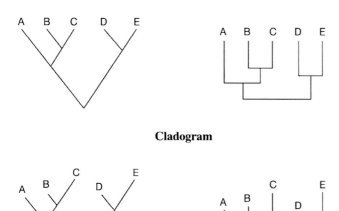

Cladogram

Phylogram

Figure 10.4: Phylogenetic trees drawn as cladograms (*top*) and phylograms (*bottom*). The branch lengths are unscaled in the cladograms and scaled in the phylograms. The trees can be drawn as angled form (*left*) or squared form (*right*).

FORMS OF TREE REPRESENTATION

The topology of branches in a tree defines the relationships between the taxa. The trees can be drawn in different ways, such as a cladogram or a phylogram (Fig. 10.4). In each of these tree representations, the branches of a tree can freely rotate without changing the relationships among the taxa.

In a *phylogram*, the branch lengths represent the amount of evolutionary divergence. Such trees are said to be scaled. The scaled trees have the advantage of showing both the evolutionary relationships and information about the relative divergence time of the branches. In a *cladogram*, however, the external taxa line up neatly in a row or column. Their branch lengths are not proportional to the number of evolutionary changes and thus have no phylogenetic meaning. In such an unscaled tree, only the topology of the tree matters, which shows the relative ordering of the taxa.

To provide information of tree topology to computer programs without having to draw the tree itself, a special text format known as the *Newick format* is developed. In this format, trees are represented by taxa included in nested parentheses. In this linear representation, each internal node is represented by a pair of parentheses that enclose all member of a monophyletic group separated by a comma. For a tree with scaled branch lengths, the branch lengths in arbitrary units are placed immediately after the name of the taxon separated by a colon. An example of using the Newick format to describe tree topology is shown in Figure 10.5.

Sometimes a tree-building method may result in several equally optimal trees. A consensus tree can be built by showing the commonly resolved bifurcating portions

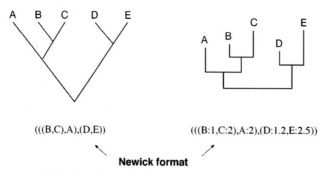

$(((B,C),A),(D,E))$ $(((B:1,C:2),A:2),(D:1.2,E:2.5))$

Newick format

Figure 10.5: Newick format of tree representation that employs a linear form of nested parentheses within which taxa are separated by commas. If the tree is scaled, branch lengths are indicated immediately after the taxon name. The numbers are relative units that represent divergent times.

and collapsing the ones that disagree among the trees, which results in a polytomy. Combining the nodes can be done either by strict consensus or by majority rule. In a strict consensus tree, all conflicting nodes are collapsed into polytomies. In a consensus tree based on a majority rule, among the conflicting nodes, those that agree by more than 50% of the nodes are retained whereas the remaining nodes are collapsed into multifurcation (Fig. 10.6).

WHY FINDING A TRUE TREE IS DIFFICULT

The main objective of molecular phylogenetics is to correctly reconstruct the evolutionary history based on the observed sequence divergence between organisms. That means finding a correct tree topology with correct branch lengths. However, the search for a correct tree topology can sometimes be extremely difficult and computationally demanding. The reason is that the number of potential tree topologies can

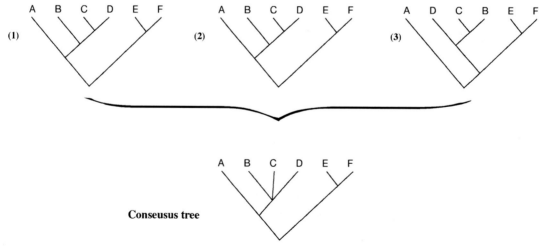

Figure 10.6: A consensus tree is derived from three individual inferred trees based on a majority rule. Conflicting nodes are represented by a multifurcating node in the consensus tree.

be enormously large even with a moderate number of taxa. The increase of possible tree topologies follows an exponential function. The number of rooted trees (N_R) for n taxa is determined by the following formula:

$$N_R = (2n - 3)!/2^{n-2}(n - 2)! \qquad \text{(Eq. 10.1)}$$

In this formula, $(2n-3)!$ is a mathematical expression of factorial, which is the product of positive integers from 1 to $2n - 3$. For example, $5! = 1 \times 2 \times 3 \times 4 \times 5 = 120$.

For unrooted trees, the number of unrooted tree topologies (N_U) is:

$$N_U = (2n - 5)!/2^{n-3}(n - 3)! \qquad \text{(Eq. 10.2)}$$

An example of all possible rooted and unrooted tree topologies for three and four taxa is shown in Figure 10.7. For three taxa, there is only one possible unrooted tree but three different rooted trees. For four taxa, one can construct three possible unrooted trees and fifteen rooted ones. The number of possible topologies increases extremely rapidly with the number of taxa. According to Equation 10.1 and Equation 10.2, for six taxa, there are 105 unrooted trees and 945 rooted trees. If there are ten taxa, there can be 2,027,025 unrooted trees and 34,459,425 rooted ones. The exponential relationship between the number of tree topologies and the number of taxa is clearly represented in Figure 10.8. There can be an explosive increase in the possible tree topologies as the number of taxa increases. Therefore, it can be computationally very demanding to find a true phylogenetic tree when the number of sequences is large. Because the number of rooted topologies is much larger than that for unrooted ones, the search for a true phylogenetic tree can be simplified by calculating the unrooted trees first. Once an optimal tree is found, rooting the tree can be performed by designating a number of taxa in the data set as an outgroup based on external information to produce a rooted tree.

PROCEDURE

Molecular phylogenetic tree construction can be divided into five steps: (1) choosing molecular markers; (2) performing multiple sequence alignment; (3) choosing a model of evolution; (4) determining a tree building method; and (5) assessing tree reliability. Each of first three steps is discussed herein; steps 4 and 5 are discussed in Chapter 11.

Choice of Molecular Markers

For constructing molecular phylogenetic trees, one can use either nucleotide or protein sequence data. The choice of molecular markers is an important matter because it can make a major difference in obtaining a correct tree. The decision to use nucleotide or protein sequences depends on the properties of the sequences and the purposes of the study. For studying very closely related organisms, nucleotide sequences, which evolve more rapidly than proteins, can be used. For example, for evolutionary analysis of different individuals within a population, noncoding regions of mitochondrial

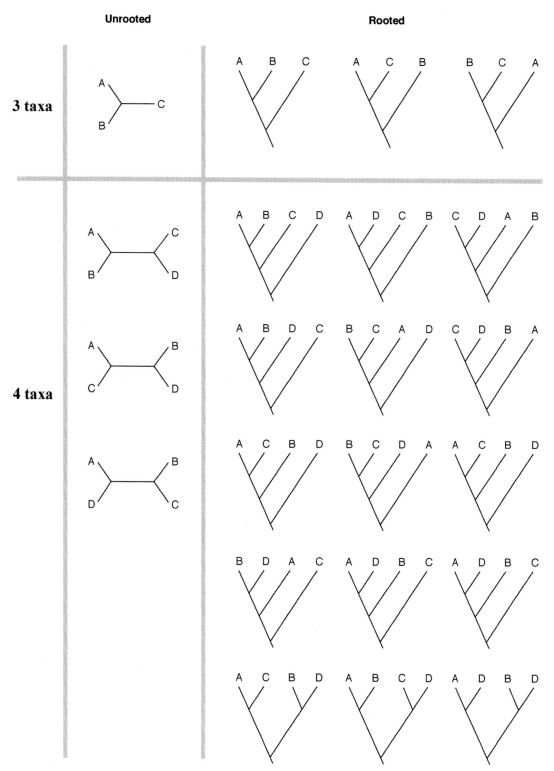

Figure 10.7: All possible tree topologies for three and four taxa. For three taxa, there are one unrooted and three rooted trees. For four taxa, there are three unrooted and fifteen rooted trees.

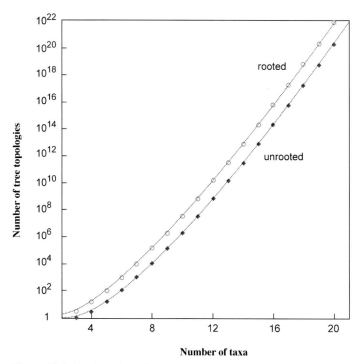

Figure 10.8: Total number of rooted (○) and unrooted (◆) tree topologies as a function of the number of taxa. The values in the *y*-axis are plotted in the log scale.

DNA are often used. For studying the evolution of more widely divergent groups of organisms, one may choose either slowly evolving nucleotide sequences, such as ribosomal RNA or protein sequences. If the phylogenetic relationships to be delineated are at the deepest level, such as between bacteria and eukaryotes, using conserved protein sequences makes more sense than using nucleotide sequences. The reason is explained in more detail next.

In many cases, protein sequences are preferable to nucleotide sequences because protein sequences are relatively more conserved as a result of the degeneracy of the genetic code in which sixty-one codons encode for twenty amino acids, meaning thereby a change in a codon may not result in a change in amino acid. Thus, protein sequences can remain the same while the corresponding DNA sequences have more room for variation, especially at the third codon position. The significant difference in evolutionary rates among the three nucleotide positions also violates one of the assumptions of tree-building. In contrast, the protein sequences do not suffer from this problem, even for divergent sequences.

DNA sequences are sometimes more biased than protein sequences because of preferential codon usage in different organisms. In this case, different codons for the same amino acid are used at different frequencies, leading to sequence variations not attributable to evolution. In addition, the genetic code of mitochondria varies from the standard genetic code. Therefore, for comparison of mitochondria protein-coding genes, it is necessary to translate the DNA sequences into protein sequences.

As mentioned in Chapter 4, protein sequences allow more sensitive alignment than DNA sequences because the former has twenty characters versus four in the latter. It has been shown that two randomly related DNA sequences can result in up to 50% sequence identity when gaps are allowed compared to only 10% for protein sequences. For moderately divergent sequences, it is almost impossible to use DNA sequences to obtain correct alignment. In addition, to align protein-coding DNA sequences, when gaps are introduced to maximize alignment scores, they almost always cause frameshift errors, making the alignment biologically meaningless. Protein sequences clearly have a higher signal-to-noise ratio when it comes to alignment and phylogenetic analysis. Thus, protein-based phylogeny in most cases may be more appropriate than DNA-based phylogeny.

Despite the advantages of using protein sequences in phylogenetic inference, DNA sequences can still be very informative in some cases, such as those for closely related sequences. In this case, faster evolutionary rates at the DNA level become an advantage. In addition, DNA sequences depict synonymous and nonsynonymous substitutions, which can be useful for revealing evidence of positive or negative selection events.

To understand positive or negative selection, it is necessary to make a distinction between synonymous substitutions and nonsynonymous substitutions. *Synonymous substitutions* are nucleotide changes in the coding sequence that do not result in amino acid sequence changes for the encoded protein. *Nonsynonymous substitutions* are nucleotide changes that result in alterations in the amino acid sequences.

Comparing the two types of substitution rates helps to understand an evolutionary process of a sequence. For example, if the nonsynonymous substitution rate is found to be significantly greater than the synonymous substitution rate, this means that certain parts of the protein are undergoing active mutations that may contribute to the evolution of new functions. This is described as *positive selection* or *adaptive evolution*. On the other hand, if the synonymous substitution rate is greater than the nonsynonymous substitution rate, this causes only neutral changes at the amino acid level, suggesting that the protein sequence is critical enough that changes at the amino acid sequence level are not tolerated. In this case, the sequence is said to be under *negative* or *purifying selection*.

Alignment

The second step in phylogenetic analysis is to construct sequence alignment. This is probably the most critical step in the procedure because it establishes positional correspondence in evolution. Only the correct alignment produces correct phylogenetic inference because aligned positions are assumed to be genealogically related. Incorrect alignment leads to systematic errors in the final tree or even a completely wrong tree. For that reason, it is essential that the sequences are correctly aligned. Multiple state-of-the-art alignment programs such as T-Coffee should be used. The alignment results from multiple sources should be inspected and compared carefully to identify the most reasonable one. Automatic sequence alignments almost always contain errors and should be further edited or refined if necessary.

Manual editing is often critical in ensuring alignment quality. However, there is no firm rule on how to modify a sequence alignment. As a general guideline, a correct alignment should ensure the matching of key cofactor residues and residues of similar physicochemical properties. If secondary structure elements are known or can be predicted (see Chapter 14), they can serve to guide the alignment. One of the few alignment programs that incorporates protein secondary structure information is Praline (see Chapter 5).

It is also often necessary to decide whether to use the full alignment or to extract parts of it. Truly ambiguously aligned regions have to be removed from consideration prior to phylogenetic analysis. Which part of the alignment to remove is often at the discretion of the researcher. It is a rather subjective process. In extreme cases, some researchers like to remove all insertions and deletions (indels) and only use positions that are shared by all sequences in the dataset. The clear drawback of this practice is that many phylogenetic signals are lost. In fact, gap regions often belong to *signature indels* unique to identification of a subgroup of sequences and should to be retained for treeing purposes.

In addition, there is an automatic approach in improving alignment quality. Rascal and NorMD (see Chapter 5) can help to improve alignment by correcting alignment errors and removing potentially unrelated or highly divergent sequences. Furthermore, the program Gblocks (http://woody.embl-heidelberg.de/phylo/) can help to detect and eliminate the poorly aligned positions and divergent regions so to make the alignment more suitable for phylogenetic analysis.

Multiple Substitutions

A simple measure of the divergence between two sequences is to count the number of substitutions in an alignment. The proportion of substitutions defines the observed distance between the two sequences. However, the observed number of substitutions may not represent the true evolutionary events that actually occurred. When a mutation is observed as A replaced by C, the nucleotide may have actually undergone a number of intermediate steps to become C, such as $A \rightarrow T \rightarrow G \rightarrow C$. Similarly, a back mutation could have occurred when a mutated nucleotide reverted back to the original nucleotide. This means that when the same nucleotide is observed, mutations like $G \rightarrow C \rightarrow G$ may have actually occurred. Moreover, an identical nucleotide observed in the alignment could be due to parallel mutations when both sequences mutate into T, for instance.

Such multiple substitutions and convergence at individual positions obscure the estimation of the true evolutionary distances between sequences. This effect is known as *homoplasy*, which, if not corrected, can lead to the generation of incorrect trees. To correct homoplasy, statistical models are needed to infer the true evolutionary distances between sequences.

Choosing Substitution Models

The statistical models used to correct homoplasy are called *substitution models* or *evolutionary models*. For constructing DNA phylogenies, there are a number of nucleotide substitution models available. These models differ in how multiple substitutions of each nucleotide are treated. The caveat of using these models is that if there are too many multiple substitutions at a particular position, which is often true for very divergent sequences, the position may become saturated. This means that the evolutionary divergence is beyond the ability of the statistical models to correct. In this case, true evolutionary distances cannot be derived. Therefore, only reasonably similar sequences are to be used in phylogenetic comparisons.

Jukes–Cantor Model

The simplest nucleotide substitution model is the Jukes–Cantor model, which assumes that all nucleotides are substituted with equal probability. A formula for deriving evolutionary distances that include hidden changes is introduced by using a logarithmic function.

$$d_{AB} = -(3/4)\ \ln[1 - (4/3)\,p_{AB}] \tag{Eq. 10.3}$$

where d_{AB} is the evolutionary distance between sequences A and B and p_{AB} is the observed sequence distance measured by the proportion of substitutions over the entire length of the alignment.

For example, if an alignment of sequences A and B is twenty nucleotides long and six pairs are found to be different, the sequences differ by 30%, or have an observed distance 0.3. To correct for multiple substitutions using the Jukes–Cantor model, the corrected evolutionary distance based on Equation 10.3 is:

$$d_{AB} = -3/4\ \ln[1 - (4/3 \times 0.3)] = 0.38$$

The Jukes–Cantor model can only handle reasonably closely related sequences. According to Equation 10.3, the normalized distance increases as the actual observed distance increases. For distantly related sequences, the correction can become too large to be reliable. If two DNA sequences have 25% similarity, p_{AB} is 0.75. This leads the log value to be infinitely large.

Kimura Model

Another model to correct evolutionary distances is called the Kimura two-parameter model. This is a more sophisticated model in which mutation rates for transitions and transversion are assumed to be different, which is more realistic. According to this model, transitions occur more frequently than transversions, which, therefore, provides a more realistic estimate of evolutionary distances. The Kimura model uses the following formula:

$$d_{AB} = -(1/2)\ \ln(1 - 2p_{ti} - p_{tv}) - (1/4)\ \ln(1 - 2p_{tv}) \tag{Eq. 10.4}$$

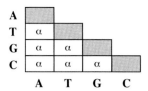   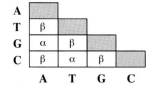

Jukes-Cantor model   **Kimura model**

Figure 10.9: The Jukes–Cantor and Kimura models for DNA substitutions. In the Jukes–Cantor model, all nucleotides have equal substitution rates (α). In the Kimura model, there are unequal rates of transitions (α) and transversions (β). The probability values for identical matches are shaded because evolutionary distances only count different residue positions.

where d_{AB} is the evolutionary distance between sequences A and B, p_{ti} is the observed frequency for transition, and p_{tv} the frequency of transversion. Comparison of the Jukes–Cantor model and the Kimura model is graphically illustrated in Figure 10.9.

An example of using the Kimura model can be illustrated by the comparison of sequences A and B that differ by 30%. If 20% of changes are a result of transitions and 10% of changes are a result of transversions, the evolutionary distance can be calculated using Equation 10.4:

$$d_{AB} = -1/2 \ln(1 - 2 \times 0.2 - 0.1) - 1/4 \ln(1 - 2 \times 0.1) = 0.40$$

In addition to these models, there are more complex models, such as TN93, HKY, and GTR, that take many more parameters into consideration. However, these more complex models are normally not used in practice because the calculations are too complicated and the variance levels resulting from the formula are too high.

For protein sequences, the evolutionary distances from an alignment can be corrected using a PAM or JTT amino acid substitution matrix whose construction already takes into account the multiple substitutions (see Chapter 3). Alternatively, protein equivalents of Jukes–Cantor and Kimura models can be used to correct evolutionary distances. For example, the Kimura model for correcting multiple substitutions in protein distances is:

$$d = -\ln(1 - p - 0.2p^2) \qquad \text{(Eq. 10.5)}$$

whereas p is the observed pairwise distance between two sequences.

Among-Site Variations

In all these calculations, different positions in a sequence are assumed to be evolving at the same rate. However, this assumption may not hold up in reality. For example, in DNA sequences, the rates of substitution differ for different codon positions. The third codon mutates much faster than the other two. For protein sequences, some amino acids change much more rarely than others owing to functional constraints. This variation in evolutionary rates is the so-called among-site rate heterogeneity, which can also cause artifacts in tree construction.

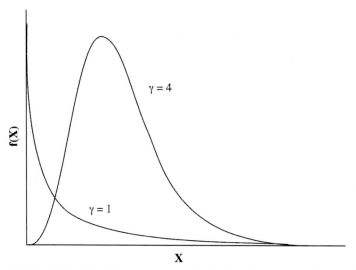

Figure 10.10: Probability curves of γ distribution. The mathematical function of the distribution is $f(x) + (x^{\gamma-1} e^{-x})/\Gamma(\gamma)$. The curves assume different shapes depending on the γ-shape parameter (γ).

It has been shown that there are always a proportion of positions in a sequence dataset that have invariant rates and a proportion that have more variable rates. The distribution of variant sites follows a γ distribution pattern. The γ distribution is a general probability function that has distribution curves with variable shapes depending on the values of the γ shape parameter (Fig. 10.10). Therefore, to account for site-dependent rate variation, a γ correction factor can be used. For the Jukes–Cantor model, the evolution distance can be adjusted with the following formula:

$$d_{AB} = (3/4)\alpha[(1 - 4/3\,p_{AB})^{-1/\alpha} - 1 \qquad \text{(Eq. 10.6)}$$

where α is the γ correction factor. For the Kimura model, the evolutionary distance with γ correction factor becomes

$$d_{AB} = (\alpha/2)[1 - 2p_{ti} - p_{tv})^{-1/\alpha} - (1/2)(1 - 2p_{tv})^{-1/\alpha} - 1/2] \qquad \text{(Eq. 10.7)}$$

Estimation of the value of the γ correction factor (α) is implemented in a number of tree-building programs.

SUMMARY

Molecular phylogenetics is the study of evolutionary relationships among living organisms using molecular data such as DNA and protein sequences. It operates on the basis of a number of assumptions – e. g., an evolutionary tree is always binary and all sequence positions evolve independently. The branches of a tree define its topology. The number of possible tree topologies depends on the number of taxa and increases extremely rapidly as the number taxa goes up. A tree based on gene sequences does not always correlate with the evolution of the species. Caution is needed in extrapolation of phylogenetic results. A phylogenetic tree can be rooted or unrooted. The best way to root a tree is to use an outgroup, the selection of which relies on external knowledge. The first step in phylogenetic construction is to decide whether to use DNA sequences or protein sequences, each having merits and limitations. Protein sequences are preferable in most cases. However, for studying very recent evolution, DNA is the marker of choice.

The second step is to perform multiple sequence alignment. Obtaining accurate alignment is critical for phylogenetic tree construction. The unique aspect of multiple alignment for phylogenetic analysis is that it often requires manual truncation of ambiguously aligned regions. The next step is to select a proper substitution model that provides estimates of the true evolutionary event by taking into account multiple substitution events. Corrected evolutionary distances are used both in distance-based and likelihood-based tree-building methods. The commonly used nucleotide substitution models are the Jukes–Cantor and Kimura models. The commonly used amino acid substitution models are the PAM and JTT models. Other adjustments to improve the estimation of true evolutionary distances include the incorporation of rate heterogeneity among sites.

FURTHER READING

Graur, D., and Li, W.-H. 2000. *Fundamentals of Molecular Evolution.* Sunderland, MA: Sinauer Associates.

Higgins, D. G. 2000. Amino acid-based phylogeny and alignment. *Adv. Protein Chem.* 54:99–135.

Nei, M., and Kumar, S. 2000. *Molecular Evolution and Phylogenetics.* New York: Oxford University Press.

Salemi, M., and Vandamme, A.-M. 2003. *The Phylogenetics Handbook – A Practical Approach to DNA and Protein Phylogeny.* Cambridge, UK: Cambridge University Press.

Thornton, J. W., and DeSalle, R. 2000. Gene family evolution and homology: Genomics meets phylogenetics. *Annu. Rev. Genomics Hum. Genet.* 1:41–73.

Whelan, S., Lio, P., and Goldman, N. 2001. Molecular phylogenetics: State of the art methods for looking into the past. *Trends Genet.* 17:262–72.

CHAPTER ELEVEN

Phylogenetic Tree Construction Methods and Programs

To continue discussion of molecular phylogenetics from Chapter 10, this chapter introduces the theory behind various phylogenetic tree construction methods along with the strategies used for executing the tree construction.

There are currently two main categories of tree-building methods, each having advantages and limitations. The first category is based on discrete characters, which are molecular sequences from individual taxa. The basic assumption is that characters at corresponding positions in a multiple sequence alignment are homologous among the sequences involved. Therefore, the character states of the common ancestor can be traced from this dataset. Another assumption is that each character evolves independently and is therefore treated as an individual evolutionary unit. The second category of phylogenetic methods is based on distance, which is the amount of dissimilarity between pairs of sequences, computed on the basis of sequence alignment. The distance-based methods assume that all sequences involved are homologous and that tree branches are additive, meaning that the distance between two taxa equals the sum of all branch lengths connecting them. More details on procedures and assumptions for each type of phylogenetic method are described.

DISTANCE-BASED METHODS

As mentioned in Chapter 10, true evolutionary distances between sequences can be calculated from observed distances after correction using a variety of evolutionary models. The computed evolutionary distances can be used to construct a matrix of distances between all individual pairs of taxa. Based on the pairwise distance scores in the matrix, a phylogenetic tree can be constructed for all the taxa involved. The algorithms for the distance-based tree-building method can be subdivided into either clustering based or optimality based. The clustering-type algorithms compute a tree based on a distance matrix starting from the most similar sequence pairs. These algorithms include an unweighted pair group method using arithmetic average (UPGMA) and neighbor joining. The optimality-based algorithms compare many alternative tree topologies and select one that has the best fit between estimated distances in the tree and the actual evolutionary distances. This category includes the Fitch–Margoliash and minimum evolution algorithms.

Clustering-Based Methods

Unweighted Pair Group Method Using Arithmetic Average

The simplest clustering method is UPGMA, which builds a tree by a sequential clustering method. Given a distance matrix, it starts by grouping two taxa with the smallest pairwise distance in the distance matrix. A node is placed at the midpoint or half distance between them. It then creates a reduced matrix by treating the new cluster as a single taxon. The distances between this new composite taxon and all remaining taxa are calculated to create a reduced matrix. The same grouping process is repeated and another newly reduced matrix is created. The iteration continues until all taxa are placed on the tree (see Box 11.1). The last taxon added is considered the outgroup producing a rooted tree.

The basic assumption of the UPGMA method is that all taxa evolve at a constant rate and that they are equally distant from the root, implying that a molecular clock (see Chapter 10) is in effect. However, real data rarely meet this assumption. Thus, UPGMA often produces erroneous tree topologies. However, owing to its fast speed of calculation, it has found extensive usage in clustering analysis of DNA microarray data (see Chapter 17).

Neighbor Joining

The UPGMA method uses unweighted distances and assumes that all taxa have constant evolutionary rates. Since this molecular clock assumption is often not met in biological sequences, to build a more accurate phylogenetic trees, the neighbor-joining (NJ) method can be used, which is somewhat similar to UPGMA in that it builds a tree by using stepwise reduced distance matrices. However, the NJ method does not assume the taxa to be equidistant from the root. It corrects for unequal evolutionary rates between sequences by using a conversion step. This conversion requires the calculations of "r-values" and "transformed r-values" using the following formula:

$$d'_{AB} = d_{AB} - 1/2 \times (r_A + r_B) \tag{Eq. 11.1}$$

where d'_{AB} is the converted distance between A and B and d_{AB} is the actual evolutionary distance between A and B. The value of r_A (or r_B) is the sum of distances of A (or B) to all other taxa. A generalized expression of the r-value is r_i calculated based on the following formula:

$$r_i = \Sigma d_{ij} \tag{Eq. 11.2}$$

where i and j are two different taxa. The r-values are needed to create a modified distance matrix. The transformed r-values (r') are used to determine the distances of an individual taxon to the nearest node.

$$r'_i = r_i/n - 2 \tag{Eq. 11.3}$$

Box 11.1 An Example of Phylogenetic Tree Construction Using the UPGMA Method

	A	B	C
B	0.40		
C	0.35	0.45	
D	0.60	0.70	0.55

1. Using a distance matrix involving four taxa, A, B, C, and D, the UPGMA method first joins two closest taxa together which are A and C (**0.35** in grey). Because all taxa are equidistant from the node, the branch length for A to the node is AC/2 = 0.35/2 = 0.175.

0.175

A

C

0.175

2. Because A and C are joined into a cluster, they are treated as one new composite taxon, which is used to create a reduced matrix. The distance of A-C cluster to every other taxa is one half of a taxon to A and C, respectively. That means that the distance of B to A-C is (AB + BC)/2; and that of D to A-C is (AD + CD)/2.

	A-C	B
B	$\frac{0.4 + 0.45}{2} = 0.425$	
D	$\frac{0.55 + 0.6}{2} = 0.575$	0.70

3. In the newly reduced-distance matrix, the smallest distance is between B and A-C (in grey), which allows the grouping of B and A-C to create a three-taxon cluster. The branch length for the B is one half of B to the A-C cluster.

0.175

A

C

0.175

B

0.425/2 = 0.212

4. When B and A-C are grouped and treated as a single taxon, this allows the matrix to reduce further into only two taxa, D and B-A-C. The distance of D to the composite taxon is the average of D to every single component which is (BD + AD + CD)/3.

	B-A-C
D	$\frac{0.7 + 0.6 + 0.55}{3} = 0.617$

5. D is the last branch to add to the tree, whose branch length is one half of D to B-A-C.

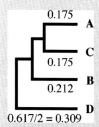

6. Because distance trees allow branches to be additive, the resulting distances between taxa from the tree path can be used to create a distance matrix. Obviously, the estimated distances do not match the actual evolutionary distances shown, which illustrates the failure of UPGMA to precisely reflect the experimental observation.

	A	B	C
B	0.42		
C	0.35	0.42	
D	0.62	0.62	0.62

where n is the total number of taxa. For example, assuming A and B form a node called U, the distance A to U is determined by the following formula:

$$d_{AU} = [d_{AB} + (r'_A - r'_B)]/2 \qquad \text{(Eq. 11.4)}$$

An example of this distance conversion and NJ tree building is shown in Box 11.2. The tree construction process is somewhat opposite to that used UPGMA. Rather than building trees from the closest pair of branches and progressing to the entire tree, the NJ tree method begins with a completely unresolved star tree by joining all taxa onto a single node and progressively decomposes the tree by selecting pairs of taxa based on the above modified pairwise distances. This allows the taxa with the shortest corrected distances to be joined first as a node. After the first node is constructed, the newly created cluster reduces the matrix by one taxon and allows the next most closely related taxon to be joined next to the first node. The cycle is repeated until all internal nodes are resolved. This process is called *star decomposition*. Unlike UPGMA, NJ and most other phylogenetic methods produce unrooted trees. The outgroup has to be determined based on external knowledge (see Chapter 10).

Generalized Neighbor Joining

One of the disadvantages of the NJ method is that it generates only one tree and does not test other possible tree topologies. This can be problematic because, in many cases, in the initial step of NJ, there may be more than one equally close pair

Box 11.2 Phylogenetic Tree Construction Using the Neighbor Joining Method

	A	B	C
B	0.40		
C	0.35	0.45	
D	0.60	0.70	0.55

1. The NJ method is similar to UPGMA, but uses an evolutionary rate correction step before tree building. Using the same distance matrix as in the UPGMA tree building (see Box 11.1), the first step of the NJ method is r-value and r'-value calculation. According to Eq. 11.1 and 11.2, r and r' for each taxon are calculated as follows:

$r_A = AB+AC+AD = 0.4+0.35+0.6 = 1.35$

$r'_A = r_A/(4-2) = 1.35/2 = 0.675$

$r_B = BA+BC+BD = 0.4+0.45+0.7 = 1.55$

$r'_B = r_B/(4-2) = 1.55/2 = 0.775$

$r_C = CA+CB+CD = 0.35+0.45+0.55 = 1.35$

$r'_C = r_C/(4-2) = 1.35/2 = 0.675$

$r_D = DA+DB+DC = 0.6+0.7+0.55 = 1.85$

$r'_D = r_D/(4-2) = 1.85/2 = 0.925$

2. Based on Eq. 11.4 and the above r-values, the corrected distances are obtained as follows:

$d'_{AB} = d_{AB} - 1/2 * (r_A+r_B) = 0.4 - (1.35+1.55)/2 = -1.05$

$d'_{AC} = d_{AC} - 1/2 * (r_A+r_C) = 0.35 - (1.35+1.35)/2 = -1$

$d'_{AD} = d_{AD} - 1/2 * (r_A+r_D) = 0.6 - (1.35+1.85)/2 = -1$

$d'_{BC} = d_{BC} - 1/2 * (r_B+r_C) = 0.45 - (1.55+1.35)/2 = -1$

$d'_{BD} = d_{BD} - 1/2 * (r_B+r_D) = 0.7 - (1.55+1.85)/2 = -1$

$d'_{CD} = d_{CD} - 1/2 * (r_C+r_D) = 0.55 - (1.35+1.85)/2 = -1.05$

3. The rate-corrected distances allow the construction of a new distance matrix.

	A	B	C
B	-1.05		
C	-1	-1	
D	-1	-1	-1.05

4. Before tree construction, all possible nodes are collapsed into a star tree. The pair of taxa with the shortest distances in the new

matrix are separated from the star tree first, according to the corrected distances. In this case, A and B as well as C and D are the
shortest (−1.05, in grey). Therefore, the first node to be built can be
either A-B or C-D. Choosing either pair first will give the same result.
Let's choose A and B first and name the node U.

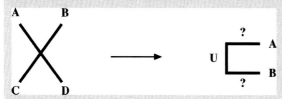

5. The branch lengths for A and B to the node U are calculated according
to Eq. 11.4.

$$d_{AU} = [d_{AB} + (r'_A - r'_B)]/2 = [0.4 + (0.675 - 0.775)]/2 = 0.15$$
$$d_{BU} = [d_{AB} + (r'_B - r'_A)]/2 = [0.4 + (0.775 - 0.675)]/2 = 0.25$$

6. The new cluster allows the construction of a reduced matrix. This
starts with actual distances. Unlike in UPGMA, the distance from a taxon
to a node is the average of the original distances to each of the components of the composite taxon, subtracted from the inferred branch
lengths.

$$d_{CU} = [(d_{AC} - d_{UA}) + (d_{BC} - d_{UB})]/2 = [(0.35 - 0.15) + (0.45 - 0.25)]/2 = 0.2$$
$$d_{DU} = [(d_{AD} - d_{UA}) + (d_{BD} - d_{UB})]/2 = [(0.6 - 0.15) + (0.7 - 0.25)]/2 = 0.45$$

	U	B
C	0.20	
D	0.45	0.55

7. Based on the reduced distance matrix, a new set of r- and r'-values
are calculated.

$$r_C = CU + CD = 0.2 + 0.55 = 0.75$$
$$r'_C = r_C/(3 - 2) = 0.75/1 = 0.75$$
$$r_D = DU + CD = 0.45 + 0.55 = 1$$
$$r'_C = r_C/(3 - 2) = 1/1 = 1$$
$$r_U = CU + DU = 0.2 + 0.45 = 0.65$$
$$r'_U = r_U/(3 - 2) = 0.65/1 = 0.65$$

Box 11.2 (continued)

8. The new r- and r'-values allow construction of the corrected distance matrix.

$d'_{CU} = d_{CU} - 1/2 * (r_C + r_U) = 0.2 - (0.75 + 0.65)/2 = -0.5$

$d'_{DU} = d_{DU} - 1/2 * (r_D + r_U) = 0.45 - (1 + 0.65)/2 = -0.375$

$d'_{CD} = d_{CD} - 1/2 * (r_C + r_D) = 0.55 - (0.75 + 1)/2 = -0.325$

	U	B
C	-0.5	
D	-0.375	-0.325

9. In the corrected distance matrix, C to node U has the shortest distance (-0.5, in grey). This allows creation of the second node named V. The branch length is calculated as in step 5.

$d_{CV} = [d_{CU} + (r'_C - r'_U)]/2 = [0.2 + (0.75 - 0.65)]/2 = 0.15$

$d_{UV} = [d_{CU} + (r'_U - r'_C)]/2 = [0.2 + (0.65 - 0.75)]/2 = 0.05$

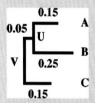

10. Because D is the last branch to be decomposed from the star tree, there is no need to convert to r and r' because r' is infinitely large when $n - 2 = 0$. Its branch length is calculated as one half of the sum of D to node V and D to C, subtracted from respective branch lengths.

$d_D = [(d_{DU} - d_{UV}) + (d_{DC} - d_{CV})]/2 = [(0.45 - 0.05) + (0.55 - 0.15)]/2 = 0.4$

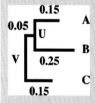

11. When the overall branch lengths are compiled into a distance matrix, which is used to compare with the original distance matrix, it is clear that the estimated distances completely match the actual evolutionary distances, indicating that this treeing method is able to satisfy the constraint of the experimental observation in this case.

	U	B
C	-0.5	
D	-0.375	-0.325

of neighbors to join, leading to multiple trees. Ignoring these multiple options may yield a suboptimal tree. To overcome the limitations, a generalized NJ method has been developed, in which multiple NJ trees with different initial taxon groupings are generated. A best tree is then selected from a pool of regular NJ trees that best fit the actual evolutionary distances. This more extensive tree search means that this approach has a better chance of finding the correct tree.

Optimality-Based Methods

The clustering-based methods produce a single tree as output. However, there is no criterion in judging how this tree is compared to other alternative trees. In contrast, optimality-based methods have a well-defined algorithm to compare all possible tree topologies and select a tree that best fits the actual evolutionary distance matrix. Based on the differences in optimality criteria, there are two types of algorithms, Fitch–Margoliash and minimum evolution, that are described next. The exhaustive search for an optimal tree necessitates a slow computation, which is a clear drawback especially when the dataset is large.

Fitch–Margoliash

The Fitch–Margoliash (FM) method selects a best tree among all possible trees based on minimal deviation between the distances calculated in the overall branches in the tree and the distances in the original dataset. It starts by randomly clustering two taxa in a node and creating three equations to describe the distances, and then solving the three algebraic equations for unknown branch lengths. The clustering of the two taxa helps to create a newly reduced matrix. This process is iterated until a tree is completely resolved. The method searches for all tree topologies and selects the one that has the lowest squared deviation of actual distances and calculated tree branch lengths. The optimality criterion is expressed in the following formula:

$$E = \sum_{i=1}^{T-1} \sum_{j=j+1}^{T} \frac{(d_{ij} - p_{ij})^2}{d_{ij}^2} \qquad \text{(Eq. 11.5)}$$

where E is the error of the estimated tree fitting the original data, T is the number of taxa, d_{ij} is the pairwise distance between ith and jth taxa in the original dataset, and p_{ij} is the corresponding tree branch length.

Minimum Evolution

Minimum evolution (ME) constructs a tree with a similar procedure, but uses a different optimality criterion that finds a tree among all possible trees with a minimum overall branch length. The optimality criterion relies on the formula:

$$S = \Sigma b_i \qquad \text{(Eq. 11.6)}$$

where b_i is the ith branch length. Searching for the minimum total branch length is an indirect approach to achieving the best fit of the branch lengths with the original dataset. Analysis has shown that minimum evolution in fact slightly outperforms the least square-based FM method.

Pros and Cons

The most frequently used distance methods are clustering based. The major advantage is that they are computationally fast and are therefore capable of handling datasets that are deemed to be too large for any other phylogenetic method. The methods, however, are not guaranteed to find the best tree. Exhaustive tree-searching algorithms such as FM and ME have better accuracies overall. However, they can be computationally prohibitive to use when the number of taxa is large (e.g., >12), because the overall number of tree topologies becomes too large to handle. A compromise between the two types of algorithm is a hybrid approach such as the generalized NJ, with a performance similar to that of ME but computationally much faster.

The overall advantage of all distance-based methods is the ability to make use of a large number of substitution models to correct distances. The drawback is that the actual sequence information is lost when all the sequence variation is reduced to a single value. Hence, ancestral sequences at internal nodes cannot be inferred.

CHARACTER-BASED METHODS

Character-based methods (also called *discrete methods*) are based directly on the sequence characters rather than on pairwise distances. They count mutational events accumulated on the sequences and may therefore avoid the loss of information when characters are converted to distances. This preservation of character information means that evolutionary dynamics of each character can be studied. Ancestral sequences can also be inferred. The two most popular character-based approaches are the maximum parsimony (MP) and maximum likelihood (ML) methods.

Maximum Parsimony

The parsimony method chooses a tree that has the fewest evolutionary changes or shortest overall branch lengths. It is based on a principle related to a medieval philosophy called *Occam's razor*. The theory was formulated by William of Occam in the thirteenth century and states that the simplest explanation is probably the correct one. This is because the simplest explanation requires the fewest assumptions and the fewest leaps of logic. In dealing with problems that may have an infinite number of possible solutions, choosing the simplest model may help to "shave off" those variables that are not really necessary to explain the phenomenon. By doing this, model development may become easier, and there may be less chance of introducing inconsistencies, ambiguities, and redundancies, hence, the name Occam's razor.

For phylogenetic analysis, parsimony seems a good assumption. By this principle, a tree with the least number of substitutions is probably the best to explain the differences among the taxa under study. This view is justified by the fact that evolutionary

sites taxa	1	2	3	4	5	6	7	8
I	A	A	T	T	A	G	C	T
II	G	G	T	C	G	T	A	G
III	A	A	T	G	C	G	C	T
IV	A	G	T	A	A	G	C	A
V	A	C	T	T	C	G	C	G
VI	A	C	A	T	G	G	C	A

Figure 11.1: Example of identification of informative sites that are used in parsimony analysis. Sites 2, 5, and 8 (*grey boxes*) are informative sites. Other sites are noninformative sites, which are either constant or having characters occurring only once.

changes are relatively rare within a reasonably short time frame. This implies that a tree with minimal changes is likely to be a good estimate of the true tree. By minimizing the changes, the method minimizes the phylogenetic noise owing to homoplasy and independent evolution. The MP approach is in principle similar to the ME approach albeit the latter is distance based instead of character based.

How Does MP Tree Building Work?

Parsimony tree building works by searching for all possible tree topologies and reconstructing ancestral sequences that require the minimum number of changes to evolve to the current sequences. To save computing time, only a small number of sites that have the richest phylogenetic information are used in tree determination. These sites are the so-called informative sites, which are defined as sites that have at least two different kinds of characters, each occurring at least twice (Fig. 11.1). Informative sites are the ones that can often be explained by a unique tree topology. Other sites are *noninformative*, which are constant sites or sites that have changes occurring only once. *Constant sites* have the same state in all taxa and are obviously useless in evaluating the various topologies. The sites that have changes occurring only once are not very useful either for constructing parsimony trees because they can be explained by multiple tree topologies. The noninformative sites are thus discarded in parsimony tree construction.

Once the informative sites are identified and the noninformative sites discarded, the minimum number of substitutions at each informative site is computed for a given tree topology. The total number of changes at all informative sites are summed up for each possible tree topology. The tree that has the smallest number of changes is chosen as the best tree.

The key to counting a minimum number of substitutions for a particular site is to determine the ancestral character states at internal nodes. Because these ancestral character states are not known directly, multiple possible solutions may exist. In this case, the parsimony principle applies to choose the character states that result in a minimum number of substitutions. The inference of an ancestral sequence is made by first going from the leaves to internal nodes and to the common root to determine all possible ancestral character states and then going back from the common root to the leaves to assign ancestral sequences that require the minimum number of substitutions. An example of predicting ancestral sequences at internal nodes is given

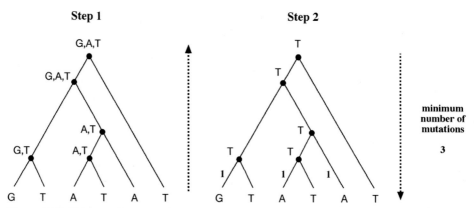

Figure 11.2: Using parsimony to infer ancestral characters at internal nodes involves a two-step procedure. The first step involves going from the leaves to the root and counting all possible ancestral characters at the internal nodes. The second step goes from the root to the leaves and assigns ancestral characters that involve minimum number of mutations. In this example, the total number of mutations is three if T is at the root, whereas other possible character states increase that number.

in Figure 11.2. It needs to be emphasized that, in reality, the ancestral node sequence cannot always be determined unambiguously. Sometimes, there may be more than one character that gives a total minimum number for a given tree topology. It is also possible that there may be two or more topologies that have the same minimum number of total substitutions. In that case, equally parsimonious trees are produced. A consensus tree has to be built that represents all the parsimonious trees (see Chapter 10).

Weighted Parsimony

The parsimony method discussed is unweighted because it treats all mutations as equivalent. This may be an oversimplification; mutations of some sites are known to occur less frequently than others, for example, transversions versus transitions, functionally important sites versus neutral sites. Therefore, a weighting scheme that takes into account the different kinds of mutations helps to select tree topologies more accurately. The MP method that incorporates a weighting scheme is called *weighted parsimony*. In the example shown in Figure 11.3, different branch lengths are obtained using weighted parsimony compared with using unweighted parsimony. In some cases, the weighting scheme may result in different tree topologies.

Tree-Searching Methods

As mentioned, the parsimony method examines all possible tree topologies to find the maximally parsimonious tree. This is an exhaustive search method. It starts by building a three taxa unrooted tree, for which only one topology is available. The choice of the first three taxa can be random. The next step is to add a fourth taxon to the existing branches, producing three possible topologies. The remaining taxa are progressively added to form all possible tree topologies (Fig. 11.4). Obviously, this brute-force approach only works if there are relatively few sequences. The exponential increase in possible tree topologies with the number of taxa means that this exhaustive

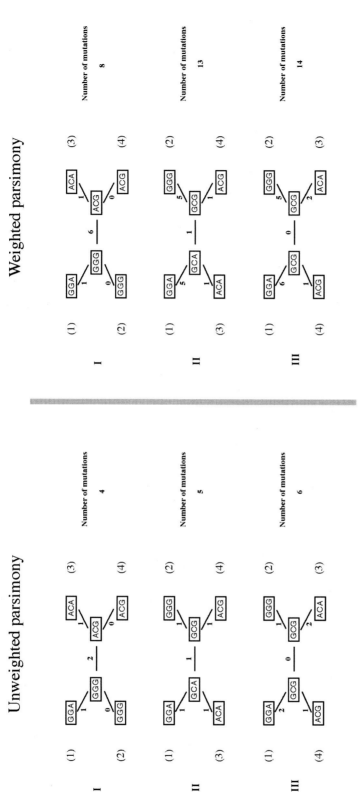

Figure 11.3: Comparison of unweighted and weighted parsimony. In the latter, transitions are weighted as 1 and transversions are weighted as 5.

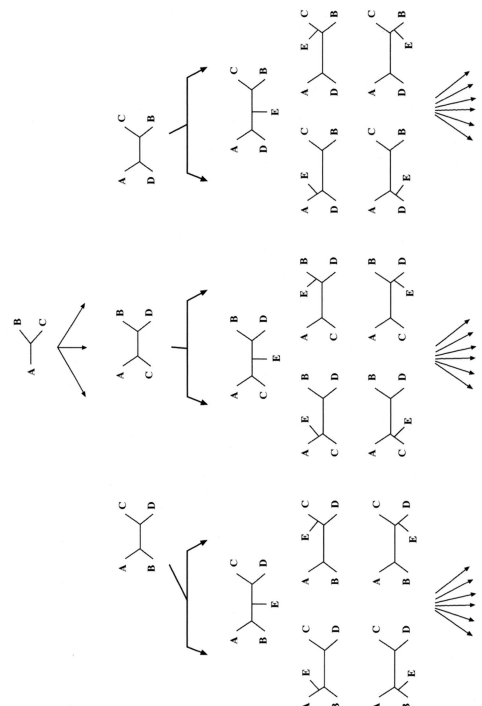

Figure 11.4: Schematic of exhaustive tree construction in the MP procedure. The tree starts with three taxa with one topology. One taxon is then added at a time in an progressive manner, during which the total branch lengths of all possible topologies are calculated.

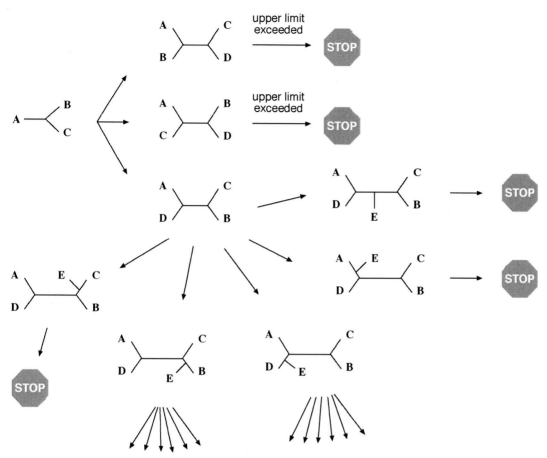

Figure 11.5: Schematic illustration of the branch-and-bound algorithm. Tree building starts with a step-wise addition of taxa of all possible topologies. Whenever the total branch length for a given topology exceeds the upper bound, the tree search in that direction stops, thereby reducing the total computing time.

method is computationally too demanding to use when the number of taxa is more than ten. When this is the case, some simplified steps have to be introduced to reduce the complexity of the search.

One such simplified method is called branch-and-bound, which uses a shortcut to find an MP tree by establishing an upper limit (or upper bound) for the number of allowed sequence variations. It starts by building a distance tree for all taxa involved using either NJ or UPGMA and then computing the minimum number of substitutions for this tree. The resulting number defines the upper bound to which any other trees are compared. The rationale is that a maximally parsimonious tree must be equal to or shorter than the distance-based tree.

The branch-and-bound method starts building trees in a similar way as in the exhaustive method. The difference is that the previously established upper bound limits the tree growth. Whenever the overall tree length at every single stage exceeds the upper bound, the topology search toward a particular direction aborts (Fig. 11.5).

By doing so, it dramatically reduces the number of trees considered hence the computing time while at the same time guaranteeing to find the most parsimonious tree.

When the number of taxa exceeds twenty, even the branch-and-bound method becomes computationally unfeasible. A more heuristic search method must be used. As a reminder, a computer heuristic procedure is an approximation strategy to find an empirical solution for a complicated problem (see Chapter 4). This strategy generates quick answers, but not necessarily the best answer. In a heuristic tree search, only a small subset of all possible trees is examined. This method starts by carrying out a quick initial approximation, which is to build an NJ tree and subsequently modifying it slightly into a different topology to see whether that leads to a shorter tree.

The modification includes cutting a branch or subtree and regrafting it to other parts of the tree (Fig. 11.6). The total branch length for the new tree is recomputed. If the tree is found to be shorter through rearrangement, it is used as a starting point for another round of rearrangement. The iteration continues until no shorter trees are found. This method is very fast, but does not guarantee to find the most parsimonious tree. The commonly used branch-swapping algorithms are nearest neighbor interchange, tree bisection and reconnection, and subtree pruning and regrafting.

The pitfall with branch swapping is that the tree rearrangement tends to focus on a local area and stalls when a local branch length minimum is reached. To avoid getting stuck in a local minimum, a "global search" option is implemented in certain programs. This allows the removal of every possible subtree and its reattachment in every possible way, to increase the chance of finding the most parsimonious tree. This approach significantly increases the computing time and thus compromises the trade-off between obtaining an optimal tree and obtaining a tree within a realistic time.

Pros and Cons

The main advantage of MP is that it is intuitive – its assumptions are easily understood. In addition, the character-based method is able to provide evolutionary information about the sequence characters, such as information regarding homoplasy and ancestral states. It tends to produce more accurate trees than the distance-based methods when sequence divergence is low because this is the circumstance when the parsimony assumption of rarity in evolutionary changes holds true. However, when sequence divergence is high, or the amount of homoplasies is large, tree estimation by MP can be less effective, because the original parsimony assumption no longer holds. Estimation of branch lengths may also be erroneous because MP does not employ substitution models to correct for multiple substitutions. This drawback can become prominent when dealing with divergent sequences. In addition, MP only considers informative sites, and ignores other sites. Consequently, certain phylogenetic signals may be lost. MP is also slow compared to the distance methods, and more important, is very sensitive to the "long-branch attraction" (LBA) artifacts.

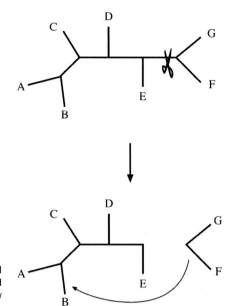

Figure 11.6: Schematic representation of a typical branch swapping process in which a branch is cut and moved to another part of the tree, generating a new topology.

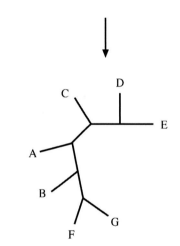

Long-Branch Attraction

LBA is a particular problem associated with parsimony methods. It refers to a phylogenetic artifact in which rapidly evolving taxa with long branches are placed together in a tree, regardless of their true positions in a tree (Fig. 11.7). This is partly due to the assumption in parsimony that all lineages evolve at the same rate and that all mutations (transitions versus transversions) contribute equally to branch lengths. It may also be partly owing to multiple substitutions at individual sites and among-site rate heterogeneity for which MP is not capable of correcting.

Figure 11.7: The LBA artifact showing taxa A and D are artifactually clustered during phylogenetic construction.

There are several possible solutions to the LBA artifact. For homoplasies that cause LBA, distance and likelihood (discussed below) methods that employ substitution models and rate heterogeneity models should be able to alleviate the problem. In addition, weighted parsimony should be more advantageous than unweighted parsimony in countering the transitional bias when transitions occur more often than transversions. Increasing the taxon sampling size may also help because introduction of intermediate taxa breaks up the long branches. A dataset with concatenated multiple genes also has less chance of LBA because the combined gene analysis may dampen the effect of a single gene having a high rate of evolution.

Maximum Likelihood Method

Another character-based approach is ML, which uses probabilistic models to choose a best tree that has the highest probability or likelihood of reproducing the observed data. It finds a tree that most likely reflects the actual evolutionary process. ML is an exhaustive method that searches every possible tree topology and considers every position in an alignment, not just informative sites. By employing a particular substitution model that has probability values of residue substitutions, ML calculates the total likelihood of ancestral sequences evolving to internal nodes and eventually to existing sequences. It sometimes also incorporates parameters that account for rate variations across sites.

How Does the Maximum Likelihood Method Work?

ML works by calculating the probability of a given evolutionary path for a particular extant sequence. The probability values are determined by a substitution model (either for nucleotides or amino acids). For example, for DNA sequences using the Jukes–Cantor model, the probability (P) that a nucleotide remains the same after time t is:

$$P(t) = 1/4 + 3/4e^{-\alpha t} \tag{Eq. 11.7}$$

where α is the nucleotide substitution rate in the Jukes–Cantor model, which is either empirically assigned or estimated from the raw datasets. In Figure 11.8, the elapsed time t from X to A can be assigned as 1 and from Z to A as 2. For a nucleotide to change into a different residue after time t, the probability value is determined by the

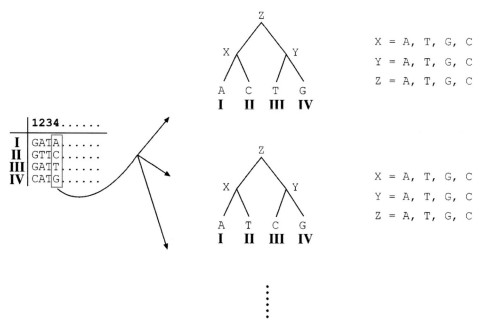

Figure 11.8: Schematic representation of the ML approach to build phylogenetic trees for four taxa, I, II, III, and IV. The ancestral character states at the internal nodes and root node are assigned X, Y, and Z, respectively. The example only shows some of the topologies derived from one of the sites in the original alignment. The method actually uses all the sites in probability calculation for all possible trees with all combinations of possible ancestral sequences at internal nodes according to a predefined substitution model.

following formula:

$$P(t) = 1/4 - 1/4e^{-\alpha t} \tag{Eq. 11.8}$$

For other substitution models, the formulas are much more complex and are not described here. For a particular site, the probability of a tree path is the product of the probability from the root to all the tips, including every intermediate branches in the tree topology. Because multiplication often results in very small values, it is computationally more convenient to express all probability values as natural log likelihood (lnL) values, which also converts multiplication into summation. Because ancestral characters at internal nodes are normally unknown, all possible scenarios of ancestral states (X, Y, and Z in Fig. 11.8) have to be computed.

After logarithmic conversion, the likelihood score for the topology is the sum of log likelihood of every single branch of the tree. After computing for all possible tree paths with different combinations of ancestral sequences, the tree path having the highest likelihood score is the final topology at the site. Because all characters are assumed to have evolved independently, the log likelihood scores are calculated for each site independently. The overall log likelihood score for a given tree path for the entire sequence is the sum of log likelihood of all individual sites. The same procedure has to be repeated for all other possible tree topologies. The tree having the highest likelihood score among all others is chosen as the best tree, which is the ML tree. This process is exhaustive in nature and therefore very time consuming.

$$L_{(4)} = \Pr(Z \to X) * \Pr(Z \to Y) * \Pr(X \to A) * \Pr(X \to C) * \Pr(Y \to T) * \Pr(Y \to G)$$
$$\ln L_{(4)} = \ln \Pr(Z \to X) + \ln \Pr(Z \to Y) + \ln \Pr(X \to A) + \ln \Pr(X \to C)$$
$$+ \ln \Pr(Y \to T) + \ln \Pr(Y \to G)$$

Pros and Cons

ML is based on well-founded statistics instead of a medieval philosophy. It is thus considered mathematically more rigorous than MP. In fact, it is the most rigorous among all approaches. ML uses the full sequence information, not just the informative sites and therefore may be more robust. ML employs substitution models and is not sensitive to LBA. Some of these strengths, however, can also be the weakness of ML depending on the context. For example, accuracy depends on the substitution model used. Choosing an unrealistic substitution model may lead to an incorrect tree. Because of the exhaustive nature of the ML method, when the number of taxa increases to a modest size, it becomes impossible to use. To overcome the problem, several heuristic or alternative approaches have been proposed. These alternative methods include quartet puzzling, genetic algorithms (GAs), and Bayesian inference, which are introduced in the following sections.

Quartet Puzzling

The most commonly used heuristic ML method is called quartet puzzling, which uses a divide-and-conquer approach. In this approach, the total number of taxa are divided into many subsets of four taxa known as *quartets*. An optimal ML tree is constructed from each of these quartets. This is a relatively easy process as there are only three possible unrooted topologies for a four-taxon tree. All the quartet trees are subsequently combined into a larger tree involving all taxa (Fig. 11.9). This process is like joining pieces in a jigsaw puzzle, hence the name. The problem in drawing a consensus is that the branching patterns in quartets with shared taxa may not agree. In this case, a majority rule is used to determine the positions of branches to be inserted to create the consensus tree.

The reason that quartet puzzling is computationally faster than exhaustive ML is because there are fewer tree topologies to search. To take four-taxon subsets out of n sequences, there are total C_n^4 combinations. Each subset has only three possible trees, and so the total number of trees that need to be computed are $3 \times C_n^4$. For instance, for twenty taxa, there are $3 \times C_{20}^4 = \frac{3 \times 20!}{(20-4)! \times 4!} = 14,535$ tree topologies to search, compared with 2×10^{20} trees if using the exhaustive search strategy. Thus, the method significantly reduces the computing time. The caveat of using the puzzling approach is that it does not necessarily return a tree with ML, but instead produces a consensus tree that is supported by the results of most quartets. Although the heuristic method is not as robust as regular ML, it has become a popular choice with many researchers because of its computational feasibility with large datasets.

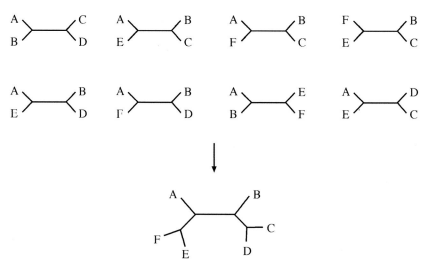

Figure 11.9: Schematic illustration of quartet puzzling in deriving a consensus tree by combining multiple quartet trees.

NJML

NJML is a hybrid algorithm combining aspects of NJ and ML. It constructs an initial tree using the NJ method with bootstrapping (which will be described). The branches with low bootstrap support are collapsed to produce multifurcating branches. The polytomy is resolved using the ML method. Although the performance of this method is not yet as good as the complete ML method, it is at least ten times faster.

Genetic Algorithm

A recent addition to fast ML search methods is the GA, a computational optimization strategy that uses biological terminology as a metaphor because the method involves "crossing" mathematical routines to generate new "offspring" routines. The algorithm works by selecting an optimal result through a mix-and-match process using a number of existing random solutions. A "fitness" measure is used to monitor the optimization process. By keeping record of the fitness scores, the process simulates the natural selection and genetic crossing processes. For instance, a subroutine that has the best score (best fit process) is selected in the first round and is used as a starting point for the next round of the optimization cycle. Again using biological metaphors, this is to generate more "offspring," which are mathematical trials with modifications from the previous ones. Different computational routines (or "chromosomes") are also allowed to combine (or "crossover") to produce a new solution. The iteration continues until an optimal solution is found.

When applying GA to phylogenetic inference, the method strongly resembles the pruning and regrafting routines used in the branch-swapping process. In GA-based tree searching, the fitness measure is the log likelihood scores. The tree search begins with a population of random trees with an arbitrary branch lengths. The tree with a highest log likelihood score is allowed to leave more "offspring" with "mutations" on the tree topology. The mutational process is essentially branch rearrangement. Mutated new trees are scored. Those that are scored higher than the parent tree are allowed to mutate more to produce even higher scored offspring, if possible. This process is repeated until no higher scored trees can be found. The advantage of this algorithm is its speed; a near optimal tree can often be obtained within a limited number of iterations.

Bayesian Analysis

Another recent development of a speedy ML method is the use of the Bayesian analysis method. The essence of Bayesian analysis is to make inference on something unobserved based on existing observations. It makes use of an important concept of known as *posterior probability*, which is defined as the probability that is revised from prior expectations, after learning something new about the data. In mathematical terms, Bayesian analysis is to calculate posterior probability of two joint events by using the prior probability and conditional probability values using the following simplified formula:

$$\text{Posterior probability} = \frac{\text{Prior probability} * \text{Conditional likelihood}}{\text{Total probability}} \qquad \text{(Eq. 11.9)}$$

Without going into much mathematical detail, it is important to know that the Bayesian method can be used to infer phylogenetic trees with maximum posterior probability. In Bayesian tree selection, the prior probability is the probability for all possible topologies before analysis. The probability for each of these topologies is equal before tree building. The conditional probability is the substitution frequency of characters observed from the sequence alignment. These two pieces of information are used as a condition by the Bayesian algorithm to search for the most probable trees that best satisfy the observations.

The tree search incorporates an iterative random sampling strategy based on the Markov chain Monte Carlo (MCMC) procedure. MCMC is designed as a "hill-climbing" procedure, seeking higher and higher likelihood scores while searching for tree topologies, although occasionally it goes downhill because of the random nature of the search. Over time, high-scoring trees are sampled more often than low-scoring trees. When MCMC reaches high scored regions, a set of near optimal trees are selected to construct a consensus tree.

In the end, the Bayesian method can achieve the same or even better performance than the complete ML method, but is much faster than regular ML and is able to handle very large datasets. The reason that the Bayesian analysis may achieve better performance than ML is that the ML method searches one single best tree, whereas the Bayesian method searches a set of best trees. The advantage of the Bayesian method can be explained by the matter of probability. Because the true tree is not known, an optimal ML tree may have, say, 90% probability of representing the reality. However, the Bayesian method produces hundreds or thousands of optimal or near-optimal trees with 88% to 90% probability to represent the reality. Thus, the latter approach has a better chance overall to guess the true tree correctly.

PHYLOGENETIC TREE EVALUATION

After phylogenetic tree construction, the next step is to statistically evaluate the reliability of the inferred phylogeny. There are two questions that need to be addressed. One is how reliable the tree or a portion of the tree is; and the second is whether this tree is significantly better than another tree. To answer the first question, we need to use analytical resampling strategies such as bootstrapping and jackknifing, which repeatedly resample data from the original dataset. For the second question, conventional statistical tests are needed.

What Is Bootstrapping?

Bootstrapping is a statistical technique that tests the sampling errors of a phylogenetic tree. It does so by repeatedly sampling trees through slightly perturbed datasets. By doing so, the robustness of the original tree can be assessed. The rationale for bootstrapping is that a newly constructed tree is possibly biased owing to incorrect alignment or chance fluctuations of distance measurements. To determine the robustness or reproducibility of the current tree, trees are repeatedly constructed with slightly perturbed alignments that have some random fluctuations introduced. A truly robust phylogenetic relationship should have enough characters to support the relationship even if the dataset is perturbed in such a way. Otherwise, the noise introduced in the resampling process is sufficient to generate different trees, indicating that the original topology may be derived from weak phylogenetic signals. Thus, this type of analysis gives an idea of the statistical confidence of the tree topology.

Parametric and Nonparametric Bootstrapping

Bootstrap resampling relies on perturbation of original sequence datasets. There are two perturbation strategies. One way to produce perturbations is through random replacement of sites. This is referred to as *nonparametric bootstrapping*. Alternatively, new datasets can be generated based on a particular sequence distribution, which is *parametric bootstrapping*. Both types of bootstrapping can be applied to the distance, parsimony, and likelihood tree construction methods.

In nonparametric bootstrapping, a new multiple sequence alignment of the same length is generated with random duplication of some of the sites (i.e., the columns in an alignment) at the expense of some other sites. In other words, certain sites are randomly replaced by other existing sites. Consequently, certain sites may appear multiple times, and other sites may not appear at all in the new alignment (Fig. 11.10). This process is repeated 100 to 1,000 times to create 100 to 1,000 new alignments that are used to reconstruct phylogenetic trees using the same method as the originally inferred tree. The new datasets with altered the nucleotide or amino acid composition

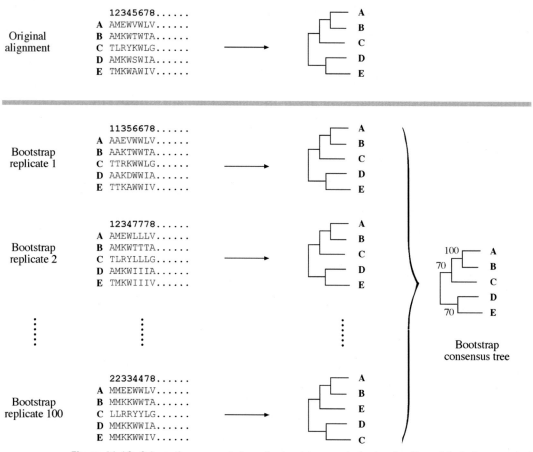

Figure 11.10: Schematic representation of a bootstrap analysis showing the original alignment and modified replicates in which certain sites are randomly replaced with other existing sites. The resulting altered replicates are used to building trees for statistical analysis at each node.

and rate heterogeneity may result in certain parts of the tree having a different topology from the original inferred tree.

All the bootstrapped trees are summarized into a consensus tree based on a majority rule. The most supported branching patterns shown at each node are labeled with bootstrap values, which are the percentage of appearance of a particular clade. Thus, the bootstrap test provides a measure for evaluating the confidence levels of the tree topology. Analysis has shown that a bootstrap value of 70% approximately corresponds to 95% statistical confidence, although the issue is still a subject of debate.

Instead of randomly duplicating sites to generate new datasets, parametric bootstrapping uses altered datasets with random sequences confined within a particular sequence distribution according to a given substitution model. For instance, for a nucleotide dataset, according to the Juke–Cantor model, all four nucleotides are identically distributed, whereas the Kimura model provides a different distribution (see Fig. 10.8). The parametric bootstrapping method may help avoid the problem of certain sites being repeated too many times as in nonparametric bootstrapping resulting in skewed sequence distribution. If a correct nucleotide/amino acid distribution model is used, parametric bootstrapping generates more reasonable replicates than random replicates. Thus, this procedure is considered more robust than nonparametric bootstrapping.

Caveats

Strictly speaking, bootstrapping does not assess the accuracy of a tree, but only indicates consistency and stability of individual clades of the tree. This means that, because of systematic errors, wrong trees can still be obtained with high bootstrap values. Therefore, bootstrap results should be interpreted with caution. Unusually high GC content in the original dataset, unusually accelerated evolutionary rates and unrealistic evolutionary models are the potential causes for generating biased trees, as well as biased bootstrap estimates, which come after the tree generation.

In addition, from a statistical point of view, a large number of bootstrap resampling steps are needed to achieve meaningful results. It is generally recommended that a phylogenetic tree should be bootstrapped 500 to 1,000 times. However, this presents a practical dilemma. In many instances, it may take hours or days to construct one ML or MP tree. So the multiplication of computing time makes bootstrapping virtually impossible to use with limited computing resources.

Jackknifing

In addition to bootstrapping, another often used resampling technique is jackknifing. In jackknifing, one half of the sites in a dataset are randomly deleted, creating datasets half as long as the original. Each new dataset is subjected to phylogenetic tree construction using the same method as the original. The advantage of jackknifing is that sites are not duplicated relative to the original dataset and that computing time is much shortened because of shorter sequences. One criticism of this approach is that the size of datasets has been changed into one half and that the datasets are no longer considered replicates. Thus, the results may not be comparable with that from bootstrapping.

Bayesian Simulation

In terms of statistical evaluation, the Bayesian method is probably the most efficient; it does not require bootstrapping because the MCMC procedure itself involves thousands or millions of steps of resampling. As a result of Bayesian tree construction, posterior probabilities are assigned at each node of a best Bayesian tree as statistical support. Because of fast computational speed of MCMC tree searching, the Bayesian method offers a practical advantage over regular ML and makes the statistical evaluation of ML trees more feasible. Unlike bootstrap values, Bayesian probabilities are normally higher because most trees are sampled near a small number of optimal trees. Therefore, they have a different statistical meaning from bootstrap.

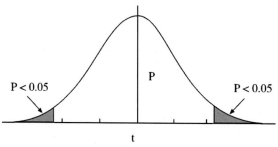

Figure 11.11: A *t*-distribution curve showing highlighted areas with margins of statistical significance.

Kishino–Hasegawa Test

In phylogenetic analysis, it is also important to test whether two competing tree topologies can be distinguished and whether one tree is significantly better than the other. The task is different from bootstrapping in that it tests the statistical significance of the entire phylogeny, not just portions of it. For that purpose, several statistical tests have been developed specifically for each of the three types of tree reconstruction methods, distance, parsimony, and likelihood. A test devised specifically for MP trees is called the Kishino–Hasegawa (KH) test.

The KH test sets out to test the null hypothesis that the two competing tree topologies are not significantly different. A paired Student *t*-test is used to assess whether the null hypothesis can be rejected at a statistically significant level. In this test, the difference of branch lengths at each informative site between the two trees is calculated. The standard deviation of the difference values can then be calculated. This in turn allows derivation of a *t*-value (see Eq. 11.10), which is used for evaluation against the *t*-distribution to see whether the value falls within the significant range (e.g., $P < .05$) to warrant the rejection of the null hypothesis (Fig. 11.11).

$$t = \frac{D_a - D_t}{S_D/\sqrt{n}}$$

(Eq. 11.10)

$$df = n - 1$$

(Eq. 11.11)

where n is the number of informative sites, df is the degree of freedom, t is the test statistical value, D_a is the average site-to-site difference between the two trees, S_D is the standard deviation, and D_t is the total difference of branch lengths of the two trees.

Shimodaira–Hasegawa Test

A frequently used statistical test for ML trees is the Shimodaira–Hasegawa (SH) test (likelihood ratio test). It tests the goodness of fit of two competing trees using the χ^2 test. For this test, log likelihood scores of two competing trees have to be obtained first. The degree of freedom used for the analysis depends on the substitution model used. It relies on the following test formula:

$$d = 2(\ln L_A - \ln L_B) = 2\ln(L_A/L_B)$$

(Eq. 11.12)

where d is the log likelihood ratio score and $\ln L_A$ and $\ln L_B$ are likelihood scores for tree A and tree B, respectively. The statistical meaning of d can be obtained from calculating the probability value from a χ^2 table.

Once the log ratio of the two scores is obtained, it is used to test against the χ^2 distribution. The resulting probability value (P-value) determines whether the difference between the two trees is significant.

PHYLOGENETIC PROGRAMS

Phylogenetic tree reconstruction is not a trivial task. Although there are numerous phylogenetic programs available, knowing the theoretical background, capabilities, and limitations of each is very important. For a list of hundreds of phylogenetic software programs, see Felsenstein's collection at: http://evolution.genetics.washington. edu/phylip/software.html. Most of these programs are freely available. Some are comprehensive packages; others are more specialized to perform a single task. Most require special efforts to learn how to use them effectively. Because this book is not intended as a computer manual, a brief introduction to several of the most commonly used programs is provided.

PAUP* (Phylogenetic analysis using parsimony and other methods, by David Swofford, http://paup.csit.fsu.edu/) is a commercial phylogenetic package. It is probably one of the most widely used phylogenetic programs available from Sinauer Publishers. It is a Macintosh program (UNIX version available in the GCG package) with a very user-friendly graphical interface. PAUP was originally developed as a parsimony program, but expanded to a comprehensive package that is capable of performing distance, parsimony, and likelihood analyses. The distance options include NJ, ME, FM, and UPGMA. For distance or ML analyses, PAUP has the option for detailed specifications of substitution models, base frequencies, and among site rate heterogeneity (γ-shape parameters, proportion of invariant sites). PAUP is also able to perform nonparametric bootstrapping, jackknifing, KH testing, and SH testing.

Phylip (Phylogenetic inference package; by Joe Felsenstein) is a free multiplatform comprehensive package containing thirty-five subprograms for performing distance, parsimony, and likelihood analysis, as well as bootstrapping for both nucleotide and amino acid sequences. It is command-line based, but relatively easy to use for each single program. The only problem is that to complete an analysis the user is required to move between different subprograms while keeping modifying names of the intermediate output files. The program package is downloadable from http://evolution. genetics.washington.edu/phylip.html. An online version is also available at http:// bioweb.pasteur.fr/seqanal/phylogeny/phylip-uk.html. A more user-friendly online version is WebPhylip available at http://sdmc.krdl.org.sg:8080/~lxzhang/phylip/.

TREE-PUZZLE is a program performing quartet puzzling. The advantage is that it allows various substitution models for likelihood score estimation and incorporates a discrete γ model for rate heterogeneity among sites (see Chapter 10). Because of the heuristic nature of the program, it allows ML analyses of large datasets. The resulting puzzle trees are automatically assigned puzzle support values to internal branches. These values are percentages of consistent quartet trees and do not have the same meaning as bootstrap values. TREE-PUZZLE version 5.0 is available for Mac, UNIX, and Windows and can be downloaded from www.tree-puzzle.de/. There is also an online version of the program available at: http://bioweb.pasteur.fr/seqanal/interfaces/Puzzle.html.

PHYML (http://atgc.lirmm.fr/phyml/) is a web-based phylogenetic program using the GA. It first builds an NJ tree and uses it as a starting tree for subsequent iterative refinement through subtree swapping. Branch lengths are simultaneously optimized during this process. The tree searching stops when the total ML score no longer increases. The main advantage of this program is the ability to build trees from very large datasets with hundreds of taxa and to complete tree searching within a relatively short time frame.

MrBayes is a Bayesian phylogenetic inference program. It randomly samples tree topologies using the MCMC procedure and infers the posterior distribution of tree topologies. It has a range of probabilistic models available to search for a set of trees with the highest posterior probability. It is fast and capable of handling large datasets. The program is available in multiplatform versions and can be downloaded from http://morphbank.ebc.uu.se/mrbayes/. A web program that also employs Bayesian inference for phylogenetic analysis is BAMBE (http://bioweb.pasteur.fr/seqanal/interfaces/bambe.html).

SUMMARY

Molecular phylogeny is a fundamental tool for understanding sequence evolution and relationships. The accuracy of the tree-building methods used for phylogenetic analysis depends on the assumption on which each the method is based. Understanding these assumptions is the first step toward efficient use of these methods. The second step is understanding how the methods actually work and what intrinsic limitations these methods have. The third step is choosing suitable phylogenetic method(s) that can give a reasonably correct picture of a phylogenetic tree.

The phylogenetic methods can be divided into distance-based and character-based methods. The distance methods include UPGMA, NJ, Fitch–Margoliash, and minimum evolution. The first two are clustering based, and are fast but not accurate; the latter two are optimality based and are accurate but not fast. Character-based approaches include the MP and ML methods. The principle of parsimony is easy to understand, but has its root in a medieval philosophy. It is slower compared to distance methods. To speed up the computation, branch-and-bound and heuristics tree searching strategies are used. The ML method is the slowest, but is based on a solid statistical foundation. To overcome the bottleneck of computation in ML, faster algorithms such as quartet puzzling, NJML, GA, and Bayesian analysis have been developed to make the method more feasible. To assess the reliability and robustness of every clade in a phylogenetic tree, bootstrapping and jackknifing are used. The KH and SH tests distinguish the overall topology of two competing trees.

It is important to realize that phylogenetic tree reconstruction is not a trivial matter, but a complicated process that often requires careful thought. Accuracy, reliability, and computational speed are all major factors for consideration when choosing a particular phylogenetic method. It is also important to realize that none of the three phylogenetic reconstruction methods are guaranteed to find the correct tree. All three methods have the potential to produce erroneous trees. To minimize phylogenetic errors, it is recommended that at least two methods be used for any phylogenetic analysis to check the consistency of tree building results obtained. Because the theories behind each of the three methods are fundamentally different, agreement in conclusion by several of these methods provides a particularly strong support for a correct phylogenetic tree. In addition, it is recommended that different rate substitution models, weighting schemes, and resampling strategies with or without exclusion of specific taxa and/or sites be applied. The same analysis should be repeated on multiple genes or proteins as well as the concatenated datasets. If more than one fundamentally different methods provide the same prediction, the confidence in the prediction is higher.

FURTHER READING

Graur, D., and Li, W.-H. 2000. *Fundamentals of Molecular Evolution.* Sunderland, MA: Sinauer Associates.

Hall, B. G. 2001. *Phylogenetic Trees Made Easy. A How-to Manual for Molecular Biologists.* Sunderland, MA: Sinauer Associates.

Huelsenbeck J. P., Ronquist, F., Nielsen R., and Bollback, J. P. 2001. Bayesian inference of phylogeny and its impact on evolutionary biology. *Science.* 294:2310–14.

Nei, M., and Kumar, S. 2000. *Molecular Evolution and Phylogenetics.* New York: Oxford University Press.

Salemi, M., and Vandamme, A.-M. 2003. *The Phylogenetics Handbook – A Practical Approach to DNA and Protein Phylogeny.* Cambridge, UK: Cambridge University Press.

Swofford, D. L., Olsen, G. J., Waddel, P. J., and Hillis, D. M. 1996. "Phylogenetic inference." In *Molecular Systematics.* 2nd ed., edited by D. M. Hillis, C. Moritz, and B. K. Mable Sunderland, MA: Sinauer Associates.

Structural Bioinformatics

CHAPTER TWELVE

Protein Structure Basics

Starting from this chapter and continuing through the next three chapters, we introduce the basics of protein structural bioinformatics. Proteins perform most essential biological and chemical functions in a cell. They play important roles in structural, enzymatic, transport, and regulatory functions. The protein functions are strictly determined by their structures. Therefore, protein structural bioinformatics is an essential element of bioinformatics. This chapter covers some basics of protein structures and associated databases, preparing the reader for discussions of more advanced topics of protein structural bioinformatics.

AMINO ACIDS

The building blocks of proteins are twenty naturally occurring amino acids, small molecules that contain a free amino group (NH_2) and a free carboxyl group (COOH). Both of these groups are linked to a central carbon ($C\alpha$), which is attached to a hydrogen and a side chain group (R) (Fig. 12.1). Amino acids differ only by the side chain R group. The chemical reactivities of the R groups determine the specific properties of the amino acids.

Amino acids can be grouped into several categories based on the chemical and physical properties of the side chains, such as size and affinity for water. According to these properties, the side chain groups can be divided into small, large, hydrophobic, and hydrophilic categories. Within the hydrophobic set of amino acids, they can be further divided into aliphatic and aromatic. *Aliphatic side chains* are linear hydrocarbon chains and *aromatic side chains* are cyclic rings. Within the hydrophilic set, amino acids can be subdivided into polar and charged. *Charged amino acids* can be either positively charged (basic) or negatively charged (acidic). Each of the twenty amino acids, their abbreviations, and main functional features once incorporated into a protein are listed in Table 12.1.

Of particular interest within the twenty amino acids are glycine and proline. Glycine, the smallest amino acid, has a hydrogen atom as the R group. It can therefore adopt more flexible conformations that are not possible for other amino acids. Proline is on the other extreme of flexibility. Its side chain forms a bond with its own backbone amino group, causing it to be cyclic. The cyclic conformation makes it very rigid, unable to occupy many of the main chain conformations adopted by other amino acids. In addition, certain amino acids are subject to modifications after

TABLE 12.1. Twenty Standard Amino Acids Grouped by Their Common Side-Chain Features

Amino Acid Group	Amino Acid Name	Three- and One-Letter Code	Main Functional Features
Small and nonpolar	Glycine Alanine Proline	Gly, G Ala, A Pro, P	Nonreactive in chemical reactions; Pro and Gly disrupt regular secondary structures
Small and polar	Cysteine Serine Threonine	Cys, C Ser, S Thr, T	Serving as posttranslational modification sites and participating in active sites of enzymes or binding metal
Large and polar	Glutamine Asparagine	Gln, Q Asn, N	Participating in hydrogen bonding or in enzyme active sites
Large and polar (basic)	Arginine Lysine Histidine	Arg, R Lys, K His, H	Found in the surface of globular proteins providing salt bridges; His participates in enzyme catalysis or metal binding
Large and polar (acidic)	Glutamate Aspartate	Glu, E Asp, D	Found in the surface of globular proteins providing salt bridges
Large and nonpolar (aliphatic)	Isoleucine Leucine Methionine Valine	Ile, I Leu, L Met, M Val, V	Nonreactive in chemical reactions; participating in hydrophobic interactions
Large and nonpolar (aromatic)	Phenylalanine Tyrosine Tryptophan	Phe, F Tyr, Y Trp, W	Providing sites for aromatic packing interactions; Tyr and Trp are weakly polar and can serve as sites for phosphorylation and hydrogen bonding

Note: Each amino acid is listed with its full name, three- and one-letter abbreviations, and main functional roles when serving as amino acid residues in a protein. Properties of some amino acid groups overlap.

a protein is translated in a cell. This is called *posttranslational modification,* and is discussed in more detail in Chapter 19.

PEPTIDE FORMATION

The peptide formation involes two amino acids covalently joined together between the carboxyl group of one amino acid and the amino group of another (Fig. 12.2). This

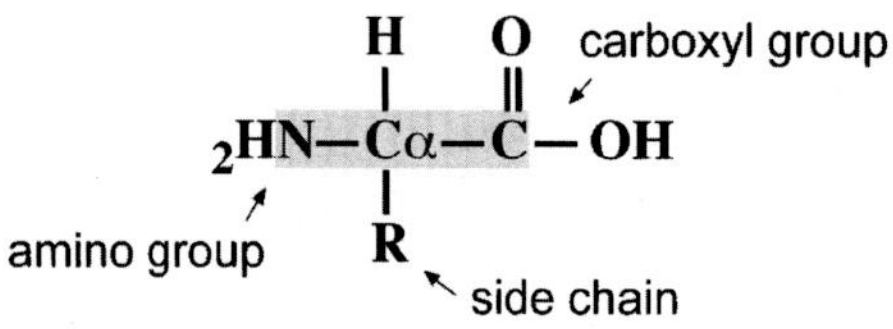

Figure 12.1: General structure of an amino acid. The main chain atoms are highlighted. The R group can be any of the twenty amino acid side chains.

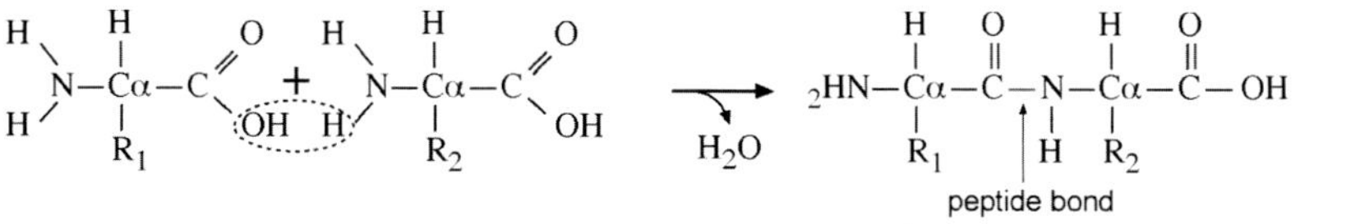

Figure 12.2: Condensation reaction between the carboxyl group of one amino acid and the amino group of another. The hydroxyl group of the carboxyl group and a hydrogen of the amino group are lost to give rise to a water molecule and a dipeptide.

reaction is a condensation reaction involving removal of elements of water from the two molecules. The resulting product is called a *dipeptide*. The newly formed covalent bond connecting the two amino acids is called a *peptide bond*. Once an amino acid is incorporated into a peptide, it becomes an amino acid residue. Multiple amino acids can be joined together to form a longer chain of amino acid polymer.

A linear polymer of more than fifty amino acid residues is referred to as a *polypeptide*. A polypeptide, also called a protein, has a well-defined three-dimensional arrangement. On the other hand, a polymer with fewer than fifty residues is usually called a peptide without a well-defined three-dimensional structure. The residues in a peptide or polypeptide are numbered beginning with the residue containing the amino group, referred to as the *N*-terminus, and ending with the residue containing the carboxyl group, known as the *C*-terminus (see Fig. 12.2). The actual sequence of amino acid residues in a polypeptide determines its ultimate structure and function.

The atoms involved in forming the peptide bond are referred to as the *backbone atoms*. They are the nitrogen of the amino group, the α carbon to which the side chain is attached and carbon of the carbonyl group.

DIHEDRAL ANGLES

A peptide bond is actually a partial double bond owing to shared electrons between O=C–N atoms. The rigid double bond structure forces atoms associated with the peptide bond to lie in the same plane, called the *peptide plane*. Because of the planar nature of the peptide bond and the size of the R groups, there are considerable restrictions on the rotational freedom by the two bonded pairs of atoms around the peptide bond. The angle of rotation about the bond is referred to as the *dihedral angle* (also called the *tortional angle*).

For a peptide unit, the atoms linked to the peptide bond can be moved to a certain extent by the rotation of two bonds flanking the peptide bond. This is measured by two dihedral angles (Fig. 12.3). One is the dihedral angle along the N–Cα bond, which is defined as phi (ϕ); and the other is the angle along the Cα–C bond, which is called psi (ψ). Various combinations of ϕ and ψ angles allow the proteins to fold in many different ways.

Ramachandran Plot

As mentioned, the rotation of ϕ and ψ is not completely free because of the planar nature of the peptide bond and the steric hindrance from the side chain R group.

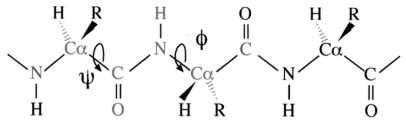

Figure 12.3: Definition of dihedral angles of ϕ and ψ. Six atoms around a peptide bond forming two peptide planes are colored in red. The ϕ angle is the rotation about the N–Cα bond, which is measured by the angle between a virtual plane formed by the C–N–Cα and the virtual plane by N–Cα–C (C in green). The ψ angle is the rotation about the Cα–C bond, which is measured by the angle between a virtual plane formed by the N–Cα–C (N in green) and the virtual plane by Cα–C–N (N in red) (see color plate section).

Consequently, there is only a limited range of peptide conformation. When ϕ and ψ angles of amino acids of a particular protein are plotted against each other, the resulting diagram is called a Ramachandran plot. This plot maps the entire conformational space of a peptide and shows sterically allowed and disallowed regions (Fig. 12.4). It can be very useful in evaluating the quality of protein models.

HIERARCHY

Protein structures can be organized into four levels of hierarchies with increasing complexity. These levels are primary structure, secondary structure, tertiary structure, and quaternary structure. A linear amino acid sequence of a protein is the primary structure. This is the simplest level with amino acid residues linked together through

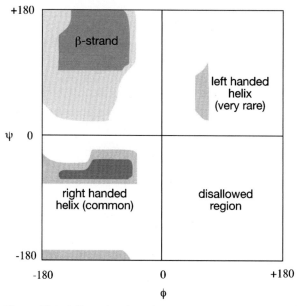

Figure 12.4: A Ramachandran plot with allowed values of ϕ and ψ in shaded areas. Regions favored by α-helices and β-strands (to be explained) are indicated.

peptide bonds. The next level up is the secondary structure, defined as the local conformation of a peptide chain. The secondary structure is characterized by highly regular and repeated arrangement of amino acid residues stabilized by hydrogen bonds between main chain atoms of the C=O group and the NH group of different residues. The level above the secondary structure is the tertiary structure, which is the three-dimensional arrangement of various secondary structural elements and connecting regions. The tertiary structure can be described as the complete three-dimensional assembly of all amino acids of a single polypeptide chain. Beyond the tertiary structure is the quaternary structure, which refers to the association of several polypeptide chains into a protein complex, which is maintained by noncovalent interactions. In such a complex, individual polypeptide chains are called *monomers* or *subunits*. Intermediate between secondary and tertiary structures, a level of supersecondary structure is often used, which is defined as two or three secondary structural elements forming a unique functional domain, a recurring structural pattern conserved in evolution.

Stabilizing Forces

Protein structures from secondary to quaternary are maintained by noncovalent forces. These include electrostatic interactions, van der Waals forces, and hydrogen bonding. Electrostatic interactions are a significant stabilizing force in a protein structure. They occur when excess negative charges in one region are neutralized by positive charges in another region. The result is the formation of salt bridges between oppositely charged residues. The electrostatic interactions can function within a relatively long range (15 Å).

Hydrogen bonds are a particular type of electrostatic interactions similar to dipole–dipole interactions involving hydrogen from one residue and oxygen from another. Hydrogen bonds can occur between main chain atoms as well as side chain atoms. Hydrogen from the hydrogen bond donor group such as the N–H group is slightly positively charged, whereas oxygen from the hydrogen bond acceptor group such as the C=O group is slightly negatively charged. When they come within a close distance (<3 Å), a partial bond is formed between them, resulting in a hydrogen bond. Hydrogen bonding patterns are a dominant factor in determining different types of protein secondary structures.

Van der Waals forces also contribute to the overall protein stability. These forces are instantaneous interactions between atoms when they become transient dipoles. A transient dipole can induce another transient dipole nearby. The dipoles of the two atoms can be reversed a moment later. The oscillating dipoles result in an attractive force. The van der Waals interactions are weaker than electrostatic and hydrogen bonds and thus only have a secondary effect on the protein structure.

In addition to these common stabilizing forces, disulfide bridges, which are covalent bonds between the sulfur atoms of the cysteine residue, are also important in maintaining some protein structures. For certain types of proteins that contain metal ions as prosthetic groups, noncovalent interactions between amino acid residues and the metal ions may play an important structural role.

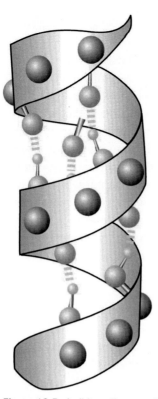

Figure 12.5: A ribbon diagram of an α-helix with main chain atoms (as grey balls) shown. Hydrogen bonds between the carbonyl oxygen (red) and the amino hydrogen (green) of two residues are shown in yellow dashed lines (see color plate section).

SECONDARY STRUCTURES

As mentioned, local structures of a protein with regular conformations are known as secondary structures. They are stabilized by hydrogen bonds formed between carbonyl oxygen and amino hydrogen of different amino acids. Chief elements of secondary structures are α-helices and β-sheets.

α-Helices

An α-helix has a main chain backbone conformation that resembles a corkscrew. Nearly all known α-helices are right handed, exhibiting a rightward spiral form. In such a helix, there are 3.6 amino acids per helical turn. The structure is stabilized by hydrogen bonds formed between the main chain atoms of residues i and $i + 4$. The hydrogen bonds are nearly parallel with the helical axis (Fig. 12.5). The average ϕ and ψ angles are 60° and 45°, respectively, and are distributed in a narrowly defined region in the lower left region of a Ramachandran plot (see Fig. 12.4). Hydrophobic residues of the helix tend to face inside and hydrophilic residues of the helix face outside. Thus, every third residue along the helix tends to be a hydrophobic residue. Ala, Gln, Leu, and Met are commonly found in an α-helix, but not Pro, Gly, and Tyr. These rules are useful in guiding the prediction of protein secondary structures.

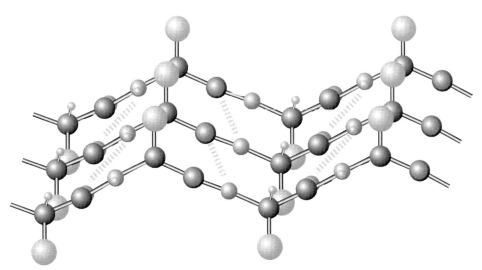

Figure 12.6: Side view of a parallel β-sheet. Hydrogen bonds between the carbonyl oxygen (red) and the amino hydrogen (green) of adjacent β-strands are shown in yellow dashed lines. R groups are shown as big balls in cyan and are positioned alternately on opposite sides of β-strands (see color plate section).

β-Sheets

A β-sheet is a fully extended configuration built up from several spatially adjacent regions of a polypeptide chain. Each region involved in forming the β-sheet is a β-strand. The β-strand conformation is pleated with main chain backbone zigzagging and side chains positioned alternately on opposite sides of the sheet. β-Strands are stabilized by hydrogen bonds between residues of adjacent strands (Fig. 12.6). β-strands near the surface of the protein tend to show an alternating pattern of hydrophobic and hydrophilic regions, whereas strands buried at the core of a protein are nearly all hydrophobic.

The β-strands can run in the same direction to form a parallel sheet or can run every other chain in reverse orientation to form an antiparallel sheet, or a mixture of both. The hydrogen bonding patterns are different in each configurations. The ϕ and ψ angles are also widely distributed in the upper left region in a Ramachandran plot (see Fig. 12.4). Because of the long-range nature of residues involved in this type of conformation, it is more difficult to predict β-sheets than α-helices.

Coils and Loops

There are also local structures that do not belong to regular secondary structures (α-helices and β-strands). The irregular structures are coils or loops. The loops are often characterized by sharp turns or hairpin-like structures. If the connecting regions are completely irregular, they belong to random coils. Residues in the loop or coil regions tend to be charged and polar and located on the surface of the protein structure. They are often the evolutionarily variable regions where mutations, deletions,

Figure 12.7: An α-helical coiled coil found in tropomyosin showing two helices wound around to form a helical bundle.

and insertions frequently occur. They can be functionally significant because these locations are often the active sites of proteins.

Coiled Coils

Coiled coils are a special type of supersecondary structure characterized by a bundle of two or more α-helices wrapping around each other (Fig. 12.7). The helices forming coiled coils have a unique pattern of hydrophobicity, which repeats every seven residues (five hydrophobic and two hydrophilic). More details on coiled coils and their structure prediction are discussed in Chapter 14.

TERTIARY STRUCTURES

The overall packing and arrangement of secondary structures form the tertiary structure of a protein. The tertiary structure can come in various forms but is generally classified as either globular or membrane proteins. The former exists in solvents through hydrophilic interactions with solvent molecules; the latter exists in membrane lipids and is stabilized through hydrophobic interactions with the lipid molecules.

Globular Proteins

Globular proteins are usually soluble and surrounded by water molecules. They tend to have an overall compact structure of spherical shape with polar or hydrophilic residues on the surface and hydrophobic residues in the core. Such an arrangement is energetically favorable because it minimizes contacts with water by hydrophobic residues in the core and maximizes interactions with water by surface polar and charged residues. Common examples of globular proteins are enzymes, myoglobins, cytokines, and protein hormones.

Integral Membrane Proteins

Membrane proteins exist in lipid bilayers of cell membranes. Because they are surrounded by lipids, the exterior of the proteins spanning the membrane must be very hydrophobic to be stable. Most typical transmembrane segments are α-helices. Occasionally, for some bacterial periplasmic membrane proteins, they are composed of β-strands. The loops connecting these segments sometimes lie in the aqueous phase, in which they can be entirely hydrophilic. Sometimes, they lie in the interface between the lipid and aqueous phases and are amphipathic in nature (containing polar residues facing the aqueous side and hydrophobic residues towards the lipid side). The amphipathic residues can also form helices which have a periodicity of three or four residues. Common examples of membrane proteins are rhodopsins, cytochrome *c* oxidase, and ion channel proteins.

DETERMINATION OF PROTEIN THREE-DIMENSIONAL STRUCTURE

Protein three-dimensional structures are obtained using two popular experimental techniques, x-ray crystallography and nuclear magnetic resonance (NMR) spectroscopy. The experimental procedures and relative merits of each method are discussed next.

X-ray Crystallography

In x-ray protein crystallography, proteins need to be grown into large crystals in which their positions are fixed in a repeated, ordered fashion. The protein crystals are then illuminated with an intense x-ray beam. The x-rays are deflected by the electron clouds surrounding the atoms in the crystal producing a regular pattern of diffraction. The diffraction pattern is composed of thousands of tiny spots recorded on a x-ray film. The diffraction pattern can be converted into an electron density map using a mathematical procedure known as Fourier transform. To interpret a three-dimensional structure from two-dimensional electron density maps requires solving the phases in the diffraction data. The phases refer to the relative timing of different diffraction waves hitting the detector. Knowing the phases can help to determine the relative positions of atoms in a crystal.

Phase solving can be carried out by two methods, molecular replacement, and multiple isomorphous replacement. Molecular replacement uses a homologous protein structure as template to derive an initial estimate of the phases. Multiple isomorphous replacement derives phases by comparing electron intensity changes in protein crystals containing heavy metal atoms and the ones without heavy metal atoms. The heavy atoms diffract x-rays with unusual intensities, which can serve as a marker for relative positions of atoms.

Once the phases are available, protein structures can be solved by modeling with amino acid residues that best fit the electron density map. The quality of the final model is measured by an R factor, which indicates how well the model reproduces the experimental electron intensity data. The R factor is expressed as a percentage of difference between theoretically reproduced diffraction data and experimentally determined diffraction data. R values can range from 0.0, which is complete agreement, to 0.59, which is complete disagreement. A major limitation of x-ray crystallography is whether suitable crystals of proteins of interest can be obtained.

Nuclear Magnetic Resonance Spectroscopy

NMR spectroscopy detects spinning patterns of atomic nuclei in a magnetic field. Protein samples are labeled with radioisotopes such as ^{13}C and ^{15}N. A radiofrequency radiation is used to induce transitions between nuclear spin states in a magnetic field. Interactions between spinning isotope pairs produce radio signal peaks that correlate with the distances between them. By interpreting the signals observed using NMR, proximity between atoms can be determined. Knowledge of distances between all labeled atoms in a protein allows a protein model to be built that satisfies all the constraints. NMR determines protein structures in solution, which has the advantage of not requiring the crystallization process. However, the proteins in solution are mobile and vibrating, reflecting the dynamic behavior of proteins. For that reason, usually a number of slightly different models (twenty to forty) have to be constructed that satisfy all the NMR distance measurements. The NMR technique obviates the need of growing protein crystals and can solve structures relatively more quickly than x-ray crystallography. The major problem associated with using NMR is the current limit of protein size (<200 residues) that can be determined. Another problem is the requirement of heavy instrumentation.

PROTEIN STRUCTURE DATABASE

Once the structure of a particular protein is solved, a table of (x, y, z) coordinates representing the spatial position of each atom of the structure is created. The coordinate information is required to be deposited in the Protein Data Bank (PDB, www.rcsb.org/pdb/) as a condition of publication of a journal paper. PDB is a worldwide central repository of structural information of biological macromolecules and is currently managed by the Research Collaboratory for Structural Bioinformatics (RCSB). In addition, the PDB website provides a number of services for structure submission and data searching and retrieval. Through its web interface, called *Structure Explorer*, a user is able to read the summary information of a protein structure, view and download structure coordinate files, search for structure neighbors of a particular protein or access related research papers through links to the NCBI PubMed database.

There are currently more than 30,000 entries in the database with the number increasing at a dramatic rate in recent years owing to large-scale structural proteomics projects being carried out. Most of the database entries are structures of proteins. However, a small portion of the database is composed of nucleic acids, carbohydrates, and theoretical models. Most protein structures are determined by x-ray crystallography and a smaller number by NMR.

Although the total number of entries in PDB is large, most of the protein structures are redundant, namely, they are structures of the same protein determined under different conditions, at different resolutions, or associated with different ligands or with single residue mutations. Sometimes, structures from very closely related proteins are determined and deposited in PDB. A small number of well-studied proteins such as hemoglobins and myoglobins have hundreds of entries. Excluding the redundant entries, there are approximately 3,000 unique protein structures represented in the database. Among the unique protein structures, there are only a limited number of protein folds available (800) compared to ~1,000,000 unique protein sequences already known, suggesting that the protein structures are much more conserved. A protein fold is a particular topological arrangement of helices, strands, and loops. Protein classification by folds is discussed in Chapter 13.

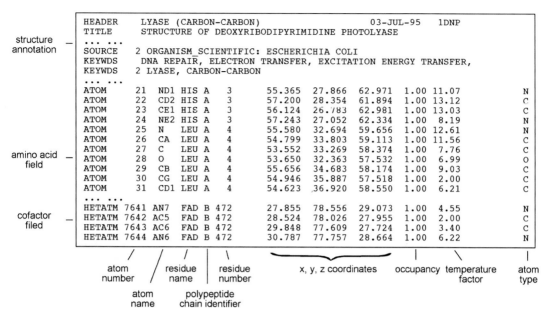

Figure 12.8: A partial PDB file of DNA photolyase (*boxed*) showing the header section and the coordinate section. The coordinate section is dissected based on individual fields.

PDB Format

A deposited set of protein coordinates becomes an entry in PDB. Each entry is given a unique code, PDBid, consisting of four characters of either letters A to Z or digits 0 to 9 such as 1LYZ and 4RCR. One can search a structure in PDB using the four-letter code or keywords related to its annotation. The identified structure can be viewed directly online or downloaded to a local computer for analysis. The PDB website provides options for retrieval, analysis, and direct viewing of macromolecular structures. The viewing can be still images or manipulable images through interactive viewing tools (see Chapter 13). It also provides links to protein structural classification results available in databases such as SCOP and CATH (see Chapter 13).

The data format in PDB was created in the early 1970s and has a rigid structure of 80 characters per line, including spaces. This format was initially designed to be compatible with FORTRAN programs. It consists of an explanatory header section followed by an atomic coordinate section (Fig. 12.8).

The header section provides an overview of the protein and the quality of the structure. It contains information about the name of the molecule, source organism, bibliographic reference, methods of structure determination, resolution, crystallographic parameters, protein sequence, cofactors, and description of structure types and locations and sometimes secondary structure information. In the structure coordinates section, there are a specified number of columns with predetermined contents. The ATOM part refers to protein atom information whereas the HETATM (for heteroatom group) part refers to atoms of cofactor or substrate molecules. Approximately ten columns of text and numbers are designated. They include information for the atom number, atom name, residue name, polypeptide chain identifier, residue number, x, y, and z Cartesian coordinates, temperature factor, and occupancy factor. The last two parameters, occupancy and temperature factors, relate to disorders of atomic positions in crystals.

The PDB format has been in existence for more than three decades. It is fairly easy to read and simple to use. However, the format is not designed for computer extraction of information from the records. Certain restrictions in the format have significantly complicated its current use. For instance, in the PDB format, only Cartesian coordinates of atoms are given without bonding information. Information such as disulfide bonds has to be interpreted by viewing programs, some of which may fail to do so. In addition, the field width for atom number is limited to five characters, meaning that the maximum number of atoms per model is 99,999. The field width for polypeptide chains is only one character in width, meaning that no more than 26 chains can be used in a multisubunit protein model. This has made many large protein complexes such as ribosomes unable to be represented by a single PDB file. They have to be divided into multiple PDB files.

mmCIF and MMDB Formats

Significant limitations of the PDB format have allowed the development of new formats to handle increasingly complicated structure data. The most popular new formats include the macromolecular crystallographic information file (mmCIF) and the molecular modeling database (MMDB) file. Both formats are highly parsable by computer software, meaning that information in each field of a record can be retrieved separately. These new formats facilitate the retrieval and organization of information from database structures.

The mmCIF format is similar to the format for a relational database (see Chapter 2) in which a set of tables are used to organize database records. Each table or field of information is explicitly assigned by a tag and linked to other fields through a special syntax. An example of an mmCIF containing multiple fields is given below. As shown in Figure 12.9, a single line of description in the header section of PDB is divided into many lines or fields with each field having explicit assignment of item names and item values. Each field starts with an underscore character followed by category name and keyword description separated by a period. The annotation in Figure 12.9 shows that the data items belong to the category of "struct" or "database." Following a keyword tag, a short text string enclosed by quotation marks is used to assign values for the keyword. Using multiple fields with tags for the same information has the advantage of providing an explicit reference to each item in a data file and ensures a one-to-one relationship between item names and item values. By presenting the data item by item, the format provides much more flexibility for information storage and retrieval.

Another new format is the MMDB format developed by the NCBI to parse and sort pieces of information in PDB. The objective is to allow the information to be more easily integrated with GenBank and Medline through Entrez (see Chapter 2).

PDB	HEADER PLANT SEED PROTEIN 11-OCT-91 1CBN

| **mmCIF** | ```
_struct.entry_id '1CBN'
_struct.title 'PLANT SEED PROTEIN'
_struct_keywords.entry_id '1CBN'
_struct_keywords.text 'plant seed protein'
_database_2.database_id 'PDB'
_database_2.database_code '1CBN'
_database_PDB_rev.rev_num 1
_database_PDB_rev.date_original '1991-10-11'
``` |
|---|---|

Figure 12.9: A comparison of PDB and mmCIF formats in two different boxes. To show the same header information in PDB, multiple fields are required in mmCIF to establish explicit relationships of item name and item values. The advantage of such format is easy parsing by computer software.

An MMDB file is written in the ASN.1 format (see Chapter 2), which has information in a record structured as a nested hierarchy. This allows faster retrieval than mmCIF and PDB. Furthermore, the MMDB format includes bond connectivity information for each molecule, called a "chemical graph," which is recorded in the ASN.1 file. The inclusion of the connectivity data allows easier drawing of structures.

SUMMARY

Proteins are considered workhorses in a cell and carry out most cellular functions. Knowledge of protein structure is essential to understand the behavior and functions of specific proteins. Proteins are polypeptides formed by joining amino acids together via peptide bonds. The folding of a polypeptide can be described by rotational angles around the main chain bonds such as ϕ and ψ angles. The degree of rotation depends on the preferred protein conformation. Allowable ϕ and ψ angles in a protein can be specified in a Ramachandran plot. There are four levels of protein structures, primary, secondary, tertiary, and quaternary. The primary structure is the sequence of amino acid residues. The secondary structure is the repeated main chain conformation, which includes α-helices and β-sheets. The tertiary structure is the overall three-dimensional conformation of a polypeptide chain. The quaternary structure is the complex arrangement of multiple polypeptide chains. Protein structures are stabilized by electrostatic interactions, hydrogen bonds, and van der Waals interactions. Proteins can be classified as being soluble globular proteins or integral membrane proteins, whose structures vary tremendously. Protein structures can be determined by x-ray crystallography and NMR spectroscopy. Both methods have advantages and disadvantages, but are clearly complementary. The solved structures are deposited in PDB, which uses a PDB format to describe structural details. However, the original PDB format has limited capacity and is difficult to be parsed by computer software.
```

To overcome the limitations, new formats such as mmCIF and MMDB have been developed.

FURTHER READING

Branden, C., and Tooze, J. 1999. *Introduction to Protein Structure*, 2nd ed. New York: Garland Publishing.

Scheeff, E. D., and Fink, J. L. 2003. "Fundamentals of protein structure." In *Structural Bioinformatics*, edited by P. E. Bourne and H. Weissig, 15–39. Hoboken, NJ: Wiley-Liss.

Westbrook, J. D., and Fitzgerald, P. M. D. 2003. "The PDB format, mmCIF and other data formats." In *Structural Bioinformatics*, edited by P. E. Bourne and H. Weissig, 161–79. Hoboken, NJ: Wiley-Liss.

Protein Structure Visualization, Comparison, and Classification

Once a protein structure has been solved, the structure has to be presented in a three-dimensional view on the basis of the solved Cartesian coordinates. Before computer visualization software was developed, molecular structures were represented by physical models of metal wires, rods, and spheres. With the development of computer hardware and software technology, sophisticated computer graphics programs have been developed for visualizing and manipulating complicated three-dimensional structures. The computer graphics help to analyze and compare protein structures to gain insight to functions of the proteins.

PROTEIN STRUCTURAL VISUALIZATION

The main feature of computer visualization programs is interactivity, which allows users to visually manipulate the structural images through a graphical user interface. At the touch of a mouse button, a user can move, rotate, and zoom an atomic model on a computer screen in real time, or examine any portion of the structure in great detail, as well as draw it in various forms in different colors. Further manipulations can include changing the conformation of a structure by protein modeling or matching a ligand to an enzyme active site through docking exercises.

Because a Protein Data Bank (PDB) data file for a protein structure contains only x, y, and z coordinates of atoms (see Chapter 12), the most basic requirement for a visualization program is to build connectivity between atoms to make a view of a molecule. The visualization program should also be able to produce molecular structures in different styles, which include wire frames, balls and sticks, space-filling spheres, and ribbons (Fig. 13.1).

A wire-frame diagram is a line drawing representing bonds between atoms. The wire frame is the simplest form of model representation and is useful for localizing positions of specific residues in a protein structure, or for displaying a skeletal form of a structure when Cα atoms of each residue are connected. Balls and sticks are solid spheres and rods, representing atoms and bonds, respectively. These diagrams can also be used to represent the backbone of a structure. In a space-filling representation (or Corey, Pauling, and Koltan [CPK]), each atom is described using large solid spheres with radii corresponding to the van der Waals radii of the atoms. Ribbon diagrams use cylinders or spiral ribbons to represent α-helices and broad, flat arrows to represent β-strands. This type of representation is very attractive in that it allows easy

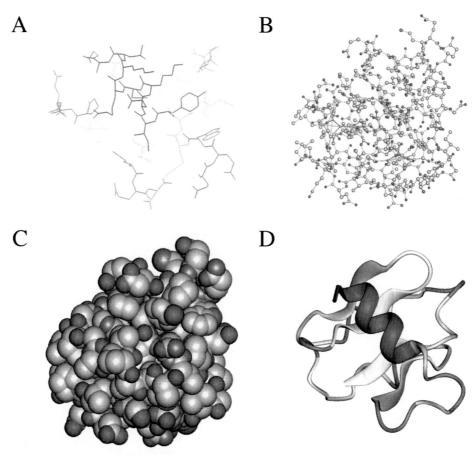

Figure 13.1: Examples of molecular structure visualization forms. **(A)** Wireframes. **(B)** Balls and sticks. **(C)** Space-filling spheres. **(D)** Ribbons (see color plate section).

identification of secondary structure elements and gives a clear view of the overall topology of the structure. The resulting images are also visually appealing.

Different representation styles can be used in combination to highlight a certain feature of a structure while deemphasizing the structures surrounding it. For example, a cofactor of an enzyme can be shown as space-filling spheres while the rest of the protein structure is shown as wire frames or ribbons. Some widely used and freely available software programs for molecular graphics are introduced next with examples of rendering provided in Figure 13.2.

RasMol (http://rutgers.rcsb.org/pdb/help-graphics.html#rasmol_download) is a command-line–based viewing program that calculates connectivity of a coordinate file and displays wireframe, cylinder, stick bonds, α-carbon trace, space-filling (CPK) spheres, and ribbons. It reads both PDB and mmCIF formats and can display a whole molecule or specific parts of it. It is available in multiple platforms: UNIX, Windows, and Mac. RasTop (www.geneinfinity.org/rastop/) is a new version of RasMol for Windows with a more enhanced user interface.

A

B

C

D

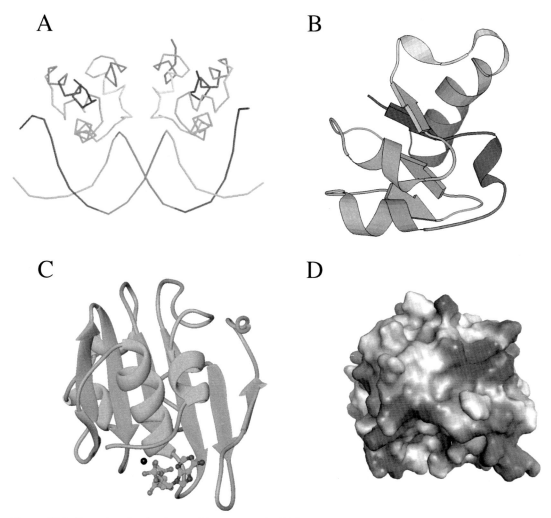

Figure 13.2: Examples of molecular graphic generated by **(A)** Rasmol, **(B)** Molscript, **(C)** Ribbons, and **(D)** Grasp (see color plate section).

Swiss-PDBViewer (www.expasy.ch/spdbv/) is a structure viewer for multiple plat-forms. It is essentially a Swiss-Army knife for structure visualization and modeling because it incorporates so many functions in a small shareware program. It is capable of structure visualization, analysis, and homology modeling. It allows display of multiple structures at the same time in different styles, by charge distribution, or by surface accessibility. It can measure distances, angles, and even mutate residues. In addition, it can calculate molecular surface, electrostatic potential, Ramachandran plot, and so on. The homology modeling part includes energy minimization and loop modeling.

Molscript (www.avatar.se/molscript/) is a UNIX program capable of generating wire-frame, space-filling, or ball-and-stick styles. In particular, secondary struc-ture elements can be drawn with solid spirals and arrows representing α-helices and β-strands, respectively. Visually appealing images can be generated that are of publication quality. The drawback is that the program is command-line–based and not very user friendly. A modified UNIX program called Bobscript (www.strubi.ox. ac.uk/bobscript/) is available with enhanced features.

Ribbons (http://sgce.cbse.uab.edu/ribbons/) another UNIX program similar to Molscript, generates ribbon diagrams depicting protein secondary structures. Aesthetically appealing images can be produced that are of publication quality. However, the program, which is also command-line-based, is extremely difficult to use.

Grasp (http://trantor.bioc.columbia.edu/grasp/) is a UNIX program that generates solid molecular surface images and uses a gradated coloring scheme to display electrostatic charges on the surface.

There are also a number of web-based visualization tools that use Java applets. These programs tend to have limited molecular display features and low-quality images. However, the advantage is that the user does not have to download, compile, and install the programs locally, but simply view the structures on a web browser using any kind of computer operating system. In fact, the PDB also attempts to simplify the database structure display for end users. It has incorporated a number of light-weight Java-based structure viewers in the PDB web site (see Chapter 12).

WebMol (www.cmpharm.ucsf.edu/cgi-bin/webmol.pl) is a web-based program built based on a modified RasMol code and thus shares many similarities with RasMol. It runs directly on a browser of any type as an applet and is able to display simple line drawing models of protein structures. It also has a feature of interactively displaying Ramachandran plots for structure model evaluation.

Chime (www.mdlchime.com/chime/) is a plug-in for web browsers; it is not a standalone program and has to be invoked in a web browser. The program is also derived from RasMol and allows interactive display of graphics of protein structures inside a web browser.

Cn3D (www.ncbi.nlm.nih.gov/Structure/CN3D/cn3d.shtml) is a helper application for web browsers to display structures in the MMDB format from the NCBI's structural database. It can be used on- or offline as a stand-alone program. It is able to render three-dimensional molecular models and display secondary structure cartoons. The drawback is that it does not recognize the PDB format.

PROTEIN STRUCTURE COMPARISON

With the visualization and computer graphics tools available, it becomes easy to observe and compare protein structures. To compare protein structures is to analyze two or more protein structures for similarity. The comparative analysis often, but not always, involves the direct alignment and superimposition of structures in a three-dimensional space to reveal which part of structure is conserved and which part is different at the three-dimensional level.

This structure comparison is one of the fundamental techniques in protein structure analysis. The comparative approach is important in finding remote protein homologs. Because protein structures have a much higher degree of conservation than the sequences, proteins can share common structures even without sequence similarity. Thus, structure comparison can often reveal distant evolutionary relationships between proteins, which is not feasible using the sequence-based alignment approach alone. In addition, protein structure comparison is a prerequisite for protein structural classification into different fold classes. It is also useful in evaluating protein prediction methods by comparing theoretically predicted structures with experimentally determined ones.

One can always compare structures manually or by eye, which is often practiced. However, the best approach is to use computer algorithms to automate the task and thereby get more accurate results. Structure comparison algorithms all employ scoring schemes to measure structural similarities and to maximize the structural similarities measured using various criteria. The algorithmic approaches to comparing protein geometric properties can be divided into three categories: the first superposes protein structures by minimizing intermolecular distances; the second relies on measuring intramolecular distances of a structure; and the third includes algorithms that combine both intermolecular and intramolecular approaches.

Intermolecular Method

The intermolecular approach is normally applied to relatively similar structures. To compare and superpose two protein structures, one of the structures has to be moved with respect to the other in such a way that the two structures have a maximum overlap in a three-dimensional space. This procedure starts with identifying equivalent residues or atoms. After residue–residue correspondence is established, one of the structures is moved laterally and vertically toward the other structure, a process known as *translation,* to allow the two structures to be in the same location (or same coordinate frame). The structures are further rotated relative to each other around the three-dimensional axes, during which process the distances between equivalent positions are constantly measured (Fig. 13.3). The rotation continues until the shortest intermolecular distance is reached. At this point, an optimal superimposition of the two structures is reached. After superimposition, equivalent residue pairs can be identified, which helps to quantitate the fitting between the two structures.

An important measurement of the structure fit during superposition is the distance between equivalent positions on the protein structures. This requires using a least-square-fitting function called *root mean square deviation* (RMSD), which is the square root of the averaged sum of the squared differences of the atomic distances.

$$\text{RMSD} = \sqrt{\sum_{i=1}^{N} D_i^2 / N} \qquad \qquad \text{(Eq. 13.1)}$$

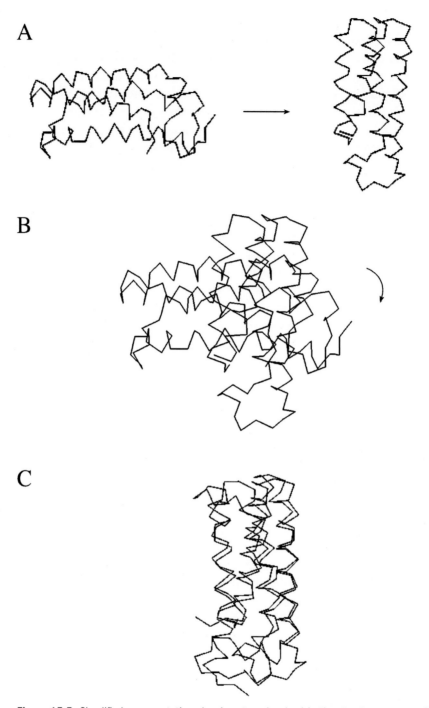

Figure 13.3: Simplified representation showing steps involved in the structure superposition of two protein molecules. **(A)** Two protein structures are positioned in different places in a three dimensional space. Equivalent positions are identified using a sequence based alignment approach. **(B)** To superimpose the two structures, the first step is to move one structure (*left*) relative to the other (*right*) through lateral and vertical movement, which is called translation. **(C)** The left structure is then rotated relative to the reference structure until such a point that the relative distances between equivalent positions are minimal.

where D is the distance between coordinate data points and N is the total number of corresponding residue pairs.

In practice, only the distances between Cα carbons of corresponding residues are measured. The goal of structural comparison is to achieve a minimum RMSD. However, the problem with RMSD is that it depends on the size of the proteins being compared. For the same degree of sequence identity, large proteins tend to have higher RMSD values than small proteins when an optimal alignment is reached. Recently, a logarithmic factor has been proposed to correct this size-dependency problem. This new measure is called $RMSD_{100}$ and is determined by the following formula:

$$RMSD_{100} = \frac{RMSD}{-1.3 + 0.5\ln(N)} \qquad \text{(Eq. 13.2)}$$

where N is the total number of corresponding atoms.

Although this corrected RMSD is more reliable than the raw RMSD for structure superposition, a low RMSD value by no means guarantees a correct alignment or an alignment with biological meaning. Careful scrutiny of the automatic alignment results is always recommended.

The most challenging part of using the intermolecular method is to identify equivalent residues in the first place, which often resorts to sequence alignment methods. Obviously, this restricts the usefulness of structural comparison between distant homologs.

A number of solutions have been proposed to compare more distantly related structures. One approach that has been proposed is to delete sequence variable regions outside secondary structure elements to reduce the search time required to find an optimum superposition. However, this method does not guarantee an optimal alignment. Another approach adopted by some researchers is to divide the proteins into small fragments (e.g., every six to nine residues). Matching of similar regions at the three-dimensional level is then done fragment by fragment. After finding the best fitting fragments, a joint superposition for the entire structure is performed. The third approach is termed *iterative optimization*, during which the two sequences are first aligned using dynamic programming. Identified equivalent residues are used to guide a first round of superposition. After superposition, more residues are identified to be in close proximity at the three-dimensional level and considered as equivalent residues. Based on the newly identified equivalent residues, a new round of superposition is generated to refine from the previous alignment. This procedure is repeated until the RMSD values cannot be further improved.

Intramolecular Method

The intramolecular approach relies on structural internal statistics and therefore does not depend on sequence similarity between the proteins to be compared. In addition, this method does not generate a physical superposition of structures, but instead provides a quantitative evaluation of the structural similarity between corresponding residue pairs.

The method works by generating a distance matrix between residues of the same protein. In comparing two protein structures, the distance matrices from the two structures are moved relative to each other to achieve maximum overlaps. By overlaying two distance matrices, similar intramolecular distance patterns representing similar structure folding regions can be identified. For the ease of comparison, each matrix is decomposed into smaller submatrices consisting of hexapeptide fragments. To maximize the similarity regions between two structures, a Monte Carlo procedure is used. By reducing three-dimensional information into two-dimensional information, this strategy identifies overall structural resemblances and common structure cores.

Combined Method

A recent development in structure comparison involves combining both inter- and intramolecular approaches. In the hybrid approach, corresponding residues can be identified using the intramolecular method. Subsequent structure superposition can be performed based on residue equivalent relationships. In addition to using RMSD as a measure during alignment, additional structural properties such as secondary structure types, torsion angles, accessibility, and local hydrogen bonding environment can be used. Dynamic programming is often employed to maximize overlaps in both inter- and intramolecular comparisons.

Multiple Structure Alignment

In addition to pairwise alignment, a number of algorithms can also perform multiple structure alignment. The alignment strategy is similar to the Clustal sequence alignment using a progressive approach (see Chapter 5). That is, all structures are first compared in a pairwise fashion. A distance matrix is developed based on structure similarity scores such as RMSD. This allows construction of a phylogenetic tree, which guides the subsequent clustering of the structures. The most similar two structures are then realigned. The aligned structures create a median structure that allows other structures to be progressively added for comparison based on the hierarchy described in the guide tree. When all the structures in the set are added, this eventually creates a multiple structure alignment. Several popular on-line structure comparison resources are discussed next.

DALI (www2.ebi.ac.uk/dali/) is a structure comparison web server that uses the intramolecular distance method. It works by maximizing the similarity of two distance graphs. The matrices are based on distances between all Cα atoms for each individual protein. Two distance matrices are overlaid and moved one relative to the other to identify most similar regions. DALI uses a statistical significance value called a Z-score to evaluate structural alignment. The Z-score is the number of standard deviations from the average score derived from the database background distribution. The higher the Z-score when comparing a pair of protein structures, the less likely the similarity observed is a result of random chance. Empirically, a Z-score >4 indicates a significant level of structure similarity. The web server is at the same time a database that contains Z-scores of all precomputed structure pairs of proteins in PDB. The user can upload a structure to compare it with all known structures, or perform a pairwise comparison of two uploaded structures.

CE (Combinatorial Extension; http://cl.sdsc.edu/ce.html) is a web-based program that also uses the intramolecular distance approach. However, unlike DALI, a type of heuristics is used. In this method, every eight residues are treated as a single residue. The $C\alpha$ distance matrices are constructed at the level of octameric "residues." In this way, the computational time required to search for the best alignment is considerably reduced, at the expense of alignment accuracy. CE also uses a Z-score as a measure of significance of an alignment. A Z-score >3.5 indicates a similar fold.

VAST (Vector Alignment Search Tool; www.ncbi.nlm.nih.gov:80/Structure/VAST/vast.shtml) is a web server that performs alignment using both the inter- and intramolecular approaches. The superposition is based on information of directionality of secondary structural elements (represented as vectors). Optimal alignment between two structures is defined by the highest degree of vector matches.

SSAP (www.biochem.ucl.ac.uk/cgi-bin/cath/GetSsapRasmol.pl) is a web server that uses an intramolecular distance–based method in which matrices are built based on the $C\beta$ distances of all residue pairs. When comparing two different matrices, a dynamic programming approach is used to find the path of residue positions with optimal scores. The dynamic programming is applied at two levels, one at a lower level in which all residue pairs between the proteins are compared and another at an upper level in which subsequently identified equivalent residue pairs are processed to refine the matching positions. This process is known as double dynamic programming. An SSAP score is reported for structural similarity. A score above 70 indicates a good structural similarity.

STAMP (www.compbio.dundee.ac.uk/Software/Stamp/stamp.html) is a UNIX program that uses the intermolecular approach to generate protein structure alignment. The main feature is the use of iterative alignment based on dynamic programming to obtain the best superposition of two or more structures.

PROTEIN STRUCTURE CLASSIFICATION

One of the applications of protein structure comparison is structural classification. The ability to compare protein structures allows classification of the structure data and identification of relationships among structures. The reason to develop a protein structure classification system is to establish hierarchical relationships among protein structures and to provide a comprehensive and evolutionary view of known structures. Once a hierarchical classification system is established, a newly obtained protein structure can find its place in a proper category. As a result, its functions can be better understood based on association with other proteins. To date, several systems have been developed, the two most popular being Structural Classification of Proteins (SCOP) and Class, Architecture, Topology and Homologous (CATH). The following introduces the basic steps in establishing the systems to classify proteins.

The first step in structure classification is to remove redundancy from databases. As mentioned in Chapter 12, among the tens of thousands of entries in PDB, the majority of the structures are redundant as they correspond to structures solved at different resolutions, or associated with different ligands or with single-residue mutations. The redundancy can be removed by selecting representatives through a sequence alignment–based approach. The second step is to separate structurally distinct domains within a structure. Because some proteins are composed of multiple domains, they must be subdivided before a sensible structural comparison can be carried out. This domain identification and separation can be done either manually or based on special algorithms for domain recognition. Once multidomain proteins are split into separate domains, structure comparison can be conducted at the domain level, either through manual inspection, or automated structural alignment, or a combination of both. The last step involves grouping proteins/domains of similar structures and clustering them based on different levels of resemblance in secondary structure composition and arrangement of the secondary structures in space.

As mentioned, the two most popular classification schemes are SCOP and CATH, both of which contain a number of hierarchical levels in their systems.

SCOP

SCOP (http://scop.mrc-lmb.cam.ac.uk/scop/) is a database for comparing and classifying protein structures. It is constructed almost entirely based on manual examination of protein structures. The proteins are grouped into hierarchies of classes, folds, superfamilies, and families. In the latest SCOP release version (v1.65, released December 2003), there are 7 classes, 800 folds, 1,294 superfamilies, and 2,327 families.

The SCOP families consist of proteins having high sequence identity (>30%). Thus, the proteins within a family clearly share close evolutionary relationships and normally have the same functionality. The protein structures at this level are also extremely similar. Superfamilies consist of families with similar structures, but weak sequence similarity. It is believed that members of the same superfamily share a common ancestral origin, although the relationships between families are considered distant. Folds consist of superfamilies with a common core structure, which is determined manually. This level describes similar overall secondary structures with similar orientation and connectivity between them. Members within the same fold do not always have evolutionary relationships. Some of the shared core structure may be a result of analogy. Classes consist of folds with similar core structures. This is at the highest level of the hierarchy, which distinguishes groups of proteins by secondary structure compositions such as all α, all β, α and β, and so on. Some classes are created based on general features such as membrane proteins, small proteins with few secondary structures and irregular proteins. Folds within the same class are essentially randomly related in evolution.

CATH

CATH (www.biochem.ucl.ac.uk/bsm/cath_new/index.html) classifies proteins based on the automatic structural alignment program SSAP as well as manual comparison. Structural domain separation is carried out also as a combined effort of a human expert and computer programs. Individual domain structures are classified at five major levels: class, architecture, fold/topology, homologous superfamily, and homologous family. In the CATH release version 2.5.1 (January 2004), there are 4 classes, 37 architectures, 813 topologies, 1,467 homologous superfamilies, and 4,036 homologous families.

The definition for class in CATH is similar to that in SCOP, and is based on secondary structure content. Architecture is a unique level in CATH, intermediate between fold and class. This level describes the overall packing and arrangement of secondary structures independent of connectivity between the elements. The topology level is equivalent to the fold level in SCOP, which describes overall orientation of secondary structures and takes into account the sequence connectivity between the secondary structure elements. The homologous superfamily and homologous family levels are equivalent to the superfamily and family levels in SCOP with similar evolutionary definitions, respectively.

Comparison of SCOP and CATH

SCOP is almost entirely based on manual comparison of structures by human experts with no quantitative criteria to group proteins. It is argued that this approach offers some flexibility in recognizing distant structural relatives, because human brains may be more adept at recognizing slightly dissimilar structures that essentially have the same architecture. However, this reliance on human expertise also renders the method subjective. The exact boundaries between levels and groups are sometimes arbitrary.

CATH is a combination of manual curation and automated procedure, which makes the process less subjective. For example, in defining domains, CATH first relies on the consensus of three different algorithms to recognize domains. When the computer programs disagree, human intervention will take place. In addition, the extra Architecture level in CATH makes the structure classification more continuous. The drawback of the systems is that the fixed thresholds in structural comparison may make assignment less accurate.

Due to the differences in classification criteria, one might expect that there would be huge differences in classification results. In fact, the classification results from both systems are quite similar. Exhaustive analysis has shown that the results from the two systems converge at about 80% of the time. In other words, only about 20% of the structure fold assignments are different. Figure 13.4 shows two examples of agreement and disagreement based on classification by the two systems.

PDB code: 4tim

	SCOP		CATH
Class	Alpha and Beta (α/β)	*Class*	Alpha Beta
		Architecture	Barrel
Fold	TIM beta/alpha-barrel	*Topology*	TIM Barrel
Superfamily	Triosephosphate isomerase	*Homologous Superfamily*	Triosephosphate isomerase
Family	Triosephosphate isomerase	*Homologous Family*	Triosephosphate isomerase

PDB code: 1lys

	SCOP		CATH
Class	Alpha and Beta (α+β)	*Class*	Mainly Alpha
		Architecture	Orthogonal Bundle
Fold	Lysozyme-like	*Topology*	Lysozyme
Superfamily	Lysozyme-like	*Homologous Superfamily*	Hydrolase (O-glycosyl)
Family	C-type lysozyme	*Homologous Family*	Hydrolase

Figure 13.4: Comparison of results of structure classification between SCOP and CATH. The classifications on the left is a case of overall agreement whereas the one on the right disagrees at the class level.

SUMMARY

A clear and concise visual representation of protein structures is the first step towards structural understanding. A number of visualization programs have been developed for that purpose. They include stand-alone programs for sophisticated manipulation of structures and light-weight web-based programs for simple structure viewing. Protein structure comparison allows recognition of distant evolutionary relationships among proteins and is helpful for structure classification and evaluation of protein structure prediction methods. The comparison algorithms fall into three categories: the intermolecular method, which involves transformation of atomic coordinates of structures to get optimal superimposition; the intramolecular method, which constructs an inter-residue distance matrix within a molecule and compares the matrix against that from a second molecule; and the combined method that uses both inter- and intramolecular approaches. Among all the structure comparison algorithms developed so far, DALI is most widely used. Protein structure classification is important for understanding protein structure, function and evolution. The most widely used classification schemes are SCOP and CATH. The two systems largely agree but differ somewhat. Each system has its own strengths and neither appears to be superior. It is thus advisable to compare the classification results from both systems in order to put a structure in the correct context.

FURTHER READING

Bourne, P. E., and Shindyalov, I. N. 2003. "Structure comparison and alignment." In *Structural Bioinformatics*, edited by P. E. Bourne and H. Weissig, 321–37. Hoboken, NJ: Wiley-Liss.

Carugo, O., and Pongor, S. 2002. Recent progress in protein 3D structure comparison. *Curr. Protein Pept. Sci.* 3:441–9.

Hadley, C., and Jones, D. T. 1999. A systematic comparison of protein structure classifications: SCOP, CATH, and FSSP. *Structure* 7:1099–112.

Jawad, Z., and Paoli, M. 2002. Novel sequences propel familiar folds. *Structure* 10:447–54.

Kinch, L. N., and Grishin, N. V. 2002. Evolution of protein structures and functions. *Curr. Opin. Struct. Biol.* 12:400–8.

Koehl, P. 2001. Protein structure similarities. *Curr. Opin. Struct. Biol.* 11:348–53.

Orengo, C. A., Pearl, F. M. G., and Thornton, J. M. 2003. "The CATH domain structure database." In *Structural Bioinformatics*, edited by P. E. Bourne and H. Weissig, 249–71. Hoboken, NJ: Wiley-Liss.

Ouzounis, C. A., Coulson, R. M., Enright, A. J., Kunin, V., and Pereira-Leal, J. B. 2003. Classification schemes for protein structure and function. *Nat. Rev. Genet.* 4:508–19.

Reddy, B. J. 2003. "Protein structure evolution and the scop database." In *Structural Bioinformatics*, edited by P. E. Bourne and H. Weissig, 239–48. Hoboken, NJ: Wiley-Liss.

Russell, R. B. 2002. Classification of protein folds. *Mol. Biotechnol.* 20:17–28.

Tate, J. 2003. "Molecular visualization." In *Structural Bioinformatics*, edited by P. E. Bourne and H. Weissig, 135–58. Hoboken, NJ: Wiley-Liss.

Protein Secondary Structure Prediction

Protein secondary structures are stable local conformations of a polypeptide chain. They are critically important in maintaining a protein three-dimensional structure. The highly regular and repeated structural elements include α-helices and β-sheets. It has been estimated that nearly 50% of residues of a protein fold into either α-helices and β-strands. As a review, an α-helix is a spiral-like structure with 3.6 amino acid residues per turn. The structure is stabilized by hydrogen bonds between residues i and $i + 4$. Prolines normally do not occur in the middle of helical segments, but can be found at the end positions of α-helices (see Chapter 12). A β-sheet consists of two or more β-strands having an extended zigzag conformation. The structure is stabilized by hydrogen bonding between residues of adjacent strands, which actually may be long-range interactions at the primary structure level. β-Strands at the protein surface show an alternating pattern of hydrophobic and hydrophilic residues; buried strands tend to contain mainly hydrophobic residues.

Protein secondary structure prediction refers to the prediction of the conformational state of each amino acid residue of a protein sequence as one of the three possible states, namely, helices, strands, or coils, denoted as H, E, and C, respectively. The prediction is based on the fact that secondary structures have a regular arrangement of amino acids, stabilized by hydrogen bonding patterns. The structural regularity serves the foundation for prediction algorithms.

Predicting protein secondary structures has a number of applications. It can be useful for the classification of proteins and for the separation of protein domains and functional motifs. Secondary structures are much more conserved than sequences during evolution. As a result, correctly identifying secondary structure elements (SSE) can help to guide sequence alignment or improve existing sequence alignment of distantly related sequences. In addition, secondary structure prediction is an intermediate step in tertiary structure prediction as in threading analysis (see Chapter 15).

Because of significant structural differences between globular proteins and transmembrane proteins, they necessitate very different approaches to predicting respective secondary structure elements. Prediction methods for each of two types of proteins are discussed herein. In addition, prediction of supersecondary structures, such as coiled coils, is also described.

SECONDARY STRUCTURE PREDICTION FOR GLOBULAR PROTEINS

Protein secondary structure prediction with high accuracy is not a trivial ask. It remained a very difficult problem for decades. This is because protein secondary structure elements are context dependent. The formation of α-helices is determined by short-range interactions, whereas the formation of β-strands is strongly influenced by long-range interactions. Prediction for long-range interactions is theoretically difficult. After more than three decades of effort, prediction accuracies have only been improved from about 50% to about 75%.

The secondary structure prediction methods can be either ab initio based, which make use of single sequence information only, or homology based, which make use of multiple sequence alignment information. The ab initio methods, which belong to early generation methods, predict secondary structures based on statistical calculations of the residues of a single query sequence. The homology-based methods do not rely on statistics of residues of a single sequence, but on common secondary structural patterns conserved among multiple homologous sequences.

Ab Initio–Based Methods

This type of method predicts the secondary structure based on a single query sequence. It measures the relative propensity of each amino acid belonging to a certain secondary structure element. The propensity scores are derived from known crystal structures. Examples of ab initio prediction are the Chou–Fasman and Garnier, Osguthorpe, Robson (GOR) methods. The ab initio methods were developed in the 1970s when protein structural data were very limited. The statistics derived from the limited data sets can therefore be rather inaccurate. However, the methods are simple enough that they are often used to illustrate the basics of secondary structure prediction.

The Chou–Fasman algorithm (http://fasta.bioch.virginia.edu/fasta/chofas.htm) determines the propensity or intrinsic tendency of each residue to be in the helix, strand, and β-turn conformation using observed frequencies found in protein crystal structures (conformational values for coils are not considered). For example, it is known that alanine, glutamic acid, and methionine are commonly found in α-helices, whereas glycine and proline are much less likely to be found in such structures.

The calculation of residue propensity scores is simple. Suppose there are n residues in all known protein structures from which m residues are helical residues. The total number of alanine residues is y of which x are in helices. The propensity for alanine to be in helix is the ratio of the proportion of alanine in helices over the proportion of alanine in overall residue population (using the formula $[x/m]/[y/n]$). If the propensity for the residue equals 1.0 for helices ($P[\alpha\text{-helix}]$), it means that the residue has an equal chance of being found in helices or elsewhere. If the propensity ratio is less than 1, it indicates that the residue has less chance of being found in helices. If the propensity is larger than 1, the residue is more favored by helices. Based on this concept, Chou

TABLE 14.1. Relative Amino Acid Propensity Values for Secondary Structure Elements Used in the Chou–Fasman Method

Amino Acid	(α-Helix)	P (β-Strand)	P (Turn)
Alanine	1.42	0.83	0.66
Arginine	0.98	0.93	0.95
Asparagine	0.67	0.89	1.56
Aspartic acid	1.01	0.54	1.46
Cysteine	0.70	1.19	1.19
Glutamic acid	1.51	0.37	0.74
Glutamine	1.11	1.11	0.98
Glycine	0.57	0.75	1.56
Histidine	1.00	0.87	0.95
Isoleucine	1.08	1.60	0.47
Leucine	1.21	1.30	0.59
Lysine	1.14	0.74	1.01
Methionine	1.45	1.05	0.60
Phenylalanine	1.13	1.38	0.60
Proline	0.57	0.55	1.52
Serine	0.77	0.75	1.43
Threonine	0.83	1.19	0.96
Tryptophan	0.83	1.19	0.96
Tyrosine	0.69	1.47	1.14
Valine	1.06	1.70	0.50

and Fasman developed a scoring table listing relative propensities of each amino acid to be in an α-helix, a β-strand, or a β-turn (Table 14.1).

Prediction with the Chou–Fasman method works by scanning through a sequence with a certain window size to find regions with a stretch of contiguous residues each having a favored SSE score to make a prediction. For α-helices, the window size is six residues, if a region has four contiguous residues each having P(α-helix) > 1.0, it is predicted as an α-helix. The helical region is extended in both directions until the P(α-helix) score becomes smaller than 1.0. That defines the boundaries of the helix. For β-strands, scanning is done with a window size of five residues to search for a stretch of at least three favored β-strand residues. If both types of secondary structure predictions overlap in a certain region, a prediction is made based on the following criterion: if $\Sigma P(\alpha) > \Sigma P(\beta)$, it is declared as an α-helix; otherwise, a β-strand.

The GOR method (http://fasta.bioch.virginia.edu/fasta_www/garnier.htm) is also based on the "propensity" of each residue to be in one of the four conformational states, helix (H), strand (E), turn (T), and coil (C). However, instead of using the propensity value from a single residue to predict a conformational state, it takes short-range interactions of neighboring residues into account. It examines a window of every seventeen residues and sums up propensity scores for all residues for each of the four states resulting in four summed values. The highest scored state defines the conformational state for the center residue in the window (ninth position). The GOR method has been shown to be more accurate than Chou–Fasman because it takes the neighboring effect of residues into consideration.

Both the Chou–Fasman and GOR methods, which are the first-generation methods developed in the 1970s, suffer from the fact that the prediction rules are somewhat arbitrary. They are based on single sequence statistics without clear relation to known protein-folding theories. The predictions solely rely on local sequence information and fail to take into account long range interactions. A Chou-Fasman–based prediction does not even consider the short-range environmental information. These reasons, combined with unreliable statistics derived from a very small structural database, limit the prediction accuracy of these methods to about 50%. This performance is considered dismal; any random prediction can have a 40% accuracy given the fact that, in globular proteins, the three-state distribution is 30% α-helix, 20% β-strands, and 50% coil.

Newer algorithms have since been developed to overcome some of these short-comings. The improvements include more refined residue statistics based on a larger number of solved protein structures and the incorporation of more local residue interactions. Examples of the improved algorithms are GOR II, GOR III, GOR IV, and SOPM. These tools can be found at http://npsa-pbil.ibcp.fr/cgi-bin/npsa_automat.pl?page=/NPSA/npsa_server.html. These are the second-generation prediction algorithms developed in the 1980s and early 1990s. They have improved accuracy over the first generation by about 10%. Although it is already significantly better than that by random prediction, the programs are still not reliable enough for routine application. Prediction errors mainly occur through missed β-strands and short-lengthed secondary structures for both helices and strands. Prediction of β-strands is still somewhat random. This may be attributed to the fact that long range interactions are not sufficiently taken into consideration in these algorithms.

Homology-Based Methods

The third generation of algorithms were developed in the late 1990s by making use of evolutionary information. This type of method combines the ab initio secondary structure prediction of individual sequences and alignment information from multiple homologous sequences (>35% identity). The idea behind this approach is that close protein homologs should adopt the same secondary and tertiary structure. When each individual sequence is predicted for secondary structure using a method similar to the GOR method, errors and variations may occur. However, evolutionary conservation dictates that there should be no major variations for their secondary structure elements. Therefore, by aligning multiple sequences, information of positional conservation is revealed. Because residues in the same aligned position are assumed to have the same secondary structure, any inconsistencies or errors in prediction of individual sequences can be corrected using a majority rule (Fig. 14.1). This homology-based method has helped improve the prediction accuracy by another 10% over the second-generation methods.

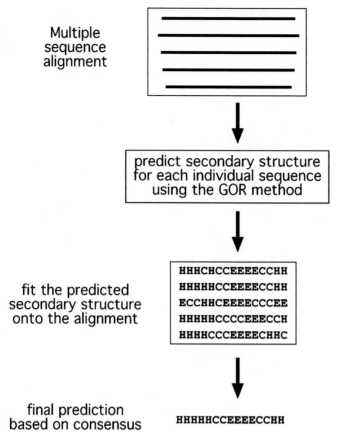

Multiple sequence alignment

predict secondary structure for each individual sequence using the GOR method

fit the predicted secondary structure onto the alignment

HHHCHCCEEEECCHH
HHHHHCCEEEECCHH
ECCHHCEEEECCCEE
HHHHHCCCCEEECCH
HHHHCCCEEEECHHC

final prediction based on consensus   HHHHHCCEEEECCHH

Figure 14.1: Schematic representation of secondary structure prediction using multiple sequence alignment information. Each individual sequence in the multiple alignment is subject to secondary structure prediction using the GOR method. If variations in predictions occur, they can be corrected by deriving a consensus of the secondary structure elements from the alignment.

Prediction with Neural Networks

The third-generation prediction algorithms also extensively apply sophisticated neural networks (see Chapter 8) to analyze substitution patterns in multiple sequence alignments. As a review, a *neural network* is a machine learning process that requires a structure of multiple layers of interconnected variables or nodes. In secondary structure prediction, the input is an amino acid sequence and the output is the probability of a residue to adopt a particular structure. Between input and output are many connected hidden layers where the machine learning takes place to adjust the mathematical weights of internal connections. The neural network has to be first trained by sequences with known structures so it can recognize the amino acid patterns and their relationships with known structures. During this process, the weight functions in hidden layers are optimized so they can relate input to output correctly. When the sufficiently trained network processes an unknown sequence, it applies the rules learned in training to recognize particular structural patterns.

When multiple sequence alignments and neural networks are combined, the result is further improved accuracy. In this situation, a neural network is trained not by a single sequence but by a sequence profile derived from the multiple sequence alignment. This combined approach has been shown to improve the accuracy to above 75%, which is a breakthrough in secondary structure prediction. The improvement mainly comes from enhanced secondary structure signals through consensus drawing. The following lists several frequently used third generation prediction algorithms available as web servers.

PHD (Profile network from Heidelberg; http://dodo.bioc.columbia.edu/predict protein/submit_def.html) is a web-based program that combines neural network with multiple sequence alignment. It first performs a BLASTP of the query sequence against a nonredundant protein sequence database to find a set of homologous sequences, which are aligned with the MAXHOM program (a weighted dynamic programming algorithm performing global alignment). The resulting alignment in the form of a profile is fed into a neural network that contains three hidden layers. The first hidden layer makes raw prediction based on the multiple sequence alignment by sliding a window of thirteen positions. As in GOR, the prediction is made for the residue in the center of the window. The second layer refines the raw prediction by sliding a window of seventeen positions, which takes into account more flanking positions. This step makes adjustments and corrections of unfeasible predictions from the previous step. The third hidden layer is called the *jury network*, and contains networks trained in various ways. It makes final filtering by deleting extremely short helices (one or two residues long) and converting them into coils (Fig. 14.2). After the correction, the highest scored state defines the conformational state of the residue.

PSIPRED (http://bioinf.cs.ucl.ac.uk/psiform.html) is a web-based program that predicts protein secondary structures using a combination of evolutionary information and neural networks. The multiple sequence alignment is derived from a PSI-BLAST database search. A profile is extracted from the multiple sequence alignment generated from three rounds of automated PSI-BLAST. The profile is then used as input for a neural network prediction similar to that in PHD, but without the jury layer. To achieve higher accuracy, a unique filtering algorithm is implemented to filter out unrelated PSI-BLAST hits during profile construction.

SSpro (http://promoter.ics.uci.edu/BRNN-PRED/) is a web-based program that combines PSI-BLAST profiles with an advanced neural network, known as *bidirectional recurrent neural networks* (BRNNs). Traditional neural networks are unidirectional, feed-forward systems with the information flowing in one direction from input to output. BRNNs are unique in that the connections of layers are designed to be able to go backward. In this process, known as *back propagation*, the weights in hidden layers are repeatedly refined. In predicting secondary structure elements, the network uses the sequence profile as input and finds residue correlations by iteratively recycling the network (recursive network). The averaged output from the iterations is given as a final residue prediction. PROTER (http://distill.ucd.ie/porter/) is a recently developed program that uses similar BRNNs and has been shown to slightly outperform SSPRO.

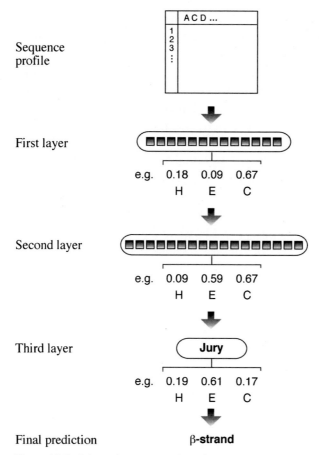

Figure 14.2: Schematic representation of secondary structure prediction in the PHD algorithm using neural networks. Multiple sequences derived from the BLAST search are used to compile a profile. The resulting profile is fed into a neural network, which contains three layers – two network layers and one jury layer. The first layer scans thirteen residues per window and makes a raw prediction, which is refined by the second layer, which scans seventeen residues per window. The third layer makes further adjustment to make a final prediction. Adjustment of prediction scores for one amino acid residue is shown.

PROF (Protein forecasting; www.aber.ac.uk/~phiwww/prof/) is an algorithm that combines PSI-BLAST profiles and a multistaged neural network, similar to that in PHD. In addition, it uses a linear discriminant function to discriminate between the three states.

HMMSTR (Hidden Markov model [HMM] for protein STRuctures; www.bioinfo. rpi.edu/~bystrc/hmmstr/server.php) uses a branched and cyclic HMM to predict secondary structures. It first breaks down the query sequence into many very short segments (three to nine residues, called I-sites) and builds profiles based on a library of known structure motifs. It then assembles these local motifs into a supersecondary structure. It further uses an HMM with a unique topology linking many smaller HMMs into a highly branched multicyclic form. This is intended to better capture the recurrent local features of secondary structure based on multiple sequence alignment.

Prediction with Multiple Methods

Because no individual methods can always predict secondary structures correctly, it is desirable to combine predictions from multiple programs with the hope of further improving the accuracy. In fact, a number of web servers have been specifically dedicated to making predictions by drawing consensus from results by multiple programs. In many cases, the consensus-based prediction method has been shown to perform slightly better than any single method.

Jpred (www.compbio.dundee.ac.uk/~www-jpred/) combines the analysis results from six prediction algorithms, including PHD, PREDATOR, DSC, NNSSP, Jnet, and ZPred. The query sequence is first used to search databases with PSI-BLAST for three iterations. Redundant sequence hits are removed. The resulting sequence homologs are used to build a multiple alignment from which a profile is extracted. The profile information is submitted to the six prediction programs. If there is sufficient agreement among the prediction programs, the majority of the prediction is taken as the structure. Where there is no majority agreement in the prediction outputs, the PHD prediction is taken.

PredictProtein (www.embl-heidelberg.de/predictprotein/predictprotein.html) is another multiple prediction server that uses Jpred, PHD, PROF, and PSIPRED, among others. The difference is that the server does not run the individual programs but sends the query to other servers which e-mail the results to the user separately. It does not generate a consensus. It is up to the user to combine multiple prediction results and derive a consensus.

Comparison of Prediction Accuracy

An important issue in protein secondary structure prediction is estimation of the prediction accuracy. The most commonly used measure for cross-validation is known as a Q_3 score, based on the three-state classification, helix (H), strand (E), and coil (C). The score is a percentage of residues of a protein that are correctly predicted. It is normally derived from the average result obtained from the testing with many proteins with known structures. For secondary structure prediction, there are well-established benchmarks for such prediction evaluation. By using these benchmarks, accuracies for several third-generation prediction algorithms have been compiled (Table 14.2).

As shown in Table 14.2, some of these best prediction methods have reached an accuracy level around 79% in the three-state prediction. Common errors include the confusion of helices and strands, incorrect start and end positions of helices and strands, and missed or wrongly assigned secondary structure elements. If a prediction is consistently 79% accurate, that means on average 21% of the residues could be predicted incorrectly.

Because different secondary structure prediction programs tend to give varied results, to maximize the accuracy of prediction, it is recommended to use several most robust prediction methods (such as Porter, PROF, and SSPRO) and draw a consensus based on the majority rule. The aforementioned metaservers provide a convenient

TABLE 14.2. Comparison of Accuracy of Some of the State-of-the-Art Secondary Structure Prediction Tools

Methods	Q_3 (%)
Porter	79.0
SSPro2	78.0
PROF	77.0
PSIPRED	76.6
Pred2ary	75.9
Jpred2	75.2
PHDpsi	75.1
Predator	74.8
HMMSTR	74.3

Note: The Q_3 score is the three-state prediction accuracy for helix, strand, and coil.

way of achieving this goal. By using the combination approach, it is possible to reach an 80% accuracy. An accuracy of 80% is an important landmark because it is equivalent to some low-resolution experimental methods to determine protein secondary structures, such as circular dichroism and Fourier transform-induced spectroscopy.

SECONDARY STRUCTURE PREDICTION FOR TRANSMEMBRANE PROTEINS

Transmembrane proteins constitute up to 30% of all cellular proteins. They are responsible for performing a wide variety of important functions in a cell, such as signal transduction, cross-membrane transport, and energy conversion. The membrane proteins are also of tremendous biomedical importance, as they often serve as drug targets for pharmaceutical development.

There are two types of integral membrane proteins: α-helical type and β-barrel type. Most transmembrane proteins contain solely α-helices, which are found in the cytoplasmic membrane. A few membrane proteins consist of β-strands forming a β-barrel topology, a cylindrical structure composed of antiparallel β-sheets. They are normally found in the outer membrane of gram-negative bacteria.

The structures of this group of proteins, however, are notoriously difficult to resolve either by x-ray crystallography or nuclear magnetic resonance (NMR) spectroscopy. Consequently, for this group of proteins, prediction of the transmembrane secondary structural elements and their organization is particularly important. Fortunately, the prediction process is somewhat easier because of the hydrophobic environment of the lipid bilayers, which restricts the transmembrane segments to be hydrophobic as well. In principle, the secondary structure prediction programs developed for soluble proteins can apply to membrane proteins as well. However, they normally do not work well in reality because the extra hydrophobicity and length requirements distort the

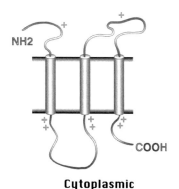

Figure 14.3: Schematic of the positive-inside rule for the orientation of membrane helices. The cylinders represent the transmembrane α–helices. There are relatively more positive charges near the helical anchor on the cytoplasmic side than on the periplasmic side.

statistical propensity of the residues. Thus, dedicated algorithms have to be used for transmembrane span predictions.

Prediction of Helical Membrane Proteins

For membrane proteins consisting of transmembrane α–helices, these transmembrane helices are predominantly hydrophobic with a specific distribution of positively charged residues. The α-helices generally run perpendicular to the membrane plane with an average length between seventeen and twenty-five residues. The hydrophobic helices are normally separated by hydrophilic loops with average lengths of fewer than sixty residues. The residues bordering the transmembrane spans are more positively charged. Another feature indicative of the presence of transmembrane segments is that residues at the cytosolic side near the hydrophobic anchor are more positively charged than those at the lumenal or periplasmic side. This is known as the *positive-inside rule* (Fig. 14.3), which allows the prediction of the orientation of the secondary structure elements. These rules form the basis for transmembrane prediction algorithms.

A number of algorithms for identifying transmembrane helices have been developed. The early algorithms based their prediction on hydrophobicity scales. They typically scan a window of seventeen to twenty-five residues and assign membrane spans based on hydrophobicity scores. Some are also able to determine the orientation of the membrane helices based on the positive-inside rule. However, predictions solely based on hydrophobicity profiles have high error rates. As with the third-generation predictions for globular proteins, applying evolutionary information with the help of neural networks or HMMs can improve the prediction accuracy significantly.

As mentioned, predicting transmembrane helices is relatively easy. The accuracy of some of the best predicting programs, such as TMHMM or HMMTOP, can exceed 70%. However, the presence of hydrophobic signal peptides can significantly compromise the prediction accuracy because the programs tend to confuse hydrophobic signal peptides with membrane helices. To minimize errors, the presence of signal peptides can be detected using a number of specialized programs (see Chapter 18) and then manually excluded.

TMHMM (www.cbs.dtu.dk/services/TMHMM/) is a web-based program based on an HMM algorithm. It is trained to recognize transmembrane helical patterns based on a training set of 160 well-characterized helical membrane proteins. When a query sequence is scanned, the probability of having an α-helical domain is given. The orientation of the α-helices is predicted based on the positive-inside rule. The prediction output returns the number of transmembrane helices, the boundaries of the helices, and a graphical representation of the helices. This program can also be used to simply distinguish between globular proteins and membrane proteins.

Phobius (http://phobius.cgb.ki.se/index.html) is a web-based program designed to overcome false positives caused by the presence of signal peptides. The program incorporates distinct HMM models for signal peptides as well as transmembrane helices. After distinguishing the putative signal peptides from the rest of the query sequence, prediction is made on the remainder of the sequence. It has been shown that the prediction accuracy can be significantly improved compared to TMHMM (94% by Phobius compared to 70% by TMHMM). In addition to the normal prediction mode, the user can also define certain sequence regions as signal peptides or other nonmembrane sequences based on external knowledge. As a further step to improve accuracy, the user can perform the "poly prediction" with the PolyPhobius module, which searches the NCBI database for homologs of the query sequence. Prediction for the multiple homologous sequences help to derive a consensus prediction. However, this option is also more time consuming.

Prediction of β-Barrel Membrane Proteins

For membrane proteins with β-strands only, the β-strands forming the transmembrane segment are amphipathic in nature. They contain ten to twenty-two residues with every second residue being hydrophobic and facing the lipid bilayers whereas the other residues facing the pore of the β-barrel are more hydrophilic. Obviously, scanning a sequence by hydrophobicity does not reveal transmembrane β-strands. These programs for predicting transmembrane α-helices are not applicable for this unique type of membrane proteins. To predict the β-barrel type of membrane proteins, a small number of algorithms have been made available based on neural networks and related techniques.

TBBpred (www.imtech.res.in/raghava/tbbpred/) is a web server for predicting transmembrane β-barrel proteins. It uses a neural network approach to predict transmembrane β-barrel regions. The network is trained with the known structural information of a limited number of transmembrane β-barrel protein structures. The algorithm contains a single hidden layer with five nodes and a single output node. In addition to neural networks, the server can also predict using a support vector machine (SVM) approach, another type of statistical learning process. Similar to

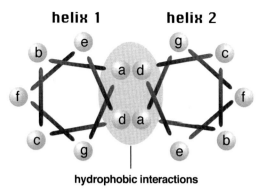

hydrophobic interactions

Figure 14.4: Cross-section view of a coiled coil structure. A coiled coil protein consisting of two interacting helical strands is viewed from top. The bars represent covalent bonds between amino acid residues. There is no covalent bond between residue *a* and *g*. The bar connecting the two actually means to connect the first residue of the next heptad. The coiled coil has a repeated seven residue motif in the form of *a-b-c-d-e-f-g*. The first and fourth positions (*a* and *d*) are hydrophobic, whose interactions with corresponding residues in another helix stabilize the structure. The positions *b, c, e, f, g* are hydrophilic and are exposed on the surface of the protein.

neural networks, in SVM the data are fed into kernels (similar to nodes), which are separated into different classes by a "hyperplane" (an abstract linear or nonlinear separator) according to a particular mathematical function. It has the advantage over neural networks in that it is faster to train and more resistant to noise. For more detailed information of SVM, see Chapter 19.

COILED COIL PREDICTION

Coiled coils are superhelical structures involving two to more interacting α-helices from the same or different proteins. The individual α-helices twist and wind around each other to form a coiled bundle structure. The coiled coil conformation is important in facilitating inter- or intraprotein interactions. Proteins possessing these structural domains are often involved in transcription regulation or in the maintenance of cytoskeletal integrity.

Coiled coils have an integral repeat of seven residues (heptads) which assume a side-chain packing geometry at facing residues (see Chapter 12). For every seven residues, the first and fourth are hydrophobic, facing the helical interface; the others are hydrophilic and exposed to the solvent (Fig. 14.4). The sequence periodicity forms the basis for designing algorithms to predict this important structural domain. As a result of the regular structural features, if the location of coiled coils can be predicted precisely, the three-dimensional structure for the coiled coil region can sometimes be built. The following lists several widely used programs for the specialized prediction.

Coils (www.ch.embnet.org/software/COILS_form.html) is a web-based program that detects coiled coil regions in proteins. It scans a window of fourteen, twenty-one, or twenty-eight residues and compares the sequence to a probability matrix compiled from known parallel two-stranded coiled coils. By comparing the similarity scores, the program calculates the probability of the sequence to adopt a coiled coil conformation. The program is accurate for solvent-exposed, left-handed coiled coils, but less sensitive for other types of coiled coil structures, such as buried or right-handed coiled coils.

Multicoil (http://jura.wi.mit.edu/cgi-bin/multicoil/multicoil.pl) is a web-based program for predicting coiled coils. The scoring matrix is constructed based on a database of known two-stranded and three-stranded coiled coils. The program is more conservative than Coils. It has been recently used in several genome-wide studies to screen for protein–protein interactions mediated by coiled coil domains.

Leucine zipper domains are a special type of coiled coils found in transcription regulatory proteins. They contain two antiparallel α-helices held together by hydrophobic interactions of leucine residues. The heptad repeat pattern is L-X(6)-L-X(6)-L-X(6)-L. This repeat pattern alone can sometimes allow the domain detection, albeit with high rates of false positives. The reason for the high false-positive rates is that the condition of the sequence region being a coiled coil conformation is not satisfied. To address this problem, algorithms have been developed that take into account both leucine repeats and coiled coil conformation to give accurate prediction.

2ZIP (http://2zip.molgen.mpg.de/) is a web-based server that predicts leucine zippers. It combines searching of the characteristic leucine repeats with coiled coil prediction using an algorithm similar to Coils to yield accurate results.

SUMMARY

Protein secondary structure prediction has a long history and is defined by three generations of development. The first generation algorithms were ab initio based, examining residue propensities that fall in the three states: helices, strands, and coils. The propensities were derived from a very small structural database. The growing structural database and use of residue local environment information allowed the development of the second-generation algorithms. A major breakthrough came from the third-generation algorithms that make use of multiple sequence alignment information, which implicitly takes the long-range intraprotein interactions into consideration. In combination with neural networks and other sophisticated algorithms, prediction efficiency has been improved significantly. To achieve high accuracy in prediction, combining results from several top-performing third-generation algorithms is recommended. Predicting secondary structures for membrane proteins is more common than for globular proteins as crystal or NMR structures are extremely difficult to obtain for the former. The prediction of transmembrane segments (mainly α-helices) involves the use of hydrophobicity, neural networks, and evolutionary information. Coiled coils are a distinct type of supersecondary structure with regular periodicity of hydrophobic residues that can be predicted using specialized algorithms.

FURTHER READING

Edwards, Y. J., and Cottage, A. 2003. Bioinformatics methods to predict protein structure and function. A practical approach. *Mol. Biotechnol.* 23:139-66.

Heringa J. 2002. Computational methods for protein secondary structure prediction using multiple sequence alignments. *Curr. Protein Pept. Sci.* 1:273-301.

Lehnert, U., Xia, Y., Royce, T. E., Goh, C. S., Liu, Y., Senes, A., Yu. H., Zhang, Z. L., Engelman, D. M, and Gerstein M. 2004. Computational analysis of membrane proteins: Genomic occurrence, structure prediction and helix interactions. *Q. Rev. Biophys.* 37:121-46.

Möller, S., Croning, M. D. R., and Apweiler, R. 2001. Evaluation of methods for the prediction of membrane spanning regions. *Bioinformatics* 17:646-53.

Przybylski, D., and Rost, B. 2002. Alignments grow, secondary structure prediction improves. *Proteins* 46:197–205.

Rost B. 2001. Review: Protein secondary structure prediction continues to rise. *J. Struct. Biol.* 134:204–18.

———. 2003. Prediction in 1D: Secondary structure, membrane helices, and accessibility. In *Structural Bioinformatics,* edited by P. E. Bourne and H. Weissig, 559–87. Hoboken, NJ: Wiley–Liss.

Protein Tertiary Structure Prediction

One of the most important scientific achievements of the twentieth century was the discovery of the DNA double helical structure by Watson and Crick in 1953. Strictly speaking, the work was the result of a three-dimensional modeling conducted partly based on data obtained from x-ray diffraction of DNA and partly based on chemical bonding information established in stereochemistry. It was clear at the time that the x-ray data obtained by their colleague Rosalind Franklin were not sufficient to resolve the DNA structure. Watson and Crick conducted one of the first-known ab initio modeling of a biological macromolecule, which has subsequently been proven to be essentially correct. Their work provided great insight into the mechanism of genetic inheritance and paved the way for a revolution in modern biology. The example demonstrates that structural prediction is a powerful tool to understand the functions of biological macromolecules at the atomic level.

We now know that the DNA structure, a double helix, is rather invariable regardless of sequence variations. Although there is little need today to determine or model DNA structures of varying sequences, there is still a real need to model protein structures individually. This is because protein structures vary depending on the sequences. Another reason is the much slower rate of structure determination by x-ray crystallography or NMR spectroscopy compared to gene sequence generation from genomic studies. Consequently, the gap between protein sequence information and protein structural information is increasing rapidly. Protein structure prediction aims to reduce this sequence–structure gap.

In contrast to sequencing techniques, experimental methods to determine protein structures are time consuming and limited in their approach. Currently, it takes 1 to 3 years to solve a protein structure. Certain proteins, especially membrane proteins, are extremely difficult to solve by x-ray or NMR techniques. There are many important proteins for which the sequence information is available, but their three-dimensional structures remain unknown. The full understanding of the biological roles of these proteins requires knowledge of their structures. Hence, the lack of such information hinders many aspects of the analysis, ranging from protein function and ligand binding to mechanisms of enzyme catalysis. Therefore, it is often necessary to obtain approximate protein structures through computer modeling.

Having a computer-generated three-dimensional model of a protein of interest has many ramifications, assuming it is reasonably correct. It may be of use for the rational design of biochemical experiments, such as site-directed mutagenesis, protein stability, or functional analysis. In addition to serving as a theoretical guide to design experiments for protein characterization, the model can help to rationalize the experimental results obtained with the protein of interest. In short, the modeling study helps to advance our understanding of protein functions.

METHODS

There are three computational approaches to protein three-dimensional structural modeling and prediction. They are homology modeling, threading, and ab initio prediction. The first two are knowledge-based methods; they predict protein structures based on knowledge of existing protein structural information in databases. Homology modeling builds an atomic model based on an experimentally determined structure that is closely related at the sequence level. Threading identifies proteins that are structurally similar, with or without detectable sequence similarities. The ab initio approach is simulation based and predicts structures based on physicochemical principles governing protein folding without the use of structural templates.

HOMOLOGY MODELING

As the name suggests, *homology modeling* predicts protein structures based on sequence homology with known structures. It is also known as *comparative modeling*. The principle behind it is that if two proteins share a high enough sequence similarity, they are likely to have very similar three-dimensional structures. If one of the protein sequences has a known structure, then the structure can be copied to the unknown protein with a high degree of confidence. Homology modeling produces an all-atom model based on alignment with template proteins.

The overall homology modeling procedure consists of six steps. The first step is template selection, which involves identification of homologous sequences in the protein structure database to be used as templates for modeling. The second step is alignment of the target and template sequences. The third step is to build a framework structure for the target protein consisting of main chain atoms. The fourth step of model building includes the addition and optimization of side chain atoms and loops. The fifth step is to refine and optimize the entire model according to energy criteria. The final step involves evaluating of the overall quality of the model obtained (Fig. 15.1). If necessary, alignment and model building are repeated until a satisfactory result is obtained.

Template Selection

The first step in protein structural modeling is to select appropriate structural templates. This forms the foundation for rest of the modeling process. The template selection involves searching the Protein Data Bank (PDB) for homologous proteins with determined structures. The search can be performed using a heuristic pairwise alignment search program such as BLAST or FASTA. However, the use of dynamic programming based search programs such as SSEARCH or ScanPS (see Chapter 4)

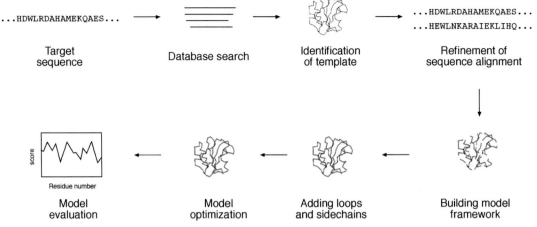

...HDWLRDAHAMEKQAES...

Target sequence

Database search

Identification of template

...HDWLRDAHAMEKQAES...
...HEWLNKARAIEKLIHQ...

Refinement of sequence alignment

score

Residue number

Model evaluation

Model optimization

Adding loops and sidechains

Building model framework

Figure 15.1: Flowchart showing steps involved in homology modeling.

can result in more sensitive search results. The relatively small size of the structural database means that the search time using the exhaustive method is still within reasonable limits, while giving a more sensitive result to ensure the best possible similarity hits.

As a rule of thumb, a database protein should have at least 30% sequence identity with the query sequence to be selected as template. Occasionally, a 20% identity level can be used as threshold as long as the identity of the sequence pair falls within the "safe zone" (see Chapter 3). Often, multiple database structures with significant similarity can be found as a result of the search. In that case, it is recommended that the structure(s) with the highest percentage identity, highest resolution, and the most appropriate cofactors is selected as a template. On the other hand, there may be a situation in which no highly similar sequences can be found in the structure database. In that instance, template selection can become difficult. Either a more sensitive profile-based PSI-BLAST method or a fold recognition method such threading can be used to identify distant homologs. Most likely, in such a scenario, only local similarities can be identified with distant homologs. Modeling can therefore only be done with the aligned domains of the target protein.

Sequence Alignment

Once the structure with the highest sequence similarity is identified as a template, the full-length sequences of the template and target proteins need to be realigned using refined alignment algorithms to obtain optimal alignment. This realignment is the most critical step in homology modeling, which directly affects the quality of the final model. This is because incorrect alignment at this stage leads to incorrect designation of homologous residues and therefore to incorrect structural models. Errors made in the alignment step cannot be corrected in the following modeling steps. Therefore, the best possible multiple alignment algorithms, such as Praline and T-Coffee (see Chapter 5), should be used for this purpose. Even alignment using the best alignment program may not be error free and should be visually inspected to ensure that conserved key residues are correctly aligned. If necessary, manual refinement of the alignment should be carried out to improve alignment quality.

Backbone Model Building

Once optimal alignment is achieved, residues in the aligned regions of the target protein can assume a similar structure as the template proteins, meaning that the coordinates of the corresponding residues of the template proteins can be simply copied onto the target protein. If the two aligned residues are identical, coordinates of the side chain atoms are copied along with the main chain atoms. If the two residues differ, only the backbone atoms can be copied. The side chain atoms are rebuilt in a subsequent procedure.

In backbone modeling, it is simplest to use only one template structure. As mentioned, the structure with the best quality and highest resolution is normally chosen if multiple options are available. This structure tends to carry the fewest errors. Occasionally, multiple template structures are available for modeling. In this situation, the template structures have to be optimally aligned and superimposed before being used as templates in model building. One can either choose to use average coordinate values of the templates or the best parts from each of the templates to model.

Loop Modeling

In the sequence alignment for modeling, there are often regions caused by insertions and deletions producing gaps in sequence alignment. The gaps cannot be directly modeled, creating "holes" in the model. Closing the gaps requires loop modeling, which is a very difficult problem in homology modeling and is also a major source of error. Loop modeling can be considered a mini–protein modeling problem by itself. Unfortunately, there are no mature methods available that can model loops reliably. Currently, there are two main techniques used to approach the problem: the database searching method and the ab initio method.

The database method involves finding "spare parts" from known protein structures in a database that fit onto the two stem regions of the target protein. The stems are defined as the main chain atoms that precede and follow the loop to be modeled. The procedure begins by measuring the orientation and distance of the anchor regions in the stems and searching PDB for segments of the same length that also match the above endpoint conformation. Usually, many different alternative segments that fit the endpoints of the stems are available. The best loop can be selected based on sequence similarity as well as minimal steric clashes with the neighboring parts of the structure. The conformation of the best matching fragments is then copied onto the anchoring points of the stems (Fig. 15.2). The ab initio method generates many random loops and searches for the one that does not clash with nearby side chains and also has reasonably low energy and ϕ and ψ angles in the allowable regions in the Ramachandran plot.

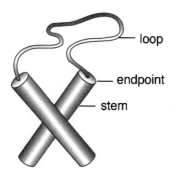

Figure 15.2: Schematic of loop modeling by fitting a loop structure onto the endpoints of existing stem structures represented by cylinders.

If the loops are relatively short (three to five residues), reasonably correct models can be built using either of the two methods. If the loops are longer, it is very difficult to achieve a reliable model. The following are specialized programs for loop modeling.

FREAD (www-cryst.bioc.cam.ac.uk/cgi-bin/coda/fread.cgi) is a web server that models loops using the database approach.

PETRA (www-cryst.bioc.cam.ac.uk/cgi-bin/coda/pet.cgi) is a web server that uses the ab initio method to model loops.

CODA (www-cryst.bioc.cam.ac.uk/~charlotte/Coda/search_coda.html) is a web server that uses a consensus method based on the prediction results from FREAD and PETRA. For loops of three to eight residues, it uses consensus conformation of both methods and for nine to thirty residues, it uses FREAD prediction only.

Side Chain Refinement

Once main chain atoms are built, the positions of side chains that are not modeled must be determined. Modeling side chain geometry is very important in evaluating protein–ligand interactions at active sites and protein–protein interactions at the contact interface.

A side chain can be built by searching every possible conformation at every torsion angle of the side chain to select the one that has the lowest interaction energy with neighboring atoms. However, this approach is computationally prohibitive in most cases. In fact, most current side chain prediction programs use the concept of *rotamers*, which are favored side chain torsion angles extracted from known protein crystal structures. A collection of preferred side chain conformations is a rotamer library in which the rotamers are ranked by their frequency of occurrence. Having a rotamer library reduces the computational time significantly because only a small number of favored torsion angles are examined. In prediction of side chain conformation, only the possible rotamers with the lowest interaction energy with nearby atoms are selected.

In many cases, even applying the rotamer library for every residue can be computationally too expensive. To reduce search time further, backbone conformation can be taken into account. It has been observed that there is a correlation of backbone conformations with certain rotamers. By using such correlations, many possible rotamers can be eliminated and the speed of conformational search can be much improved. After adding the most frequently occurring rotamers, the conformations have to be further optimized to minimize steric overlaps with the rest of the model structure.

Most modeling packages incorporate the side chain refinement function. A specialized side chain modeling program that has reasonably good performance is SCWRL (sidechain placement with a rotamer library; www.fccc.edu/research/labs/dunbrack/scwrl/), a UNIX program that works by placing side chains on a backbone template according to preferences in the backbone-dependent rotamer library. It removes rotamers that have steric clashes with main chain atoms. The final, selected set of rotamers has minimal clashes with main chain atoms and other side chains.

Model Refinement Using Energy Function

In these loop modeling and side chain modeling steps, potential energy calculations are applied to improve the model. However, this does not guarantee that the entire raw homology model is free of structural irregularities such as unfavorable bond angles, bond lengths, or close atomic contacts. These kinds of structural irregularities can be corrected by applying the energy minimization procedure on the entire model, which moves the atoms in such a way that the overall conformation has the lowest energy potential. The goal of energy minimization is to relieve steric collisions and strains without significantly altering the overall structure.

However, energy minimization has to be used with caution because excessive energy minimization often moves residues away from their correct positions. Therefore, only limited energy minimization is recommended (a few hundred iterations) to remove major errors, such as short bond distances and close atomic clashes. Key conserved residues and those involved in cofactor binding have to be restrained if necessary during the process.

Another often used structure refinement procedure is molecular dynamic simulation. This practice is derived from the concern that energy minimization only moves atoms toward a local minimum without searching for all possible conformations, often resulting in a suboptimal structure. To search for a global minimum requires moving atoms uphill as well as downhill in a rough energy landscape. This requires thermodynamic calculations of the atoms. In this process, a protein molecule is "heated" or "cooled" to simulate the uphill and downhill molecular motions. Thus, it helps overcome energy hurdles that are inaccessible to energy minimization. It is hoped that this simulation follows the protein folding process and has a better chance at finding the true structure. A more realistic simulation can include water molecules surrounding the structure. This makes the process an even more computationally expensive procedure than energy minimization, however. Furthermore, it shares a similar weakness of energy minimization: a molecular structure can be "loosened up" such that it becomes less realistic. Much caution is therefore needed in using these molecular dynamic tools.

GROMOS (www.igc.ethz.ch/gromos/) is a UNIX program for molecular dynamic simulation. It is capable of performing energy minimization and thermodynamic simulation of proteins, nucleic acids, and other biological macromolecules. The simulation can be done in vacuum or in solvents. A lightweight version of GROMOS has been incorporated in SwissPDB Viewer.

Model Evaluation

The final homology model has to be evaluated to make sure that the structural features of the model are consistent with the physicochemical rules. This involves checking anomalies in ϕ–ψ angles, bond lengths, close contacts, and so on. Another way of checking the quality of a protein model is to implicitly take these stereochemical properties into account. This is a method that detects errors by compiling statistical profiles of spatial features and interaction energy from experimentally determined structures. By comparing the statistical parameters with the constructed model, the method reveals which regions of a sequence appear to be folded normally and which regions do not. If structural irregularities are found, the region is considered to have errors and has to be further refined.

Procheck (www.biochem.ucl.ac.uk/~roman/procheck/procheck.html) is a UNIX program that is able to check general physicochemical parameters such as ϕ–ψ angles, chirality, bond lengths, bond angles, and so on. The parameters of the model are used to compare with those compiled from well-defined, high-resolution structures. If the program detects unusual features, it highlights the regions that should be checked or refined further.

WHAT IF (www.cmbi.kun.nl:1100/WIWWWI/) is a comprehensive protein analysis server that validates a protein model for chemical correctness. It has many functions, including checking of planarity, collisions with symmetry axes (close contacts), proline puckering, anomalous bond angles, and bond lengths. It also allows the generation of Ramachandran plots as an assessment of the quality of the model.

ANOLEA (Atomic Non-Local Environment Assessment; http://protein.bio.puc.cl/cardex/servers/anolea/index.html) is a web server that uses the statistical evaluation approach. It performs energy calculations for atomic interactions in a protein chain and compares these interaction energy values with those compiled from a database of protein x-ray structures. If the energy terms of certain regions deviate significantly from those of the standard crystal structures, it defines them as unfavorable regions. An example of the output from the verification of a homology model is shown in Figure 15.3A. The threshold for unfavorable residues is normally set at 5.0. Residues with scores above 5.0 are considered regions with errors.

Verify3D (www.doe-mbi.ucla.edu/Services/Verify_3D/) is another server using the statistical approach. It uses a precomputed database containing eighteen environmental profiles based on secondary structures and solvent exposure, compiled from high-resolution protein structures. To assess the quality of a protein model, the secondary structure and solvent exposure propensity of each residue are calculated. If the parameters of a residue fall within one of the profiles, it receives a high score, otherwise a low score. The result is a two-dimensional graph illustrating the folding quality of each residue of the protein structure. A verification output of the above homology

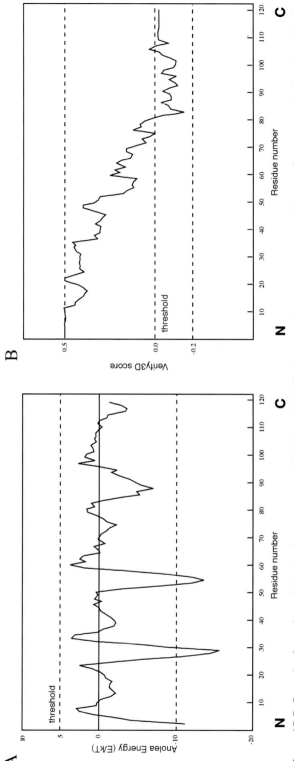

Figure 15.3: Example of protein model evaluation outputs by ANOLEA and Verify3D. The protein model was obtained from the Swiss model database (model code 1n5d). **(A)** The assessment result by the ANOLEA server. The threshold for unfavorable residues is normally set at 5.0. Residues with scores above 5.0 are considered regions with errors. **(B)** The assessment result by the Verify3D server. The threshold value is normally set at zero. The residues with the scores below zero are considered to have an unfavorable environment.

model is shown in Figure 15.3B. The threshold value is normally set at zero. Residues with scores below zero are considered to have an unfavorable environment.

The assessment results can be different using different verification programs. As shown in Figure 15.2, ANOLEA appears to be less stringent than Verify3D. Although the full-length protein chain of this model is declared favorable by ANOLEA, residues in the C-terminus of the protein are considered to be of low quality by Verify3D. Because no single method is clearly superior to any other, a good strategy is to use multiple verification methods and identify the consensus between them. It is also important to keep in mind that the evaluation tests performed by these programs only check the stereochemical correctness, regardless of the accuracy of the model, which may or may not have any biological meaning.

Comprehensive Modeling Programs

A number of comprehensive modeling programs are able to perform the complete procedure of homology modeling in an automated fashion. The automation requires assembling a pipeline that includes target selection, alignment, model generation, and model evaluation. Some freely available protein modeling programs and servers are listed.

Modeller (http://bioserv.cbs.cnrs.fr/HTML_BIO/frame_mod.html) is a web server for homology modeling. The user provides a predetermined sequence alignment of a template(s) and a target to allow the program to calculate a model containing all of the heavy atoms (nonhydrogen atoms). The program models the backbone using a homology-derived restraint method, which relies on multiple sequence alignment between target and template proteins to distinguish highly conserved residues from less conserved ones. Conserved residues are given high restraints in copying from the template structures. Less conserved residues, including loop residues, are given less or no restraints, so that their conformations can be built in a more or less ab initio fashion. The entire model is optimized by energy minimization and molecular dynamics procedures.

Swiss-Model (www.expasy.ch/swissmod/SWISS-MODEL.html) is an automated modeling server that allows a user to submit a sequence and to get back a structure automatically. The server constructs a model by automatic alignment (First Approach mode) or manual alignment (Optimize mode). In the First Approach mode, the user provides sequence input for modeling. The server performs alignment of the query with sequences in PDB using BLAST. After selection of suitable templates, a raw model is built. Refinement of the structure is done using GROMOS. Alternatively, the user can specify or upload structures as templates. The final model is sent to the user by e-mail. In the Optimize mode, the user constructs a sequence alignment in SwissPdbViewer and submits it to the server for model construction.

3D-JIGSAW (www.bmm.icnet.uk/servers/3djigsaw/) is a modeling server that works in either the automatic mode or the interactive mode. Its loop modeling relies on the database method. The interactive mode allows the user to edit alignments and select templates, loops, and side chains during modeling, whereas the automatic mode allows no human intervention and models a submitted protein sequence if it has an identity >40% with known protein structures.

Homology Model Databases

The availability of automated modeling algorithms has allowed several research groups to use the fully automated procedure to carry out large-scale modeling projects. Protein models for entire sequence databases or entire translated genomes have been generated. Databases for modeled protein structures that include nearly one third of all known proteins have been established. They provide some useful information for understanding evolution of protein structures. The large databases can also aid in target selection for drug development. However, it has also been shown that the automated procedure is unable to model moderately distant protein homologs. Automated modeling tends to be less accurate than modeling that requires human intervention because of inappropriate template selection, suboptimal alignment, and difficulties in modeling loops and side chains.

ModBase (http://alto.compbio.ucsf.edu/modbase-cgi/index.cgi) is a database of protein models generated by the Modeller program. For most sequences that have been modeled, only partial sequences or domains that share strong similarities with templates are actually modeled.

3Dcrunch (www.expasy.ch/swissmod/SWISS-MODEL.html) is another database archiving results of large-scale homology modeling projects. Models of partial sequences from the Swiss-Prot database are derived using the Swiss-Model program.

THREADING AND FOLD RECOGNITION

As discussed in Chapters 12 and 13, there are only small number of protein folds available (<1,000), compared to millions of protein sequences. This means that protein structures tend to be more conserved than protein sequences. Consequently, many proteins can share a similar fold even in the absence of sequence similarities. This allowed the development of computational methods to predict protein structures beyond sequence similarities. To determine whether a protein sequence adopts a known three-dimensional structure fold relies on threading and fold recognition methods.

By definition, *threading* or *structural fold recognition* predicts the structural fold of an unknown protein sequence by fitting the sequence into a structural database and selecting the best-fitting fold. The comparison emphasizes matching of secondary structures, which are most evolutionarily conserved. Therefore, this approach can identify structurally similar proteins even without detectable sequence similarity.

The algorithms can be classified into two categories, pairwise energy based and profile based. The pairwise energy–based method was originally referred to as *threading* and the profile-based method was originally defined as *fold recognition*. However, the two terms are now often used interchangeably without distinction in the literature.

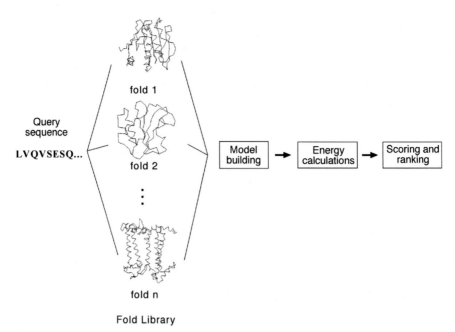

Query
sequence

LVQVSESQ...

fold 1

fold 2

⋮

fold n

Fold Library

Model
building → Energy
calculations → Scoring and
ranking

Figure 15.4: Outline of the threading method using the pairwise energy approach to predict protein structural folds from sequence. By fitting a structural fold library and assessing the energy terms of the resulting raw models, the best-fit structural fold can be selected.

Pairwise Energy Method

In the pairwise energy based method, a protein sequence is searched for in a structural fold database to find the best matching structural fold using energy-based criteria. The detailed procedure involves aligning the query sequence with each structural fold in a fold library. The alignment is performed essentially at the sequence profile level using dynamic programming or heuristic approaches. Local alignment is often adjusted to get lower energy and thus better fitting. The adjustment can be achieved using algorithms such as double-dynamic programming (see Chapter 14). The next step is to build a crude model for the target sequence by replacing aligned residues in the template structure with the corresponding residues in the query. The third step is to calculate the energy terms of the raw model, which include pairwise residue interaction energy, solvation energy, and hydrophobic energy. Finally, the models are ranked based on the energy terms to find the lowest energy fold that corresponds to the structurally most compatible fold (Fig. 15.4).

Profile Method

In the profile-based method, a profile is constructed for a group of related protein structures. The structural profile is generated by superimposition of the structures to expose corresponding residues. Statistical information from these aligned residues is then used to construct a profile. The profile contains scores that describe the propensity of each of the twenty amino acid residues to be at each profile position. The profile scores contain information for secondary structural types, the degree of solvent exposure, polarity, and hydrophobicity of the amino acids. To predict the structural fold of an unknown query sequence, the query sequence is first predicted for its secondary structure, solvent accessibility, and polarity. The predicted information is then used for comparison with propensity profiles of known structural folds to find the fold that best represents the predicted profile.

Because threading and fold recognition detect structural homologs without completely relying on sequence similarities, they have been shown to be far more sensitive than PSI-BLAST in finding distant evolutionary relationships. In many cases, they can identify more than twice as many distant homologs than PSI-BLAST. However, this high sensitivity can also be their weakness because high sensitivity is often associated with low specificity. The predictions resulting from threading and fold recognition often come with very high rates of false positives. Therefore, much caution is required in accepting the prediction results.

Threading and fold recognition assess the compatibility of an amino acid sequence with a known structure in a fold library. If the protein fold to be predicted does not exist in the fold library, the method will fail. Another disadvantage compared to homology modeling lies in the fact that threading and fold recognition do not generate fully refined atomic models for the query sequences. This is because accurate alignment between distant homologs is difficult to achieve. Instead, threading and fold recognition procedures only provide a rough approximation of the overall topology of the native structure.

A number of threading and fold recognition programs are available using either or both prediction strategies. At present, no single algorithm is always able to provide reliable fold predictions. Some algorithms work well with some types of structures, but fail with others. It is a good practice to compare results from multiple programs for consistency and judge the correctness by using external knowledge.

3D-PSSM (www.bmm.icnet.uk/~3dpssm/) is a web-based program that employs the structural profile method to identify protein folds. The profiles for each protein superfamily are constructed by combining multiple smaller profiles. First, protein structures in a superfamily based on the SCOP classification are superimposed and are used to construct a structural profile by incorporating secondary structures and solvent accessibility information for corresponding residues. In addition, each member in a protein structural superfamily has its own sequence-based PSI-BLAST profile computed. These sequence profiles are used in combination with the structure profile to form a large superfamily profile in which each position contains both sequence and structural information. For the query sequence, PSI-BLAST is performed to generate a sequence-based profile. PSI-PRED is used to predict its secondary structure. Both the sequence profile and predicted secondary structure are compared with the precomputed protein superfamily profiles, using a dynamic programming approach. The matching scores are calculated in terms of secondary structure, solvation energy, and sequence profiles and ranked to find the highest scored structure fold (Fig. 15.5).

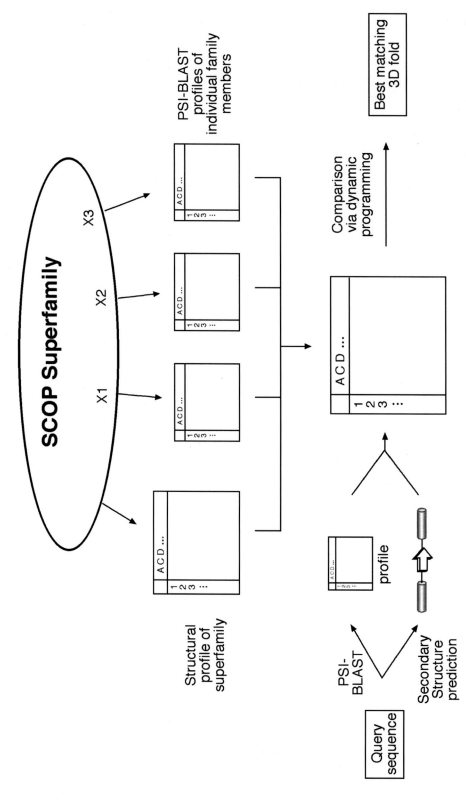

Figure 15.5: Schematic diagram of fold recognition by 3D-PSSM. A profile for protein structures in a SCOP superfamily is precomputed based on the structure profile of all members of the superfamily, as well as on PSI-BLAST sequence profiles of individual members of the superfamily. For the query sequence, a PSI-BLAST profile is constructed and its secondary structure information is predicted, which together are used to compare with the precomputed protein superfamily profile.

GenThreader (http://bioinf.cs.ucl.ac.uk/psipred/index.html) is a web-based program that uses a hybrid of the profile and pairwise energy methods. The initial step is similar to 3D-PSSM; the query protein sequence is subject to three rounds of PSI-BLAST. The resulting multiple sequence hits are used to generate a profile. Its secondary structure is predicted using PSIPRED. Both are used as input for threading computation based on a pairwise energy potential method. The threading results are evaluated using neural networks that combine energy potentials, sequence alignment scores, and length information to create a single score representing the relationship between the query and template proteins.

Fugue (www-cryst.bioc.cam.ac.uk/~fugue/prfsearch.html) is a profile-based fold recognition server. It has precomputed structural profiles compiled from multiple alignments of homologous structures, which take into account local structural environment such as secondary structure, solvent accessibility, and hydrogen bonding status. The query sequence (or a multiple sequence alignment if the user prefers) is used to scan the database of structural profiles. The comparison between the query and the structural profiles is done using global alignment or local alignment depending on sequence variability.

AB INITIO PROTEIN STRUCTURAL PREDICTION

Both homology and fold recognition approaches rely on the availability of template structures in the database to achieve predictions. If no correct structures exist in the database, the methods fail. However, proteins in nature fold on their own without checking what the structures of their homologs are in databases. Obviously, there is some information in the sequences that provides instruction for the proteins to "find" their native structures. Early biophysical studies have shown that most proteins fold spontaneously into a stable structure that has near minimum energy. This structural state is called the *native state*. This folding process appears to be nonrandom; however, its mechanism is poorly understood.

The limited knowledge of protein folding forms the basis of ab initio prediction. As the name suggests, the ab initio prediction method attempts to produce all-atom protein models based on sequence information alone without the aid of known protein structures. The perceived advantage of this method is that predictions are not restricted by known folds and that novel protein folds can be identified. However, because the physicochemical laws governing protein folding are not yet well understood, the energy functions used in the ab initio prediction are at present rather inaccurate. The folding problem remains one of the greatest challenges in bioinformatics today.

Current ab initio algorithms are not yet able to accurately simulate the protein-folding process. They work by using some type of heuristics. Because the native state of a protein structure is near energy minimum, the prediction programs are thus designed using the energy minimization principle. These algorithms search for every possible conformation to find the one with the lowest global energy. However, searching for a fold with the absolute minimum energy may not be valid in reality. This contributes to one of the fundamental flaws of this approach. In addition, searching for all possible structural conformations is not yet computationally feasible. It has been estimated that, by using one of the world's fastest supercomputers (one trillion operations per second), it takes 10^{20} years to sample all possible conformations of a 40-residue protein. Therefore, some type of heuristics must be used to reduce the conformational space to be searched. Some recent ab initio methods combine fragment search and threading to yield a model of an unknown protein. The following web program is such an example using the hybrid approach.

Rosetta (www.bioinfo.rpi.edu/~bystrc/hmmstr/server.php) is a web server that predicts protein three-dimensional conformations using the ab initio method. This in fact relies on a "mini-threading" method. The method first breaks down the query sequence into many very short segments (three to nine residues) and predicts the secondary structure of the small segments using a hidden Markov model–based program, HMMSTR (see Chapter 14). The segments with assigned secondary structures are subsequently assembled into a three-dimensional configuration. Through random combinations of the fragments, a large number of models are built and their overall energy potentials calculated. The conformation with the lowest global free energy is chosen as the best model.

It needs to be emphasized that up to now, ab initio prediction algorithms are far from mature. Their prediction accuracies are too low to be considered practically useful. Ab initio prediction of protein structures remains a fanciful goal for the future. However, with the current pace of high-throughput structural determination by the structural proteomics initiative, which aims to solve all protein folds within a decade, the time may soon come when there is little need to use the ab initio modeling approach because homology modeling and threading can provide much higher quality predictions for all possible protein folds. Regardless of the progress made in structural proteomics, exploration of protein structures using the ab initio prediction approach may still yield insight into the protein-folding process.

CASP

Discussion of protein structural prediction would not be complete without mentioning CASP (Critical Assessment of Techniques for Protein Structure Prediction). With so many protein structure prediction programs available, there is a need to know the reliability of the prediction methods. For that purpose, a common benchmark is needed to measure the accuracies of the prediction methods. To avoid letting programmers know the correct answer in the structure benchmarks in advance, already published protein structures cannot be used for testing the efficacy of new methodologies. Thus, a biannual international contest was initiated in 1994. It allows developers to predict unknown protein structures through blind testing so that the reliability of new prediction methods can be objectively evaluated. This is the experiment of CASP.

CASP contestants are given protein sequences whose structures have been solved by x-ray crystallography and NMR, but not yet published. Each contestant predicts the structures and submits the results to the CASP organizers before the structures are made publicly available. The results of the predictions are compared with the newly determined structures using structure alignment programs such as VAST, SARF, and DALI. In this way, new prediction methodologies can be evaluated without the possibility of bias. The predictions can be made at various levels of detail (secondary or tertiary structures) and in various categories (homology modeling, threading, ab initio). This experiment has been shown to provide valuable insight into the performance of prediction methods and has become the major driving force of development for protein structure prediction methods. For more information, the reader is recommended to visit the web site of the Protein Structure Prediction Center at http://predictioncenter.llnl.gov/.

SUMMARY

Protein structural prediction offers a theoretical alternative to experimental determination of structures. It is an efficient way to obtain structural information when experimental techniques are not successful. Computational prediction of protein structures is divided into three categories: homology modeling, threading, and ab initio prediction. Homology modeling, which is the most accurate prediction approach, derives models from close homologs. The process is simple in principle, but is more complicated in practice. It involves an elaborate procedure of template selection, sequence alignment correction, backbone generation, loop building, side chain modeling, model refinement, and model evaluation. Among these steps, sequence alignment is the most important step and loop modeling is the most difficult and error-prone step. Algorithms have been developed to automate the entire process and have been applied to a large-scale modeling work. However, the automated process tends to be less accurate than detailed manual modeling.

Another way to predict protein structures is through threading or fold recognition, which searches for a best fitting structure in a structural fold library by matching secondary structure and energy criteria. This approach is used when no suitable template structures can be found for homology-based modeling. The caveat is that this approach does not generate an actual model, but provide an essentially correct fold for the query protein. In addition, the protein fold of interest often does not exist in the fold library, in which case the method will fail.

The third prediction method – ab initio prediction – attempts to generate a structure without relying on templates, but by using physical rules only. It may be used when neither homology modeling nor threading can be applied. However, the ab initio approach so far has very limited success in getting correct structures. An objective evaluation platform, CASP, for protein structure prediction methodologies has been established to allow program developers to test the effectiveness of the algorithms.

FURTHER READING

Al-Lazikani, B., Jung, J., Xiang, Z., and Honig, B. 2001. Protein structure prediction. *Curr. Opin. Chem. Biol.* 5:51–6.

Baker, D., and Sali, A. 2001. Protein structure prediction and structural genomics. *Science.* 294:93–6.

Bonneau, R., and Baker, D. 2001. *Ab initio* protein structure prediction: Progress and prospects. *Annu. Rev. Biophys. Biomol. Struct.* 30:173–89.

Bourne, P. E. 2003. "CSAP and CAFASP experiments and their findings." In *Structural Bioinformatics*, edited by P. E. Bourne and H. Weissig, 501–7. Hoboken, NJ: Wiley-Liss.

Chivian, D., Robertson, T., Bonneau, R., and Baker, D. 2003. "*Ab initio* methods." In *Structural Bioinformatics*, edited by P. E. Bourne and H. Weissig, 547–56. Hoboken, NJ: Wiley-Liss.

Edwards, Y. J., and Cottage, A. 2003. Bioinformatics methods to predict protein structure and function. A practical approach. *Mol Biotechnol.* 23:139–66.

Fetrow, J. S., Giammona, A., Kolinski, A., and Skolnick, J. 2002. The protein folding problem: A biophysical enigma. *Curr. Pharm. Biotechnol.* 3:329–47.

Forster, M. J. 2000. Molecular modelling in structural biology. *Micron* 33:365–84.

Ginalski, K., Grishin, N. V., Godzik, A., and Rychlewski, L. 2005. Practical lessons from protein structure prediction. *Nucleic Acids Res.* 33:1874–91.

Godzik, A. 2003. "Fold recognition methods." In *Structural Bioinformatics*, edited by P. E. Bourne and H. Weissig, 525–46. Hoboken, NJ: Wiley-Liss.

Hardin, C., Pogorelov, T. V., and Luthey-Schulten, Z. 2002. *Ab initio* protein structure prediction. *Curr. Opin. Struct. Biol.* 12:176–81.

Krieger, E., Nabuurs, S. B., and Vriend, G. 2003. "Homology modeling." In *Structural Bioinformatics*, edited by P. E. Bourne and H. Weissig, 509–23. Hoboken, NJ: Wiley-Liss.

Marti-Renom, M. A., Stuart, A. C., Fiser, A., Sanchez, R., Melo, F., and Sali, A. 2000. Comparative protein structure modeling of genes and genomes. *Annu. Rev. Biophys. Biomol. Struct.* 29:291–325.

Xu, D., Xu, Y., and Uberbacher, E. C. 2000. Computational tools for protein modeling. *Curr. Protein Pept. Sci.* 1:1–21.

CHAPTER SIXTEEN

RNA Structure Prediction

RNA is one of the three major types of biological macromolecules. Understanding the structures of RNA provides insights into the functions of this class of molecules. Detailed structural information about RNA has significant impact on understanding the mechanisms of a vast array of cellular processes such as gene expression, viral infection, and immunity. RNA structures can be experimentally determined using x-ray crystallography or NMR techniques (see Chapter 10). However, these approaches are extremely time consuming and expensive. As a result, computational prediction has become an attractive alternative. This chapter presents the basics of RNA structures and current algorithms for RNA structure prediction, with an emphasis on secondary structure prediction.

INTRODUCTION

It is known that RNA is a carrier of genetic information and exists in three main forms. They are messenger RNA (mRNA), ribosomal RNA (rRNA), and transfer RNA (tRNA). Their main roles are as follows: mRNA is responsible for directing protein synthesis; rRNA provides structural scaffolding within ribosomes; and tRNA serves as a carrier of amino acids for polypeptide synthesis.

Recent advances in biochemistry and molecular biology have allowed the discovery of new functions of RNA molecules. For example, RNA has been shown to possess catalytic activity and is important for RNA splicing, processing, and editing. A class of small, noncoding RNA molecules, termed microRNA or miRNA, have recently been identified to regulate gene expression through interaction with mRNA molecules.

Unlike DNA, which is mainly double stranded, RNA is single stranded, although an RNA molecule can self-hybridize at certain regions to form partial double-stranded structures. Generally, mRNA is more or less linear and nonstructured, whereas rRNA and tRNA can only function by forming particular secondary and tertiary structures. Therefore, knowledge of the structures of these molecules is particularly important for understanding their functions. Difficulties in experimental determination of RNA structures make theoretical prediction a very desirable approach. In fact, computational-based analysis is a main tool in RNA-based drug design in pharmaceutical industry. In addition, knowledge of the secondary structures of rRNA is key for RNA-based phylogenetic analysis.

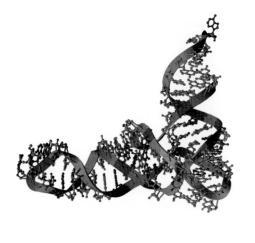

GCGGAUUUAGCUCAGUUGGGAGAGC
GCCAGACUGAAAUCUGGAGGUCCUG
UGUUCGAUCCACAGAAUUCGCACCA

Primary structure

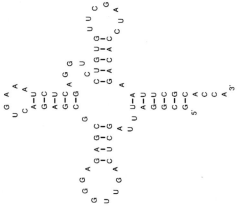

Secondary structure

Tertiary structure

Figure 16.1: The primary, secondary, and tertiary structures of a tRNA molecule illustrating the three levels of RNA structural organization.

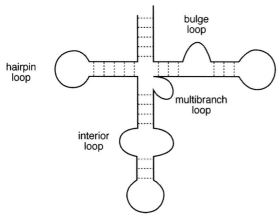

Figure 16.2: Schematic diagram of a hypothetical RNA molecular containing four basic types of RNA loops: a hairpin loop, bulge loop, interior loop, and multibranch loop. Dashed lines indicate base pairings in the helical regions of the molecule.

TYPES OF RNA STRUCTURES

RNA structures can be described at three levels as in proteins: primary, secondary, and tertiary. The primary structure is the linear sequence of RNA, consisting of four bases, adenine (A), cytosine (C), guanine (G), and uracil (U). The secondary structure refers to the planar representation that contains base-paired regions among single-stranded regions. The base pairing is mainly composed of traditional Watson–Crick base pairing, which is A–U and G–C. In addition to the canonical base pairing, there often exists noncanonical base pairing such as G and U base paring. The G–U base pair is less stable and normally occurs within a double-strand helix surrounded by Watson–Crick base pairs. Finally, the tertiary structure is the three-dimensional arrangement of bases of the RNA molecule. Examples of the three levels of RNA structural organization are illustrated in Figure 16.1.

Because the RNA tertiary structure is very difficult to predict, attention has been mainly focused on secondary structure prediction. It is therefore important to learn in more detail about RNA secondary structures. Based on the arrangement of helical base pairing in secondary structures, four main subtypes of secondary structures can be identified. They are hairpin loops, bulge loops, interior loops, and multibranch loops (Fig. 16.2).

The *hairpin loop* refers to a structure with two ends of a single-stranded region (loop) connecting a base-paired region (stem). The *bulge loop* refers to a single stranded region connecting two adjacent base-paired segments so that it "bubbles" out in the middle of a double helix on one side. The *interior loop* refers to two single-stranded regions on opposite strands connecting two adjacent base-paired segments. It can be said to "bubble" out on both sides in the middle of a double helical segment. The *multibranch loop*, also called *helical junctions*, refers to a loop that brings three or more base-paired segments in close vicinity forming a multifurcated structure.

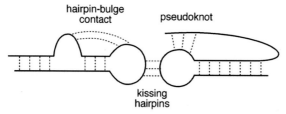

Figure 16.3: A hypothetical RNA structure containing a pseudoknot, kissing hairpin, and hairpin–bulge contact.

In addition to the traditional secondary structural elements, base pairing between loops of different secondary structural elements can result in a higher level of structures such as pseudoknots, kissing hairpins, and hairpin–bulge contact (Fig. 16.3). A *pseudoknot loop* refers to base pairing formed between loop residues within a hairpin loop and residues outside the hairpin loop. A *kissing hairpin* refers to a hydrogen bonded interaction formed between loop residues of two hairpin structures. The *hairpin–bulge* contact refers to interactions between loop residues of a hairpin loop and a bulge loop. This type of interaction forms supersecondary structures, which are relatively rare in real structures and thus are ignored by most conventional prediction algorithms.

RNA SECONDARY STRUCTURE PREDICTION METHODS

At present, there are essentially two types of method of RNA structure prediction. One is based on the calculation of the minimum free energy of the stable structure derived from a single RNA sequence. This can be considered an ab initio approach. The second is a comparative approach which infers structures based on an evolutionary comparison of multiple related RNA sequences.

AB INITIO APPROACH

This approach makes structural predictions based on a single RNA sequence. The rationale behind this method is that the structure of an RNA molecule is solely determined by its sequence. Thus, algorithms can be designed to search for a stable RNA structure with the lowest free energy. Generally, when a base pairing is formed, the energy of the molecule is lowered because of attractive interactions between the two strands. Thus, to search for a most stable structure, ab initio programs are designed to search for a structure with the maximum number of base pairs.

Free energy can be calculated based on parameters empirically derived for small molecules. G–C base pairs are more stable than A–U base pairs, which are more stable than G–U base pairs. It is also known that base-pair formation is not an independent event. The energy necessary to form individual base pairs is influenced by adjacent base pairs through helical stacking forces. This is known as *cooperativity* in helix formation. If a base pair is next to other base pairs, the base pairs tend to stabilize each other through attractive stacking interactions between aromatic rings of the base pairs. The attractive interactions lead to even lower energy. Parameters for calculating the cooperativity of the base-pair formation have been determined and can be used for structure prediction.

However, if the base pair is adjacent to loops or bulges, the neighboring loops and bulges tend to destabilize the base-pair formation. This is because there is a loss of entropy when the ends of the helical structure are constrained by unpaired loop residues. The destabilizing force to a helical structure also depends on the types of loops nearby. Parameters for calculating different destabilizing energies have also been determined and can be used as penalties for secondary structure calculations.

The scoring scheme based on the combined stabilizing and destabilizing interactions forms the foundation of the ab initio RNA secondary structure prediction method. This method works by first finding all possible base-pairing patterns from a sequence and then calculating the total energy of a potential secondary structure by taking into account all the adjacent stabilizing and destabilizing forces. If there are multiple alternative secondary structures, the method finds the conformation with the lowest energy, meaning that it is energetically most favorable.

Dot Matrices

In searching for the lowest energy form, all possible base-pair patterns have to be examined. There are several methods for finding all the possible base-paired regions from a given nucleic acid sequence. The dot matrix method and the dynamic programming method introduced in Chapter 3 can be used in detecting self-complementary regions of a sequence. A simple dot matrix can find all possible base-paring patterns of an RNA sequence when one sequence is compared with itself (Fig. 16.4). In this case, dots are placed in the matrix to represent matching complementary bases instead of identical ones.

The diagonals perpendicular to the main diagonal represent regions that can self-hybridize to form double-stranded structure with traditional A–U and G–C base pairs. In reality, the pattern detection in a dot matrix is often obscured by high noise levels. As discussed in Chapter 3, one way to reduce the noise in the matrix is to select an appropriate window size of a minimum number of contiguous base matches. Normally, only a window size of four consecutive base matches is used. If the dot plot reveals more than one feasible structures, the lowest energy one is chosen.

Dynamic Programming

The use of a dot plot can be effective in finding a single secondary structure in a small molecule (see Fig. 16.4). However, if a large molecule contains multiple secondary structure segments, choosing a combination that is energetically most stable among a large number of possibilities can be a daunting task. To overcome the problem, a quantitative approach such as dynamic programming can be used to assemble a final structure with optimal base-paired regions. In this approach, an RNA sequence

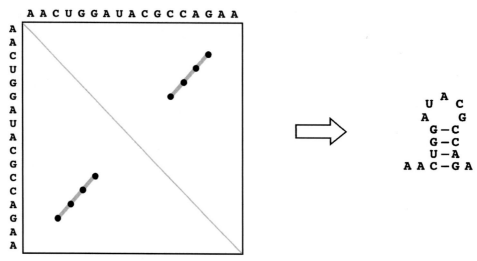

Figure 16.4: Example of a dot plot used for RNA secondary structure prediction. In this plot, an RNA sequence is compared with itself. Dots are placed for matching complementary bases when a window size of four nucleotide match is used. A main diagonal, which is perpendicular to the short diagonals, is placed for self-matching. Based on the dot plot, the predicted secondary structure for this sequence is shown on the right.

is compared with itself. A scoring scheme is applied to fill the matrix with match scores based on Watson–Crick base complementarity. Often, G–U base pairing and energy terms of the base pairing are also incorporated into the scoring process. A path with the maximal score within a scoring matrix after taking into account the entire sequence information represents the most probable secondary structure form.

The dynamic programming method produces one structure with a single best score. However, this is potentially a drawback of this approach because in reality an RNA may exist in multiple alternative forms with near minimum energy but not necessarily the one with maximum base pairs.

Partition Function

The problem of dynamic programming to select one single structure can be complemented by adding a probability distribution function, known as the *partition function*, which calculates a mathematical distribution of probable base pairs in a thermodynamic equilibrium. This function helps to select a number of suboptimal structures within a certain energy range. The following lists two well-known programs using the ab initio prediction method.

Mfold (www.bioinfo.rpi.edu/applications/mfold/) is a web-based program for RNA secondary structure prediction. It combines dynamic programming and thermodynamic calculations for identifying the most stable secondary structures with the lowest energy. It also produces dot plots coupled with energy terms. This method is reliable for short sequences, but becomes less accurate as the sequence length increases.

RNAfold (http://rna.tbi.univie.ac.at/cgi-bin/RNAfold.cgi) is one of the web programs in the Vienna package. Unlike Mfold, which only examines the energy terms of

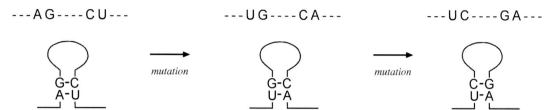

Figure 16.5: Example of covariation of residues among three homologous RNA sequences to maintain the stability of an existing secondary structure.

the optimal alignment in a dot plot, RNAfold extends the sequence alignment to the vicinity of the optimal diagonals to calculate thermodynamic stability of alternative structures. It further incorporates a partition function to select a number of statistically most probable structures. Based on both thermodynamic calculations and the partition function, a number of alternative structures that may be suboptimal are provided. The collection of the predicted structures may provide a better estimate of plausible foldings of an RNA molecule than the predictions by Mfold. Because of the much larger number of secondary structures to be computed, a more simplified energy rule has to be used to increase computational speed. Thus, the prediction results are not always guaranteed to be better than those predicted by Mfold.

COMPARATIVE APPROACH

The comparative approach uses multiple evolutionarily related RNA sequences to infer a consensus structure. This approach is based on the assumption that RNA sequences that deem to be homologous fold into the same secondary structure. By comparing related RNA sequences, an evolutionarily conserved secondary structure can be derived.

To distinguish the conserved secondary structure among multiple related RNA sequences, a concept of "covariation" is used. It is known that RNA functional motifs are structurally conserved. To maintain the secondary structures while the homologous sequences evolve, a mutation occurring in one position that is responsible for base pairing should be compensated for by a mutation in the corresponding base-pairing position so to maintain base pairing and the stability of the secondary structure (Fig. 16.5). This is the concept of *covariation*. Any lack of covariation can be deleterious to the RNA structure and functions. Based on this rule, algorithms can be written to search for the covariation patterns after a set of homologous RNA sequences are properly aligned. The detected correlated substitutions help to determine conserved base pairing in a secondary structure.

Another aspect of the comparative method is to select a common structure through consensus drawing. Because predicting secondary structures for each individual sequence may produce errors, by comparing all predicted structures of a group of aligned RNA sequences and drawing a consensus, the commonly adopted structure can be selected; many other possible structures can be eliminated in the process. The comparative-based algorithms can be further divided into two categories based on the type of input data. One requires predefined alignment and the other does not.

Algorithms That Use Prealignment

This type of algorithm requires the user to provide a pairwise or multiple alignment as input. The sequence alignment can be obtained using standard alignment programs such as T-Coffee, PRRN, or Clustal (see Chapter 5). Based on the alignment input, the prediction programs compute structurally consistent mutational patterns such as covariation and derive a consensus structure common for all the sequences. In practice, the consensus structure prediction is often combined with thermodynamic calculations to improve accuracy.

This type of program is relatively successful for reasonably conserved sequences. The requirement for using this type of program is an appropriate set of homologous sequences that have to be similar enough to allow accurate alignment, but divergent enough to allow covariations to be detected. If this condition is not met, correct structures cannot be inferred. The method also depends on the quality of the input alignment. If there are errors in the alignment, covariation signals will not be detected. The selection of one single consensus structure is also a drawback because alternative and evolutionarily unconserved structures are not predicted. The following is an example of this type of program based on predefined aligned sequences.

RNAalifold (http://rna.tbi.univie.ac.at/cgi-bin/alifold.cgi) is a program in the Vienna package. It uses a multiple sequence alignment as input to analyze covariation patterns on the sequences. A scoring matrix is created that combines minimum free energy and covariation information. Dynamic programming is used to select the structure that has the minimum energy for the whole set of aligned RNA sequences.

Algorithms That Do Not Use Prealignment

This type of algorithm simultaneously aligns multiple input sequences and infers a consensus structure. The alignment is produced using dynamic programming with a scoring scheme that incorporates sequence similarity as well as energy terms. Because the full dynamic programming for multiple alignment is computationally too demanding, currently available programs limit the input to two sequences.

Foldalign (http://foldalign.kvl.dk/server/index.html) is a web-based program for RNA alignment and structure prediction. The user provides a pair of unaligned sequences. The program uses a combination of Clustal and dynamic programming with a scoring scheme that includes covariation information to construct the alignment. A commonly conserved structure for both sequences is subsequently derived based on the alignment. To reduce computational complexity, the program ignores multibranch loops and is only suitable for handling short RNA sequences.

Dynalign (http://rna.urmc.rochester.edu/) is a UNIX program with a free source code for downloading. The user again provides two input sequences. The program calculates the possible secondary structures of each using a method similar to Mfold.

By comparing multiple alternative structures from each sequence, a lowest energy structure common to both sequences is selected that serves as the basis for sequence alignment. The unique feature of this program is that it does not require sequence similarity and therefore can handle very divergent sequences. However, because of the computation complexity, the program only predicts small RNA sequences such as tRNA with reasonable accuracy.

PERFORMANCE EVALUATION

Rigorously evaluating the performance of RNA prediction programs has traditionally been hindered by the dearth of three-dimensional structural information for RNA. The availability of recently solved crystal structures of the entire ribosome provides a wealth of structural details relating to diverse types of RNA molecules. The high-resolution structural information can then be used as a benchmark for evaluating state-of-the-art RNA structure prediction programs in all categories.

If prediction accuracy can be represented using a single parameter such as the correlation coefficient, which takes into account both sensitivity and selectivity information (see Chapter 8), the ab initio–based programs score roughly 20% to 60% depending on the length of the sequences. Generally speaking, the programs perform better for shorter RNA sequences than for longer ones. For small RNA sequences, such as tRNA, some programs may be able to produce 70% accuracy. The major limitation for performance gains of this category appears to be dependence on energy parameters alone, which may not be sufficient to distinguish different structural possibilities of the same molecule.

Based on recent benchmark comparisons, the comparative-type algorithms can reach an accuracy range of 20% to 80%. The results depend on whether a program is prealignment dependent or not. Most of the superior performance comes from prealignment-dependent programs such as RNAalifold. The prealignment-independent programs fare much worse for predicting long sequences. For small RNA sequences such as tRNA, both subtypes can achieve very high accuracy (up to 100%). This illustrates that the comparative approach is consistently more accurate than the ab initio one.

SUMMARY

Detailed understanding of RNA structures is important for understanding the functional role of RNA in the cell. The demand for structural information about RNA has motivated the development of a large number of prediction algorithms. Current RNA structure prediction is predominantly focused on secondary structures owing to the difficulty in predicting tertiary structures. The secondary structure prediction methods can be classified as either ab initio or comparative. The ab initio method is based on energetic calculations from a single query sequence. However, the accuracy of the ab initio method is limited. The comparative approach, which requires multiple sequences, is able to achieve better accuracy. However, the obvious drawback of the consensus approach is the requirement for a unique set of homologous sequences. Neither type of the prediction methods currently considers pseudoknots in the RNA structure because of the much greater computational complexity involved. To further increase prediction performance, the research and development should focus on alleviating some of the current drawbacks.

FURTHER READING

Doshi, K. J., Cannone, J. J., Cobaugh, C. W., and Gutell, R. R. 2004. Evaluation of the suitability of free-energy minimization using nearest-neighbor energy parameters for RNA secondary structure prediction. *BMC Bioinformatics* 5:105.

Doudna, J. A. 2000. Structural genomics of RNA. *Nat. Struct. Biol.* Suppl:954–6.

Gardner, P. P., and Giegerich, R. 2004. A comprehensive comparison of comparative RNA structure prediction approaches. *BMC Bioinformatics* 5:140.

Gorodkin, J. Stricklin, S. L., and Stormo, G. D. 2001. Discovering common stem-loop motifs in unaligned RNA sequences. *Nucleic Acids. Res.* 10:2135–44.

Leontis, N. B., Stombaugh, J., and Westhof, E. 2002. Motif prediction in ribosomal RNAs lessons and prospects for automated motif prediction in homologous RNA molecules. *Biochimie* 84:961–73.

Major, F., and Griffey, R. 2001. Computational methods for RNA structure determination. *Curr. Opin. Struct. Biol.* 11:282–6.

Westhof, E., Auffinger, P., and Gaspin, C. 1997. "DNA and RNA structure prediction.: In: *DNA and Protein Sequence Analysis*, edited by M. J. Bishop and C. J. Rawlings, 255–78. Oxford, UK: IRL Press.

Genomics and Proteomics

CHAPTER SEVENTEEN

Genome Mapping, Assembly, and Comparison

Genomics is the study of genomes. Genomic studies are characterized by simultaneous analysis of a large number of genes using automated data gathering tools. The topics of genomics range from genome mapping, sequencing, and functional genomic analysis to comparative genomic analysis. The advent of genomics and the ensuing explosion of sequence information are the main driving force behind the rapid development of bioinformatics today.

Genomic study can be tentatively divided into structural genomics and functional genomics. *Structural genomics* refers to the initial phase of genome analysis, which includes construction of genetic and physical maps of a genome, identification of genes, annotation of gene features, and comparison of genome structures. This is the major theme of discussion of this chapter. However, it should to be mentioned that the term *structural genomics* has already been used by a structural biology group for an initiative to determine three-dimensional structures of all proteins in a cell. Strictly speaking, the initiative of structural determination of proteins falls within the realm of *structural proteomics* and should not be confused as a subdiscipline of genomics. The structure genomics discussed herein mainly deals with structures of genome sequences. *Functional genomics* refers to the analysis of global gene expression and gene functions in a genome, which is discussed in Chapter 18.

GENOME MAPPING

The first step to understanding a genome structure is through genome mapping, which is a process of identifying relative locations of genes, mutations or traits on a chromosome. A low-resolution approach to mapping genomes is to describe the order and relative distances of genetic markers on a chromosome. *Genetic markers* are identifiable portions of a chromosome whose inheritance patterns can be followed. For many eukaryotes, genetic markers represent morphologic phenotypes. In addition to genetic linkage maps, there are also other types of genome maps such as physical maps and cytologic maps, which describe genomes at different levels of resolution. Their relations relative to the DNA sequence on a chromosome are illustrated in Figure 17.1. More details of each type of genome maps are discussed next.

Genetic linkage maps, also called *genetic maps*, identify the relative positions of genetic markers on a chromosome and are based on how frequent the markers are inherited together. The rationale behind genetic mapping is that the closer the two

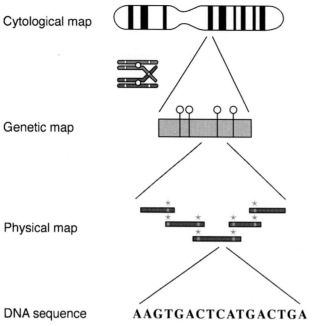

Cytological map

Genetic map

Physical map

DNA sequence   **AAGTGACTCATGACTGA**

Figure 17.1: Overview of various genome maps relative to the genomic DNA sequence. The maps represent different levels of resolution to describe a genome using genetic markers. Cytologic maps are obtained microscopically. Genetic maps (*grey bar*) are obtained through genetic crossing experiments in which chromosome recombinations are analyzed. Physical maps are obtained from overlapping clones identified by hybridizing the clone fragments (grey bars) with common probes (grey asterisks).

genetic markers are, the more likely it is that they are inherited together and are not separated in a genetic crossing event. The distance between the two genetic markers is measured in centiMorgans (cM), which is the frequency of recombination of genetic markers. One centiMorgan is defined as one percentage of the total recombination events when separation of the two genetic markers is observed in a genetic crossing experiment. One centiMorgan is approximately 1 Mb in humans and 0.5 Mb in *Drosophila*.

Physical maps are maps of locations of identifiable landmarks on a genomic DNA regardless of inheritance patterns. The distance between genetic markers is measured directly as kilobases (Kb) or megabases (Mb). Because the distance is expressed in physical units, it is more accurate and reliable than centiMorgans used in genetic maps. Physical maps are constructed by using a chromosome walking technique, which uses a number of radiolabeled probes to hybridize to a library of DNA clone fragments. By identifying overlapping clones probed by common probes, a relative order of the cloned fragments can be established.

Cytologic maps refer to banding patterns seen on stained chromosomes, which can be directly observed under a microscope. The observable light and dark bands are the visually distinct markers on a chromosome. A genetic marker can be associated with a specific chromosomal band or region. The banding patterns, however, are not always constant and are subject to change depending on the extent of chromosomal contraction. Thus, cytologic maps can be considered to be of very low resolution and hence somewhat inaccurate physical maps. The distance between two bands is expressed in relative units (Dustin units).

GENOME SEQUENCING

The highest resolution genome map is the genomic DNA sequence that can be considered as a type of physical map describing a genome at the single base-pair level. DNA sequencing is now routinely carried out using the Sanger method. This involves the use of DNA polymerases to synthesize DNA chains of varying lengths. The DNA synthesis is stopped by adding dideoxynucleotides. The dideoxynucleotides are labeled with fluorescent dyes, which terminate the DNA synthesis at positions containing all four bases, resulting in nested fragments that vary in length by a single base. When the labeled DNA is subjected to electrophoresis, the banding patterns in the gel reveal the DNA sequence.

The fluorescent traces of the DNA sequences are read by a computer program that assigns bases for each peak in a chromatogram. This process is called *base calling*. Automated base calling may generate errors and human intervention is often required to correct the sequence calls.

There are two major strategies for whole genome sequencing: the shotgun approach and the hierarchical approach. The *shotgun approach* randomly sequences clones from both ends of cloned DNA. This approach generates a large number of sequenced DNA fragments. The number of random fragments has to be very large, so large that the DNA fragments overlap sufficiently to cover the entire genome. This approach does not require knowledge of physical mapping of the clone fragments, but rather a robust computer assembly program to join the pieces of random fragments into a single, whole-genome sequence. Generally, the genome has to be redundantly sequenced in such a way that the overall length of the fragments covers the entire genome multiple times. This is designed to minimize sequencing errors and ensure correct assembly of a contiguous sequence. Overlapping sequences with an overall length of six to ten times the genome size are normally obtained for this purpose.

Despite the multiple coverage, sometimes certain genomic regions remain unsequenced, mainly owing to cloning difficulties. In such cases, the remainder gap sequences can be obtained through extending sequences from regions of known genomic sequences using a more traditional PCR technique, which requires the use of custom primers and performs genome walking in a stepwise fashion. This step of genome sequencing is also known as *finishing*, which is followed by computational assembly of all the sequence data into a final complete genome.

The hierarchical genome sequencing approach is similar to the shotgun approach, but on a smaller scale. The chromosomes are initially mapped using the physical mapping strategy. Longer fragments of genomic DNA (100 to 300 kB) are obtained and cloned into a high-capacity bacterial vector called bacterial artificial chromo-some (BAC). Based on the results of physical mapping, the locations and orders of the BAC clones on a chromosome can be determined. By successively sequencing adja-cent BAC clone fragments, the entire genome can be covered. The complete sequence of each individual BAC clone can be obtained using the shotgun approach. Overlap-ping BAC clones are subsequently assembled into an entire genome sequence. Major differences between the hierarchical and the full shotgun approaches are shown in Figure 17.2.

During the era of human genome sequencing, there was a heated debate on the merits of each of the two strategies. In fact, there are advantages and disadvantages in either. The hierarchical approach is slower and more costly than the shotgun approach because it involves an initial clone-based physical mapping step. However, once the map is generated, assembly of the whole genome becomes relatively easy and less error prone. In contrast, the whole genome shotgun approach can produce a draft sequence very rapidly because it is based on the direct sequencing approach. However, it is computationally very demanding to assemble the short random frag-ments. Although the approach has been successfully employed in sequencing small microbial genomes, for a complex eukaryotic genome that contains high levels of repetitive sequences, such as the human genome, the full shotgun approach becomes less accurate and tends to leave more "holes" in the final assembled sequence than the hierarchical approach. Current genome sequencing of large organisms often uses a combination of both approaches.

GENOME SEQUENCE ASSEMBLY

As described, initial DNA sequencing reactions generate short sequence reads from DNA clones. The average length of the reads is about 500 bases. To assemble a whole genome sequence, these short fragments are joined to form larger fragments after removing overlaps. These longer, merged sequences are termed *contigs*, which are usually 5,000 to 10,000 bases long. A number of overlapping contigs can be further merged to form scaffolds (30,000–50,000 bases, also called *supercontigs*), which are unidirectionally oriented along a physical map of a chromosome (Fig. 17.3). Overlap-ping scaffolds are then connected to create the final highest resolution map of the genome.

Correct identification of overlaps and assembly of the sequence reads into contigs are like joining jigsaw puzzles, which can be very computationally intensive when dealing with data at the whole-genome level. The major challenges in genome assem-bly are sequence errors, contamination by bacterial vectors, and repetitive sequence regions. Sequence errors can often be corrected by drawing a consensus from an align-ment of multiple overlapped sequences. Bacterial vector sequences can be removed using filtering programs prior to assembly. To overcome the problem of sequence repeats, programs such as RepeatMasker (see Chapter 4) can be used to detect and

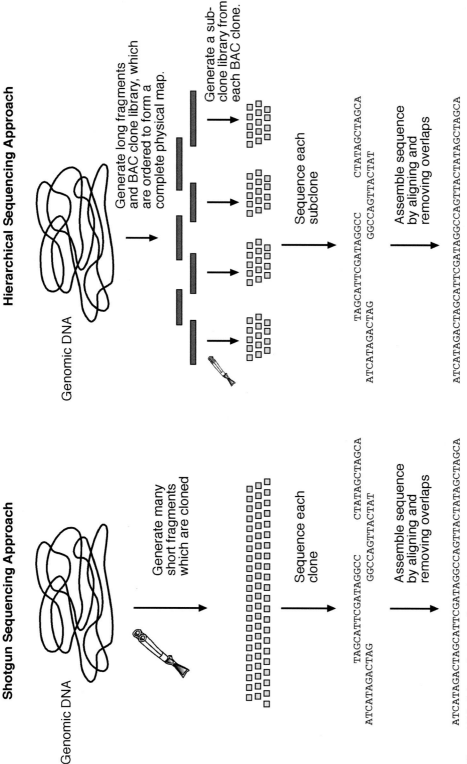

Figure 17.2: Schematic comparison of the two whole genome sequencing approaches. The full shotgun approach cuts DNA into ~2 kB fragments, which are cloned into small vectors and sequenced individually. The sequenced fragments are then put together into a final sequence in one step. The hierarchical approach cuts DNA into intermediate size fragments (~150 kB). The DNA fragments are cloned into BACs. A physical map has to be built based on the BAC clones. Each BAC clone is then subject to the shotgun approach.

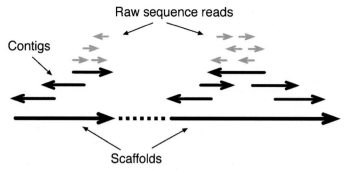

Figure 17.3: Schematic diagram showing three different levels of sequence assembly. Contigs are formed by combining raw sequence reads of various orientations after removing overlaps. Scaffolds are assembled from contigs and oriented unidirectionally on a chromosome. Because sequence fragments generated can be in either of the DNA strands, arrows are used to represent directionality of the sequences written in 5′ → 3′ orientation.

mask repeats. Additional constraints on the sequence reads can be applied to avoid misassembly caused by repeat sequences.

A commonly used constraint to avoid errors caused by sequence repeats is the so-called forward–reverse constraint. When a sequence is generated from both ends of a single clone, the distance between the two opposing fragments of a clone is fixed to a certain range, meaning that they are always separated by a distance defined by a clone length (normally 1,000 to 9,000 bases). When the constraint is applied, even when one of the fragments has a perfect match with a repetitive element outside the range, it is not able to be moved to that location to cause missassembly. An example of assembly with or without applying the forward–reverse constraints is shown in Figure 17.4.

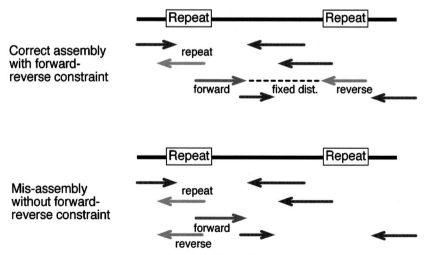

Figure 17.4: Example of sequence assembly with or without applying forward–reverse constraint, which fixes the sequence distance from both ends of a subclone. Without the restraint, the red fragment is misassembled due to matches of repetitive element in the middle of a fragment (see color plate section).

Base Calling and Assembly Programs

The first step toward genome assembly is to derive base calls and assign associated quality scores. The next step is to assemble the sequence reads into contiguous sequences. This step includes identifying overlaps between sequence fragments, assigning the order of the fragments and deriving a consensus of an overall sequence. Assembling all shotgun fragments into a full genome is a computationally very challenging step. There are a variety of programs available for processing the raw sequence data. The following is a selection of base calling and assembly programs commonly used in genome sequencing projects.

Phred (www.phrap.org/) is a UNIX program for base calling. It uses a Fourier analysis to resolve fluorescence traces and predict actual peak locations of bases. It also gives a probability score for each base call that may be attributable to error. The commonly accepted score threshold is twenty, which corresponds to a 1% chance of error. The higher the score, the better the quality of the sequence reads. If the score value falls below the threshold, human intervention is required.

Phrap (www.phrap.org/) is a UNIX program for sequence assembly. It takes Phred base-call files with quality scores as input and aligns individual fragments in a pairwise fashion using the Smith–Waterman algorithm. The base quality information is taken into account during the pairwise alignment. After all the pairwise sequence similarity is identified, the program performs assembly by progressively merging sequence pairs with decreasing similarity scores while removing overlapped regions. Consensus contigs are derived after joining all possible overlapped reads.

VecScreen (www.ncbi.nlm.nih.gov/VecScreen/VecScreen.html) is a web-based program that helps detect contaminating bacterial vector sequences. It scans an input nucleotide sequence and compares it with a database of known vector sequences by using the BLAST program.

TIGR Assembler (www.tigr.org/) is a UNIX program from TIGR for assembly of large shotgun sequence fragments. It treats the sequence input as clean reads without consideration of the sequence quality. A main feature of the program is the application of the forward–reverse constraints to avoid misassembly caused by sequence repeats. The sequence alignment in the assembly stage is performed using the Smith–Waterman algorithm.

ARACHNE (www-genome.wi.mit.edu/wga/) is a free UNIX program for the assembly of whole-genome shotgun reads. Its unique features include using a heuristic approach similar to FASTA to align overlapping fragments, evaluating alignments using statistical scores, correcting sequencing errors based on multiple sequence alignment, and using forward–reverse constraints. It accepts base calls with associated quality scores assigned by Phred as input and produces scaffolds or a fully assembled genome.

EULER (http://nbcr.sdsc.edu/euler/) is an assembly algorithm that uses a Eulerian Superpath approach, which is a polynomial algorithm for solving puzzles such as the famous "traveling salesman problem": finding the shortest path of visiting a given number of cities exactly once and returning to the starting point. In this approach, a sequence fragment is broken down to tuples of twenty nucleotides. The tuples are distributed in a diagram with numerous nodes that are all interconnected. The tuples are converted to binary vectors in the nodes. By using a Viterbi algorithm (see Chapter 6), the shortest path among the vectors can be found, which is the best way to connect the tuples into a full sequence. Because this approach does not directly rely on detecting overlaps, it may be advantageous in assembling sequences with repeat motifs.

GENOME ANNOTATION

Before the assembled sequence is deposited into a database, it has to be analyzed for useful biological features. The genome annotation process provides comments for the features. This involves two steps: gene prediction and functional assignment. Some examples of finished gene annotations in GenBank have been described in the Biological Database section (see Chapter 2). The following example illustrates the overall process employed in annotating the human genome.

As a real-world example, gene annotation of the human genome employs a combination of theoretical prediction and experimental verification. Gene structures are first predicted by ab initio exon prediction programs such as GenScan or FgenesH (see Chapter 8). The predictions are verified by BLAST searches against a sequence database. The predicted genes are further compared with experimentally determined cDNA and EST sequences using the pairwise alignment programs such as GeneWise, Spidey, SIM4, and EST2Genome. All predictions are manually checked by human curators. Once open reading frames are determined, functional assignment of the encoded proteins is carried out by homology searching using BLAST searches against a protein database. Further functional descriptions are added by searching protein motif and domain databases such as Pfam and InterPro (see Chapter 7) as well as by relying on published literature.

Gene Ontology

A problem arises when using existing literature because the description of a gene function uses natural language, which is often ambiguous and imprecise. Researchers working on different organisms tend to apply different terms to the same type of genes or proteins. Alternatively, the same terminology used in different organisms may actually refer to different genes or proteins. Therefore, there is a need to standardize protein functional descriptions. This demand has spurred the development of the gene ontology (GO) project, which uses a limited vocabulary to describe molecular functions, biological processes, and cellular components. The controlled vocabulary is organized such that a protein function is linked to the cellular function through a hierarchy of descriptions with increasing specificity. The top of the hierarchy provides an overall picture of the functional class, whereas the lower level

CYTOCHROME C OXIDASE

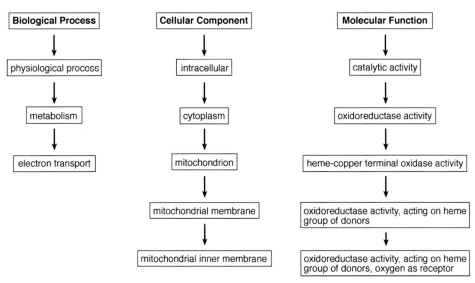

Figure 17.5: Example of GO annotation for cytochrome *c* oxidase. The functional and structural terms are arranged in three categories with a number of hierarchies indicating the levels of conceptual associations of protein functions.

in the hierarchy specifies more precisely the functional role. This way, protein functionality can be defined in a standardized and unambiguous way.

A GO description of a protein provides three sets of information: *biological process*, *cellular component*, and *molecular function*, each of which uses a unique set of nonoverlapping vocabularies. The standardization of the names, activities, and associated pathways provides consistency in describing overall protein functions and facilitates grouping of proteins of related functions. A database searching using GO for a particular protein can easily bring up other proteins of related functions in much the same way as using a thesaurus. Using GO, a genome annotator can assign functional properties of a gene product at different hierarchical levels, depending on how much is known about the gene product.

At present, the GO databases have been developed for a number of model organisms by an international consortium, in which each gene is associated with a hierarchy of GO terms. These have greatly facilitated genome annotation efforts. A good introduction of gene ontology can be found at www.geneontology.org. An example of GO annotation for cytochrome *c* oxidase is shown in Figure 17.5.

Automated Genome Annotation

With the genome sequence data being generated at an exponential rate, there is a need to develop fast and automated methods to annotate the genomic sequences. The automated approach relies on homology detection, which is essentially heuristic sequence similarity searching. If a newly sequenced gene or its gene product has significant matches with a database sequence beyond a certain threshold, a transfer of functional assignment is taking place. In addition to sequence matching at the full length, detection of conserved motifs often offers additional functional clues.

Because using a single database searching method is often incomplete and error prone, automated methods have to mimic the manual process, which takes into consideration multiple lines of evidence in assigning a gene function, to minimize errors. The following algorithm is an example that goes a step beyond examining sequence similarity and provides functional annotations based on multiple protein characteristics.

GeneQuiz (http://jura.ebi.ac.uk:8765/ext-genequiz/) is a web server for protein sequence annotation. The program compares a query sequence against databases using BLAST and FASTA to identify homologs with high similarities. In addition, it performs domain analysis using the PROSITE and Blocks databases (see Chapter 7) as well as analysis of secondary structures and supersecondary structures that includes prediction of coiled coils and transmembrane helices. Multiple search and analysis results are compiled to produce a summary of protein function with an assigned confidence level (clear, tentative, marginal, and negligible).

Annotation of Hypothetical Proteins

Although a large number of genes and proteins can be assigned functions by the sequence similarity based approach, about 40% of the genes from newly sequenced genomes have no known functions and can only be annotated as genes encoding "hypothetical proteins." Experimental discovery of the functions of these genes and proteins is often time consuming and difficult because of lack of hypotheses to design experiments. In this situation, more advanced tools can be used for functional predictions by searching for remote homologs.

One way to obtain functional hints of genes encoding hypothetical proteins is by searching for remote homologs in databases. Detecting remote homologs typically involves combined searches of protein motifs and domains and prediction for secondary and tertiary structures. Conserved functional sites can be identified by profile and hidden Markov model–based motif and domain search tools such as SMART and InterPro (see Chapter 7). The prediction can also be performed using structure-based approaches such as threading and fold recognition (see Chapter 15). If the distant homologs detected using the structural approach are linked with well-defined functions, a broad functional class of the query protein if not the precise function of the protein can be inferred. In addition, prediction results for subcellular localization, protein–protein interactions can provide further functional hints (see Chapter 19).

These suggestions do not guarantee to provide correct annotations for the "hypothetical proteins," but they may provide critical hypotheses of the protein function that can be tested in the laboratory. The remote homology detection helps to shed light on the possible functions of the proteins that previously have no functional information at all. Thus, the bioinformatic analysis can spur an important advance in knowledge in many cases. Some hypothetical proteins, because of their novel

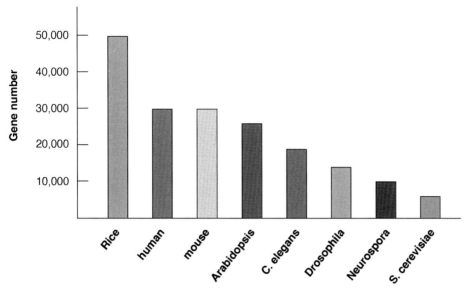

Figure 17.6: Gene numbers estimated from several sequenced eukaryotic genomes. (Data from Integrated Genomics Online Database http://ergo.integratedgenomics.com/GOLD/.)

structural folds, still cannot be predicted even with the advanced bioinformatics approaches and remain challenges for both experimental and computational work.

How Many Genes in a Genome?

One of the main tasks of genome annotation is to try to give a precise account of the total number of genes in a genome. This may be more feasible for prokaryotes as their gene structures are relatively simple. However, the number of genes in eukaryotic genomes, in particular the human genome, has been a subject of debate. This is mainly because of the complex structures of these genomes, which obscure gene prediction. Before the human genome sequencing was completed, the estimated gene numbers ranged from 20,000 to 120,000. Since the completion of the sequencing of the human genome, with the use of more sophisticated gene finding programs, the total number of human genes now dropped to close to 25,000 to 30,000. Although no exact number is agreed upon by all researchers, it is now widely believed that the total number of human genes will be no more than 30,000. This compares to estimates of 50,000 in rice, 30,000 in mouse, 26,000 in *Arabidopsis*, 18,400 in *C. elegans*, and 6,200 in yeast (Fig. 17.6).

The discovery of the low gene count in humans may be ego defeating to some as they realize that humans are only five times more complex than baker's yeast and apparently equally as complex as the mouse. What is worse, the food in their rice bowls has twice as many genes. The finding seriously challenges the view that humans are a superior species on Earth. As in many discoveries in scientific history, such as Darwin's evolutionary theory suggesting that humans arose from a "subhuman" ancestor, recent genomic discoveries have moved humans further away from this exalted status. However, before we are overwhelmed by the humble realization, we should also realize that the complexity of an organism simply cannot be represented by gene numbers. As will soon become clear, gene expression and regulation, protein expression, modification, and interactions all contribute to the overall complexity of an organism.

Genome Economy

One level of genetic complexity is manifested at the protein expression level in which there are often more expressed proteins than genes available to code for them. For example, in humans, there are more than 100,000 proteins expressed based on EST analysis (see Chapter 18) compared to no more than 30,000 genes. If the "one gene, one protein" paradigm holds true, how could this discrepancy exist? Where does the extra coding power come from?

The answer lies in "genome economy," a phenomenon of synthesizing more proteins from fewer genes. This is a major strategy that eukaryotic organisms use to achieve a myriad of phenotypic diversities. There are many underlying genetic mechanisms to help account for genome economy. A major mechanism responsible for the protein diversity is *alternative splicing*, which refers to the splicing event that joins different exons from a single gene to form different transcripts. A related mechanism, known as *exon shuffling*, which joins exons from different genes to generate more transcripts, is also common in eukaryotes. It is known that, in humans, about two thirds of the genes exhibit alternative splicing and exon shuffling during expression, generating 90% of the total proteins. In *Drosophila*, the *DSCAM* gene contains 115 exons that can be alternatively spliced to produce 38,000 different proteins. This remarkable ability to generate protein diversity and new functions highlights the true complexity of a genome. It also illustrates the evolutionary significance of introns in eukaryotic genes, which serve as spacers that make the molecular recombination possible.

There are more surprising mechanisms responsible for genome economy. For example, trans-splicing can occur between RNAs produced from both DNA strands. In the *Drosophila mdg4* mutant, RNA transcribed from four exons in the sense strand and two exons in the antisense strand are joined to form a single mRNA. With different exon combinations, four different proteins can be produced. In some circumstances, one mRNA transcript can lead to the translation of more than one protein. For example, human dentin phosphoprotein and dentin sialoprotein are proteins involved in tooth formation. An mRNA transcript that includes coding regions from both proteins is translated into a precursor protein that is cleaved to produce two different mature proteins. Another situation, called "gene within gene," can be found in a gene for human prostate-specific antigen (PSA). In addition to regular PSA, humans can produce a similar protein, called PSA-LM, that functions antagonistically to PSA and is important for prostate cancer diagnosis. PSA-LM turns out to be encoded by the fourth intron of the PSA gene.

These are just a few known mechanisms of condensing the coding potential of genomic DNA to achieve increased protein diversity. From a bioinformatics point of view, this makes gene prediction based on computational approaches all the more complicated. It also highlights one of the challenges that faces software program developers today. A number of databases have recently been established to archive alternatively spliced forms of eukaryotic genes. The following is one such example for human genes.

ProSplicer (http://prosplicer.mbc.nctu.edu.tw/) is a web-based database of human alternative spliced transcripts. The spliced variants are identified by aligning each known human protein, mRNA, and EST sequence against the genomic sequence using the SIM4 and TBLASTN program. The three sets of alignment are compiled to derive alternative splice forms. The database organizes data by tissue types and can be searched using keywords.

COMPARATIVE GENOMICS

Comparison of whole genomes from different organisms is comparative genomics, which includes comparison of gene number, gene location, and gene content from these genomes. The comparison helps to reveal the extent of conservation among genomes, which will provide insights into the mechanism of genome evolution and gene transfer among genomes. It helps to understand the pattern of acquisition of foreign genes through lateral gene transfer. It also helps to reveal the core set of genes common among different genomes, which should correspond to the genes that are crucial for survival. This knowledge can be potentially useful in future metabolic pathway engineering.

As alluded to previously, the main themes of comparative genomics include whole genome alignment, comparing gene order between genomes, constructing minimal genomes, and lateral gene transfer among genomes, each of which is discussed in more detail.

Whole Genome Alignment

With an ever-increasing number of genome sequences available, it becomes imperative to understand sequence conservation between genomes, which often helps to reveal the presence of conserved functional elements. This can be accomplished through direct genome comparison or genome alignment. The alignment at the genome level is fundamentally no different from the basic sequence alignment described in Chapters 3, 4, and 5. However, alignment of extremely large sequences presents new complexities owing to the sheer size of the sequences. Regular alignment programs tend to be error prone and inefficient when dealing with long stretches of DNA containing hundreds or thousands of genes. Another challenge of genome alignment is effective visualization of alignment results. Because it is obviously difficult to sift through and make sense of the extremely large alignments, a graphical representation is a must for interpretation of the result. Therefore, specific alignment algorithms are needed to deal with the unique challenges of whole genome alignment. A number of alignment programs for "super-long" DNA sequences are described next.

MUMmer (Maximal Unique Match, www.tigr.org/tigr-scripts/CMR2/webmum/mumplot) is a free UNIX program from TIGR for alignment of two entire genome sequences and comparison of the locations of orthologs. The program is essentially a modified BLAST, which, in the seeding step (see Chapter 4), finds the longest approximate matches that include mismatches instead of finding exact *k*-mer matches as in regular BLAST. The result of the alignment of whole genomes is shown as a dot plot with lines of connected dots to indicate collinearity of genes. It is optimized for pairwise comparison of closely related microbial genomes.

BLASTZ (http://bio.cse.psu.edu/) is a UNIX program modified from BLAST to do pairwise alignment of very large genomic DNA sequences. The modified BLAST program first masks repetitive sequences and searches for closely matched "words," which are defined as twelve identical matches within a stretch of nineteen nucleotides. The words serve as seeds for extension of alignment in both directions until the scores drop below a certain threshold. Nearby aligned regions are joined by using a weighted scheme that employs a unique gap penalty scheme that tolerates minor variations such as transitions in the seeding step of the alignment construction to increase its sensitivity.

LAGAN (Limited Area Global Alignment of Nucleotides; http://lagan.stanford.edu/) is a web-based program designed for pairwise alignment of large genomes. It first finds anchors between two genomic sequences using an algorithm that identifies short, exactly matching words. Regions that have high density of words are selected as anchors. The alignments around the anchors are built using the Needleman–Wunsch global alignment algorithm. Nearby aligned regions are further connected using the same algorithm. The unique feature of this program is that it is able to take into account degeneracy of the genetic codes and is therefore able to handle more distantly related genomes. Multi-LAGAN, an extension of LAGAN, available from the same website, performs multiple alignment of genomes using a progressive approach similar to that used in Clustal (see Chapter 5).

PipMaker (http://bio.cse.psu.edu/cgi-bin/pipmaker?basic) is a web server using the BLASTZ heuristic method to find similar regions in two DNA sequences. It produces a textual output of the alignment result and also a graphical output that presents the alignment as a percent identity plot as well as a dot plot. For comparing multiple genomes, MultiPipMaker is available from the same site.

MAVID (http://baboon.math.berkeley.edu/mavid/) is a web-based program for aligning multiple large DNA sequences. MAVID is based on a progressive alignment algorithm similar to Clustal. It produces an NJ tree as a guide tree. The sequences are aligned recursively using a heuristic pairwise alignment program called AVID. AVID works by first selecting anchors using the Smith–Waterman algorithm and then building alignments for the sequences between nearby anchors. Connected alignments are treated as new anchors for building longer alignments. The process is repeated iteratively until the entire sequence pair including weakly conserved regions are aligned.

GenomeVista (http://pipeline.lbl.gov/cgi-bin/GenomeVista) is a database searching program that searches against the human, mouse, rat, or *Drosophila* genomes using a large piece of DNA as query. It uses a program called BLAT to find anchors and extends the alignment from the anchors using AVID. (BLAT is a fast local alignment algorithm that aligns short sequences of forty bases with more than 95% similarity.) It produces a graphical output that shows the sequence percent identity.

Finding a Minimal Genome

One of the goals of genome comparison is to understand what constitutes a minimal genome, which is a minimal set of genes required for maintaining a free-living cellular organism. Finding minimal genomes helps provide an understanding of genes constituting key metabolic pathways, which are critical for a cell's survival. This analysis involves identification of orthologous genes shared between a number of divergent genomes.

Coregenes (http://pasteur.atcc.org:8050/CoreGenes1.0//) is a web-based program that determines a core set of genes based on comparison of four small genomes. The user supplies NCBI accession numbers for the genomes of interest. The program performs an iterative BLAST comparison to find orthologous genes by using one genome as a reference and another as a query. This pairwise comparison is performed for all four genomes. As a result, the common genes are compiled as a core set of genes from the genomes.

Lateral Gene Transfer

Lateral gene transfer (or *horizontal gene transfer*) is defined as the exchange of genetic materials between species in a way that is incongruent with commonly accepted vertical evolutionary pathway. Lateral gene transfer mainly occurs among prokaryotic organisms when foreign genes are acquired through mechanisms such as transformation (direct uptake of foreign DNA from environment), conjugation (gene uptake through mating behavior), and transduction (gene uptake mediated by infecting viruses). The transmission of genes between organisms can occur relatively recently or as a more ancient event.

If lateral transfer events occurred relatively recently, one would expect to discover traces of the transfer by detecting regions of genomic sequence with unusual properties compared to surrounding regions. The unusual characteristics to be examined include nucleotide composition, codon usage, and amino acid composition. This can be considered a "within-genome" approach. Another way to discern lateral gene transfer is through phylogenetic analysis (see Chapters 10 and 11), referred to as an "among-genome" approach, which can be used to discover both recent and ancient lateral gene transfer events. Abnormal groupings in phylogenetic trees are often interpreted as the possibility of lateral gene transfer events. Because phylogenetic analyses have been described in detail in previous chapters, the following introduces basic tools for identifying genomic regions that may be a result of lateral gene transfer events using the within-genome approach.

Within-Genome Approach

This approach is to identify regions within a genome with unusual compositions. Single or oligonucleotide statistics, such as G–C composition, codon bias, and

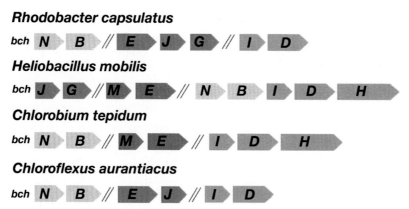

Figure 17.7: Schematic diagram showing a conserved linkage pattern of photosynthesis genes among four divergent photosynthetic bacterial groups. The synteny reveals potential physical interactions of encoded proteins, some of which have been experimentally verified. All the genes shown (*bch*) are involved in the pathway of bacteriochlorophyll biosynthesis. Intergenic regions of unspecified lengths are indicated by forward slashes (//). (*Source:* from Xiong et al., 2000; reproduced with permission from *Science*).

oligonucleotide frequencies are used. Unusual nucleotide statistics in certain genomic regions versus the rest of the genome may help to identify "foreign" genes in a genome. A commonly used parameter is GC skew $((G - C)/(G + C))$, which is compositional bias for G in a DNA sequence and is a commonly used indicator for newly acquired genetic elements.

ACT (Artemis Comparison Tool; www.sanger.ac.uk/Software/ACT) is a pairwise genomic DNA sequence comparison program (written in Java and run on UNIX, Macintosh, and Windows) for detecting gene insertions and deletions among related genomes. The pairwise sequence alignment is conducted using BLAST. The display feature includes showing collinear as well as noncollinear (rearrangement) regions between two genomes. It also calculates GC biases to indicate nucleotide patterns. However, it is up to the genome annotators to determine whether the observations constitute evidence for lateral gene transfer, as this requires combining evidence from multiple approaches.

Swaap (http://www.bacteriamuseum.org/SWAAP/SwaapPage.htm) is a Windows program that is able to distinguish coding versus noncoding regions and measure GC skews, oligonucleotide frequencies in a genomic sequence.

Gene Order Comparison

Another aspect of comparative genomics is the comparison of gene order. When the order of a number of linked genes is conserved between genomes, it is called *synteny*. Generally speaking, gene order is much less conserved compared with gene sequences. Gene order conservation is in fact rarely observed among divergent species. Therefore, comparison of syntenic relationships is normally carried out between relatively close lineages. However, if syntenic relationships for certain genes are indeed observed among divergent prokaryotes, they often provide important clues to functional relationships of the genes of interest. For example, genes involved in the same metabolic pathway tend to be clustered among phylogenetically diverse organisms. The preservation of the gene order is a result of the selective pressure to allow the genes to be coregulated and function as an operon. Furthermore, the synteny of genes from divergent groups often associates with physical interactions of the encoded gene products. The use of conserved gene neighbors as predictors of protein interactions is discussed in Chapter 18. An example of synteny of bacterial photosynthesis genes coupled with protein interactions is illustrated in Figure 17.7.

GeneOrder (http://pumpkins.ib3.gmu.edu:8080/geneorder/) is a web-based program that allows direct comparison of a pair of genomic sequences of less than 2 Mb. It displays a dot plot with diagonal lines denoting collinearity of genes and lines off the diagonal indicating inversions or rearrangements in the genomes.

SUMMARY

Genome mapping using relative positions of genetic markers without knowledge of sequence data is a low-resolution approach to describing genome structures. A genome can be described at the highest resolution by a complete genome sequence. Whole-genome sequencing can be carried out using full shotgun or hierarchical approaches. The former requires more extensive computational power in the assembly step, and the latter is inefficient because of the physical mapping process required. Among the genome sequence assembly programs, ARACHNE and EULER are the best performers. Genome annotation includes gene finding and assignment of function to these genes. Functional assignment depends on homology searching and literature information. GO projects aim to facilitate automated annotation by standardizing the descriptions used for gene functions. The exact number of genes in the human genome is unknown, but is likely to be in the same range as most other eukaryotes. The gene number, however, does not dictate complexities of a genome. One example is exhibited in protein expression in which a larger number of proteins are produced than genes available to code for them. This is the so-called genome economy. The main mechanisms responsible for genome economy are alternative splicing and exon shuffling. Genomes can be compared on the basis of their gene content and gene order. Many specialized genome comparison programs for cross-genome alignment have been developed. Among them, BLASTZ and LAGAN may be the best in terms of speed and accuracy. Gene order comparison across genomes often helps to discover potential operons and assign putative functions. Conserved gene order among prokaryotes is often indicative of protein physical interactions.

FURTHER READING

Bennetzen, J. 2002. Opening the door to comparative plant biology. *Science* 296:60–3.

Chain, P., Kurtz, S., Ohlebusch, E., and Slezak, T. 2003. An applications-focused review of comparative genomics tools: Capabilities, limitations and future challenges. *Brief. Bioinform.* 4:105–23.

Dandekar, T, Snel, B., Huynen, M., and Bork, P. 1998. Conservation of gene order: A fingerprint of proteins that physically interact. *Trends Biochem. Sci.* 23:324–8.

Frazer, K. A., Elnitski, L., Church, D. M., Dubchak, I., and Hardison, R. C. 2003. Cross-species sequence comparisons: A review of methods and available resources. *Genome Res.* 13:1–12.

Gibson, G., and Muse, S. V. 2002. *A Primer of Genome Science.* Sunderland, MA: Sinauer Associates.

Karlin, S., Mrazek, J., and Gentles, A. J. 2003. Genome comparisons and analysis. *Curr. Opin. Struct. Biol.* 13:344–52.

Lewis, R., and Palevitz, B. A. 2001. Genome economy. *The Scientist* 15:21.

Lio, P. 2003. Statistical bioinformatic methods in microbial genome analysis. *BioEssays* 25:266–73.

Michalovich, D., Overington, J., and Fagan, R. 2002. Protein sequence analysis in silico: Application of structure-based bioinformatics to genomic initiatives. *Curr. Opin. Pharmacol.* 2:574–80.

Pennacchio, L. A., and Rubin, E. M. 2003. Comparative genomic tools and databases: Providing insights into the human genome. *J. Clin. Invest.* 111:1099–106.

Primrose, S. B., and Twyman, R. M. 2003. *Principles of Genome Analysis and Genomics*, 3rd ed. Oxford, UK: Blackwell.

Stein, L. 2001. Genome annotation: From sequence to biology. *Nat. Rev. Genetics* 2:493–503.

Sterky, F., and Lundeberg, J. 2000. Sequence analysis of gene and genomes. *J. Biotechnol.* 76:1–31.

Ureta-Vidal, A., Ettwiller, L., and Birney, E. 2003. Comparative genomics: Genome-wide analysis of metazoan eukaryotes. *Nature Rev. Genetics* 4:251–62.

Waterston, R. H., Lindblad-Toh, K., Birney, E., Rogers, J., Abril, J. F., Agarwal, P., Agarwala, R., et al. 2002. Initial sequencing and comparative analysis of the mouse genome. *Nature* 420:520–62.

Wei, L., Liu, Y., Dubchak, I., Shon, J., and Park, J. 2002. Comparative genomics approaches to study organism similarities and differences. *J. Biomed. Inform.* 35:142–50.

Xiong, J., Fischer, W. M., Inoue, K., Nakahara, M., and Bauer, C. E. 2000. Molecular evidence for the early evolution of photosynthesis. *Science* 289:1724–30.

Functional Genomics

The field of genomics encompasses two main areas, structural genomics and functional genomics (see Chapter 17). The former mainly deals with genome structures with a focus on the study of genome mapping and assembly as well as genome annotation and comparison; the latter is largely experiment based with a focus on gene functions at the whole genome level using high throughput approaches. The emphasis here is on "high throughput," which is simultaneous analysis of all genes in a genome. This feature is in fact what separates genomics from traditional molecular biology, which studies only one gene at a time.

The high throughput analysis of all expressed genes is also termed *transcriptome analysis*, which is the expression analysis of the full set of RNA molecules produced by a cell under a given set of conditions. In practice, messenger RNA (mRNA) is the only RNA species being studied. Transcriptome analysis facilitates our understanding of how sets of genes work together to form metabolic, regulatory, and signaling pathways within the cell. It reveals patterns of coexpressed and coregulated genes and allows determination of the functions of genes that were previously uncharacterized. In short, functional genomics provides insight into the biological functions of the whole genome through automated high throughput expression analysis. This chapter mainly discusses the bioinformatics aspect of the transcriptome analysis that can be conducted using either sequence- or microarray-based approaches.

SEQUENCE-BASED APPROACHES

Expressed Sequence Tags

One of the high throughput approaches to genome-wide profiling of gene expression is sequencing expressed sequence tags (ESTs). ESTs are short sequences obtained from cDNA clones and serve as short identifiers of full-length genes. ESTs are typically in the range of 200 to 400 nucleotides in length obtained from either the 5′ end or 3′ end of cDNA inserts. Libraries of cDNA clones are prepared through reverse transcription of isolated mRNA populations by using oligo(dT) primers that hybridize with the poly(A) tail of mRNAs and ligation of the cDNAs to cloning vectors. To generate EST data, clones in the cDNA library are randomly selected for sequencing from either end of the inserts.

The EST data are able to provide a rough estimate of genes that are actively expressed in a genome under a particular physiological condition. This is because the frequencies for particular ESTs reflect the abundance of the corresponding mRNA in a cell, which corresponds to the levels of gene expression at that condition. Another potential benefit of EST sampling is that, by randomly sequencing cDNA clones, it is possible to discover new genes.

However, there are also many drawbacks of using ESTs for expression profile analysis. EST sequences are often of low quality because they are automatically generated without verification and thus contain high error rates. Many bases are ambiguously determined, represented by N's. Common errors also include frameshift errors and artifactual stop codons, resulting in failures of translating the sequences. In addition, there is often contamination by vector sequence, introns (from unspliced RNAs), ribosomal RNA (rRNA), mitochondrial RNA, among others. ESTs represent only partial sequences of genes. Gene sequences at the 3′ end tend to be more heavily represented than those at the 5′ end because reverse transcription is primed with oligo(dT) primers. Unfortunately, the sequences from the 3′ end are also most error prone because of the low base-call quality at the start of sequence reads. Another problem of ESTs is the presence of chimeric clones owing to cloning artifacts in library construction, in which more than one transcript is ligated in a clone resulting in the 5′ end of a sequence representing one gene and the 3′ end another gene. It has been estimated that up to 11% of cDNA clones may be chimeric. Another fundamental problem with EST profiling is that it predominantly represents highly expressed, abundant transcripts. Weakly expressed genes are hardly found in a EST sequencing survey.

Despite these limitations, EST technology is still widely used. This is because EST libraries can be easily generated from various cell lines, tissues, organs, and at various developmental stages. ESTs can also facilitate the unique identification of a gene from a cDNA library; a short tag can lead to a cDNA clone. Although individual ESTs are prone to error, an entire collection of ESTs contains valuable information. Often, after consolidation of multiple EST sequences, a full-length cDNA can be derived. By searching a nonredundant EST collection, one can identify potential genes of interest.

The rapid accumulation of EST sequences has prompted the establishment of public and private databases to archive the data. For example, GenBank has a special EST database, dbEST (www.ncbi.nlm.nih.gov/dbEST/) that contains EST collections for a large number of organisms (>250). The database is regularly updated to reflect the progress of various EST sequencing projects. Each newly submitted EST sequence is subject to a database search. If a strong similarity to a known gene is found, it is annotated accordingly.

EST Index Construction

One of the goals of the EST databases is to organize and consolidate the largely redundant EST data to improve the quality of the sequence information so the data can be used to extract full-length cDNAs. The process includes a preprocessing step that removes vector contaminants and masks repeats. Vecscreen, introduced in

Chapter 17, can be used to screen out bacterial vector sequences. This is followed by a clustering step that associates EST sequences with unique genes. The next step is to derive consensus sequences by fusing redundant, overlapping ESTs and to correct errors, especially frameshift errors. This step results in longer EST contigs. The procedure is somewhat similar to the genome assembly of shotgun sequence reads (see Chapter 17) . Finally, the coding regions are defined through the use of HMM-based gene-finding algorithms (see Chapter 8). This helps to exclude the potential intron and 3′-untranslated sequences. Once the coding sequence is identified, it can be annotated by translating it into protein sequences for database similarity searching. To go another step further, compiled ESTs can be used to align with the genomic sequence if available to identify the genome locus of the expressed gene as well as intron–exon boundaries of the gene. This is usually performed using the program SIM4 (http://pbil.univ-lyon1.fr/sim4.php).

The clustering process that reduces the EST redundancy and produces a collection of nonredundant and annotated EST sequences is known as *gene index construction*. The following lists a couple of major databases that index EST sequences.

UniGene (www.ncbi.nlm.nih.gov/UniGene/) is an NCBI EST cluster database. Each cluster is a set of overlapping EST sequences that are computationally processed to represent a single expressed gene. The database is constructed based on combined information from dbEST, GenBank mRNA database, and "electronically spliced" genomic DNA. Only ESTs with 3′ poly-A ends are clustered to minimize the the problem of chimerism. The resulting 3′ EST sequences provide more unique representation of the transcripts. The next step is to remove contaminant sequences that include bacterial vectors and linker sequences. The cleaned ESTs are used to search against a database of known unique genes (EGAD database) with the BLAST program. The compiling step identifies sequence overlaps and derives sequence consensus using the CAP3 program. During this step, errors in individual ESTs are corrected; the sequences are then partitioned into clusters and assembled into contigs. The final result is a set of nonredundant, gene-oriented clusters known as *UniGene clusters*. Each UniGene cluster represents a unique gene and is further annotated for putative function and its gene locus information, as well as information related to the tissue type where the gene has been expressed. The entire clustering procedure is outlined in Figure 18.1.

TIGR Gene Indices (www.tigr.org/tdb/tgi.shtml) is an EST database that uses a different clustering method from UniGene (Fig. 18.2). It compiles data from dbEST, GenBank mRNA and genomic DNA data, and TIGR's own sequence database. Sequences are only clustered if they are more than 95% identical for over a forty-nucleotide region in pairwise comparisons. BLAST and FASTA are used to identify sequence overlaps. In the sequence assembly stage, both TIGR Assembler (see Chapter 17) and CAP3 are used to construct contigs, producing a so-called tentative consensus (TC). To prevent chimerism, transcripts are clustered only if they match fully with known genes. Functional assignment is then given to the TC that relies most heavily on BLAST searches against protein databases. The TIGR gene indices serve as an

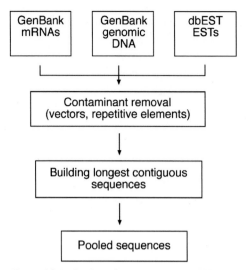

Figure 18.1: Outline of steps to process EST sequences for construction of the UniGene database.

alternative to the UniGene clusters with the resulting gene indices showing compiled EST sequences, functional annotation, and database similarity search results.

SAGE

Serial analysis of gene expression (SAGE) is another high throughput, sequence-based approach for global gene expression profile analysis. Unlike EST sampling, SAGE is more quantitative in determining mRNA expression in a cell. In this method, short fragments of DNA (usually 15 base pairs [bp]) are excised from cDNA sequences and used as unique markers of the gene transcripts. The sequence fragments are termed *tags*. They are subsequently concatenated (linked together), cloned, and sequenced. The transcript analysis is carried out computationally in a serial manner. Once gene tags are unambiguously identified, their frequency indicates the level

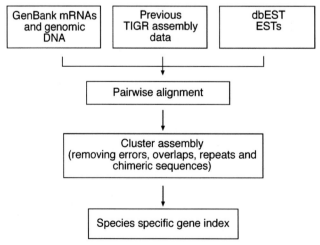

Figure 18.2: Outline of construction for TIGR gene indices.

of gene expression. This approach is much more efficient than the EST analysis in that it uses a short nucleotide tag to define a gene transcript and allows sequencing of multiple tags in a single clone. If an average clone has a size of 700 bp, it can contain up to 50 sequence tags (15 bp each), which means that the SAGE method can be at least fifty times more efficient than the brute force EST sequencing and counting. Therefore, the SAGE analysis has a better chance of detecting weakly expressed genes.

The detailed SAGE procedure (Fig. 18.3) involves the generation of short unique sequence tags (15 bp in length) by cleaving cDNA with a restriction enzyme (e.g., *Nla* III with a restriction site ↑CATG) that has a relatively high cutting frequency (*Nla* III cuts every 256 bp on average (4^4)). The *Nla* III restriction digestion produces a 4-bp overhang, which is complementary to that of a premade linker. The cleaved cDNA is divided into two pools that are ligated to different linkers, which have complementary 4-bp overhangs. The unique linker contains a restriction site for a "reach and grab" type of enzyme that cuts outside its recognition site by a specific number of base pairs downstream. For example, *Bsm*F I has a restriction site GGGAC(N_{10})↑ for the forward strand and ↑(N_{14})GTCCC for the reverse strand. When the linker with *Nla* III sticky ends is allowed to ligate with *Nla* III–treated cDNA, this creates the fusion product of linker and cDNA. This is then subject to *Bsm*F I digestion, which generates a digested product with a staggered end. The product is "blunt ended" by T4 DNA polymerase, which fills in the overhang to produce the 11-bp sequence downstream of the *Nla* III site (labeled with *X*s or *Y*s in Fig. 18.3). This sample is then allowed to ligate to the other pool of cDNA ligated to a different linker to produce a linked sequence "ditag." The linkers and the ditag are amplified using polymerase chain reaction (PCR) with primers specific to each linker. The linker sequences are then removed using *Nla* III. The ditag with sticky ends is then allowed to be concatenated with more ditags to form long serial molecules that can be cloned and sequenced. When a large number of clones with linked tags are sequenced, the frequency of occurrence of each tag is counted to obtain an accurate picture of gene expression patterns.

In a SAGE experiment, sequencing is the most costly and time-consuming step. It is difficult to know how many tags need to be sequenced to get a good coverage of the entire transcriptome. It is generally determined on a case-by-case basis. As a rule of thumb, 10,000 clones representing approximately 500,000 tags from each sample are sequenced. The scale and cost of the sequencing required for SAGE analysis are prohibitive for most laboratories. Only large sequencing centers can afford to carry out SAGE analysis routinely.

Another obvious drawback with this approach is the sensitivity to sequencing errors owing to the small size of oligonucleotide tags for transcript representation. One or two sequencing errors in the tag sequence can lead to ambiguous or erroneous tag identification. Another fundamental problem with SAGE is that a correctly sequenced SAGE tag sometimes may correspond to several genes or no gene at all. To improve the sensitivity and specificity of SAGE detection, the lengths of the tags need to be increased for the technique. The following list contains some comprehensive software tools for SAGE analysis.

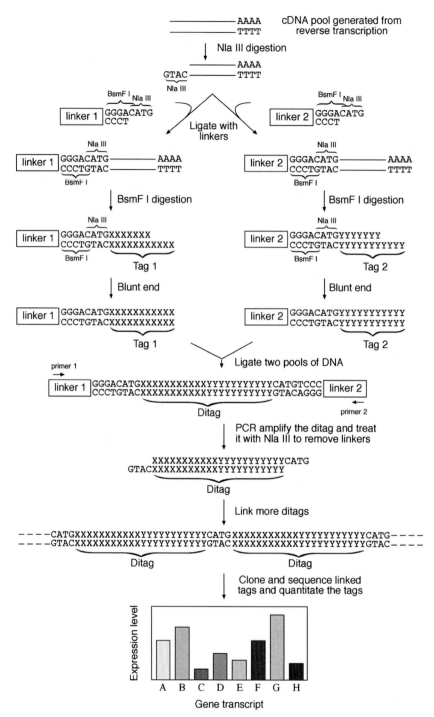

Figure 18.3: Outline of the SAGE experimental procedure.

SAGEmap (www.ncbi.nlm.nih.gov/SAGE/) is a SAGE database created by NCBI. Given a cDNA sequence, one can search SAGE libraries for possible SAGE tags and perform "virtual" Northern blots that indicate the relative abundance of a tag in a SAGE library. Each output is hyperlinked to a particular UniGene entry with sequence annotation.

SAGE xProfiler (www.ncbi.nlm.nih.gov/SAGE/sagexpsetup.cgi) is a web-based program that allows a "virtual subtraction" of an expression profile of one library (e.g., normal tissue) from another (e.g., diseased tissue). Comparison of the two libraries can provide information about overexpressed or silenced genes in normal versus diseased tissues.

SAGE Genie (http://cgap.nci.nih.gov/SAGE) is another NCBI web-based program that allows matching of experimentally obtained SAGE tags to known genes. It provides an interface for visualizing human gene expression. It has a filtering function that filters out linker sequences from experimentally obtained SAGE tags and allows expression pattern comparison between normal and diseased human tissues. The data output can be presented using subprograms such as the Anatomic Viewer, Digital Northern, and Digital Gene Expression Display.

MICROARRAY-BASED APPROACHES

The most commonly used global gene expression profiling method in current genomics research is the DNA microarray-based approach. A microarray (or gene chip) is a slide attached with a high-density array of immobilized DNA oligomers (sometimes cDNAs) representing the entire genome of the species under study. Each oligomer is spotted on the slide and serves as a probe for binding to a unique, complementary cDNA. The entire cDNA population, labeled with fluorescent dyes or radioisotopes, is allowed to hybridize with the oligo probes on the chip. The amount of fluorescent or radiolabels at each spot position reflects the amount of corresponding mRNA in the cell. Using this analysis, patterns of global gene expression in a cell can be examined. Sets of genes involved in the same regulatory or metabolic pathways can potentially be identified.

A typical DNA microarray experiment involves a multistep procedure: fabrication of microarrays by fixing properly designed oligonucleotides representing specific genes; hybridization of cDNA populations onto the microarray; scanning hybridization signals and image analysis; transformation and normalization of data; and analyzing data to identify differentially expressed genes as well as sets of genes that are coregulated (Fig. 18.4).

Oligonucleotide Design

DNA microarrays are generated by fixing oligonucleotides onto a solid support such as a glass slide using a robotic device. The oligonucleotide array slide represents thousands of preselected genes from an organism. The length of oligonucleotides is typically in the range of twenty-five to seventy bases long. The oligonucleotides are

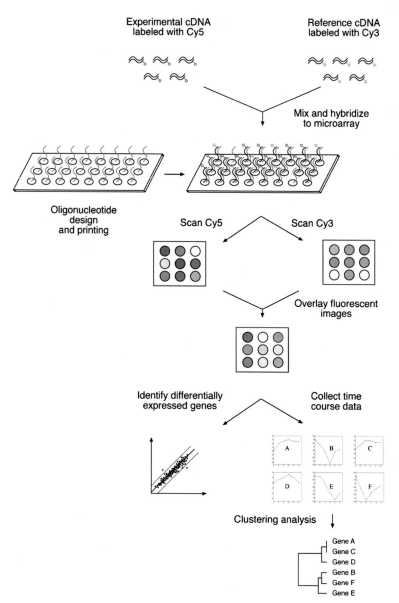

Figure 18.4: Schematic of a multistep procedure of a DNA microarray assay experiment and subsequent data analysis (see color plate section).

called probes that hybridize to labeled cDNA samples. Shorter oligo probes tend to be more specific in hybridization because they are better at discriminating perfect complementary sequences from sequences containing mismatches. However, longer oligos can be more sensitive in binding cDNAs. Sometimes, multiple distinct oligonucleotide probes hybridizing different regions of the same transcript can be used to increase the signal-to-noise ratio. To design optimal oligonucleotide sequences for microarrays, the following criteria are used.

The probes should be specific enough to minimize cross-hybridization with non-specific genes. This requires BLAST searches against genome databases to find sequence regions with least sequence similarity with nontarget genes. The probes should be sensitive and devoid of low-complexity regions (a string of identical nucleotides; see Chapter 4). The filtering program RepeatMasker (see Chapter 4) is often used in the BLAST search. The oligonucleotide sequences should not form stable internal secondary structures, such as a hairpin structure, which could interfere with the hybridization reaction. DNA/RNA folding programs such as Mfold can help to detect secondary structures. The oligo design should be close to the 3' end of the gene because the cDNA collection is often biased to the 3' end. In addition, for operational convenience, all the probes should have an approximately equal melting temperature (T_m) and a GC content of 45% to 65%. A number of programs have been developed that use these rules in designing probe sequences for microarrays spotting.

OligoWiz (www.cbs.dtu.dk/services/OligoWiz/) is a Java program that runs locally but allows the user to connect to the server to perform analysis via a graphic user interface. It designs oligonucleotides by incorporating multiple criteria including homology, T_m, low complexity, and relative position within a transcript.

OligoArray (http://berry.engin.umich.edu/oligoarray2/) is also a Java client-server program that computes oligonucleotides for microarray construction. It uses the normal criteria with an emphasis on gene specificity and secondary structure for oligonucleotides. The secondary structures and related thermodynamic parameters are calculated using Mfold.

Data Collection

The expression of genes is measured via the signals from cDNAs hybridizing with the specific oligonucleotide probes on the microarray. The cDNAs are obtained by extracting total RNA or mRNA from tissues or cells and incorporating fluorescent dyes in the DNA strands during the cDNA biosynthesis. The most common type of microarray protocol is the two-color microarray, which involves labeling one set of cDNA from an experimental condition with one dye (Cy5, red fluorescence) and another set of cDNA from a reference condition (the controls) with another dye (Cy3, green fluorescence). When the two differently labeled cDNA samples are mixed in equal quantity and allowed to hybridize with the DNA probes on the chips, gene expression patterns of both samples can be measured simultaneously.

The image of the hybridized array is captured using a laser scanner that scans every spot on the microarray. Two wavelengths of the laser beam are used to excite the red and green fluorescent dyes to produce red and green fluorescence, which is detected using a photomultiplier tube. Thus, for each spot on the microarray, red and green fluorescence signals are recorded. The two fluorescence images from the scanner are then overlaid to create a composite image, which indicates the relative expression levels of each gene. Thus, the measurement from the composite image reflects the ratio of the two color intensities. If a gene is expressed at a higher level in the experimental condition (red) than in the control (green), the spot displays a reddish color. If the gene is expressed at a lower level than the control, the spot appears greenish. Unchanged gene expression, having equal amount of green and red fluorescence, results in a yellow spot. The colored image is stored as a computer file (in TIFF format) for further processing.

Image Processing

Image processing is to locate and quantitate hybridization spots and to separate true hybridization signals from background noise. The background noise and artifacts produced in this step include nonspecific hybridization, unevenness of the slide surface, and the presence of contaminants such as dust on the surface of the slide. In addition, there are also geometric variations of hybridization spots resulting in some spots being of irregular shapes. Computer programs are used to correctly locate the boundaries of the spots and measure the intensities of the spot images after subtracting the background pixels.

After subtracting the background noise, the array signals are converted into numbers and reported as ratios between Cy5 and Cy3 for each spot. This ratio represents relative expression changes and reflects the fold change in mRNA quantity in experimental versus control conditions. The data are often presented as false colors of different intensities of red and green colors depending on whether the ratios are above 1 or below 1, respectively. Where there is an equal quantity of experimental and control mRNA (yellow in raw data), black is shown. The false color images are presented in squares in a matrix of genes versus conditions so that differentially expressed genes can be more easily analyzed (Box 18.1).

Manufacturers of microarray scanners normally provide software programs to specifically perform microarray image analysis. There are, however, also a small number of free image-processing software programs available on the Internet.

ArrayDB (http://genome.nhgri.nih.gov/arraydb/) is a web interface program that allows the user to upload data for graphical viewing. The user can present histograms, select actual microarray slide images, and display detailed information of each spot which is linked to functional annotation of the corresponding gene in the UniGene, Entrez, dbEST, and KEGG databases. This can help to provide a synopsis of gene function when interpreting the microarray data.

ScanAlyze (http://rana.lbl.gov/EisenSoftware.htm) is a Windows program for microarray fluorescent image analysis. It features semiautomatic spot definition and multichannel pixel and spot analyses.

TIGR Spotfinder (http://www.tigr.org/softlab/) is another Windows program for microarray image processing using the TIFF image format. It uses an adaptive threshold algorithm, which resolves the boundaries of spots according to their shapes. The algorithm determines the intensity of irregular spots more accurately than most other similar programs. It also interfaces with a gene expression database.

Data Transformation and Normalization

Following image processing, the digitized gene expression data need to be further processed before differentially expressed genes can be identified. This processing is

Box 18.1 Outline of the Procedure for Microarray Data Analysis

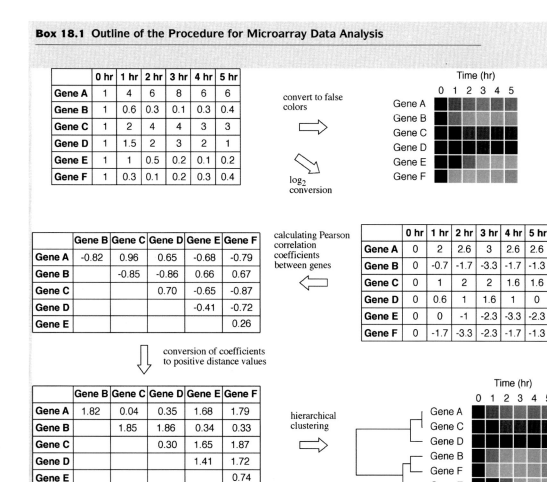

The example involves the use of six hypothetic genes whose expression is measured over a time course of 5 hours. The microarray raw data in the form of Cy5/Cy3 ratios are converted to false colors image in red, green and black. The data matrix is subjected to logarithmic transformation. The distances between genes are calculated using Pearson correlation coefficients. After conversion to a positive distance matrix, further classification analysis using the hierarchical clustering approach produces a tree showing the relationships of coexpressed genes (see color plate section).

referred to as *data normalization* and is designed to correct bias owing to variations in microarray data collection rather than intrinsic biological differences.

When the raw fluorescence intensity Cy5 is plotted against Cy3, most of the data are clustered near the bottom left of the plot, showing a non-normal distribution of the raw data (Fig. 18.5A). This is thought to be a result of the imbalance of red and green intensities during spot sampling, resulting in ineffective discrimination of differentially expressed genes. One way to improve the data discrimination is to transform

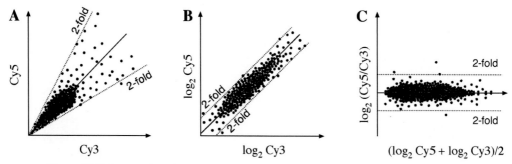

Figure 18.5: Scatter plot of gene expression analysis showing the process of data normalization. The solid line indicates linear regression of the data points; dashed lines show the cutoff for a twofold change in expression. **(A)** Plot of raw fluorescence signal intensities of Cy5 versus Cy3. **(B)** Plot of the same data after log transformation to the base of 2. **(C)** Plot of mean log intensity versus log ratio of the two fluorescence intensities, which shifts the data points to around the horizontal axis, making them easier to visualize.

raw Cy5 and Cy3 values by taking the logarithm to the base of 2. The transformation produces a more uniform distribution of data and has the advantage to display upregulated and downregulated genes more symetrically. As shown in Figure 18.5B, the data become more evenly distributed within a certain range, and assume a normal distribution pattern. By taking this transformation, the data for up-regulation and down-regulation can be more comparable.

There are many ways to further normalize the data. One way is to plot the data points horizontally. This requires plotting the log ratios (Cy5/Cy3) against the average log intensities (Fig. 18.5C). In this representation, the data are roughly symmetrically distributed about the horizontal axis. The differentially expressed genes can then be more easily visualized. This form of representation is also called *intensity-ratio plot*. In all these instances, linear regression is used.

Sometimes, the data do not conform to a linear relationship owing to systematic sampling errors. In this case, a nonlinear regression may produce a better fitting and help to eliminate the bias. The most frequently used regression type is known as Lowess (*locally weighted scatter plot smoother*) regression. This method performs a locally weighted linear fitting of the intensity-ratio data and calculates the differences between the curve-fitted values and experimental values. The algorithm further "corrects" the experimental data points by depressing large difference values more than small ones with respect to a reference. As a result, a new distribution of intensity-ratio data that conforms a linear relationship can be produced. After normalization of the data, the true outliers, which represent genes that are significantly up-regulated or down-regulated, can be more easily identified. The following two software programs that are freely available are specialized in image analysis and data normalization.

Arrayplot (www.biologie.ens.fr/fr/genetiqu/puces/publications/arrayplot/index. html) is a Windows program that allows visualization, filtering, and normalization of raw microarray data. It has an interface to view significantly up-regulated or down-regulated genes. It calculates normalization factors based on the overall median signal intensity.

SNOMAD (http://pevsnerlab.kennedykrieger.org/snomadinput.html) is a web server for microarray data normalization. It provides scatter plots based on raw signal intensities and performs log-transformation and linear regression as well as Lowess regression analysis of the data.

Statistical Analysis to Identify Differentially Expressed Genes

To separate genes that are differentially expressed, many published studies use a normalization cutoff of twofold as a criterion. However, this is an arbitrary cutoff value, which could be considered to be either too high or too low depending on the data variability. In addition, the inherent data variability is not taken into account. A data point above or below the cutoff line could simply be there by chance or because of error. The only way to ensure that a gene that appears to be differentially expressed is truly differentially expressed is to perform multiple replicate experiments and to perform statistical testing. The repeat experiments provide replicate data points that offer information about the variability of the expression data at a particular condition. The information on the distribution for the data points under particular conditions can help answer the question whether a given fold difference is significant. The main hindrance to obtaining multiple replicate datasets is often the cost: microarray experiments are extremely expensive for regular research laboratories.

If replicated datasets are available, rigorous statistical tests such as t-test and analysis of variance (ANOVA) can be performed to test the null hypothesis that a given data point is not significantly different from the mean of the data distribution. For such tests, it is common to use a P-value cutoff of .05, which means a confidence level of 95% to distinguish the data groups. This level also corresponds to a gene expression level with two standard deviations from the mean of distribution. It is noticeable that the number of standard deviations is only meaningful if the data are approximately normally distributed, which makes the previous normalization step more valuable.

MA-ANOVA (www.jax.org/staff/churchill/labsite/software/anova/) is a statistical program for Windows and UNIX that uses ANOVA to analyze microarray data. It calculates log ratios, displays ratio-intensity plots, and performs permutation tests and bootstrapping of confidence values.

Cyber-T (http://visitor.ics.uci.edu/genex/cybert/) is a web server that performs t-tests on observed changes of replicate gene expression measurements to identify significantly differentially expressed genes. It also contains a computational method for estimating false-positive and false-negative levels in experimental data based on modeling of P-value distributions.

Microarray Data Classification

One of the key features of DNA microarray analysis is to study the expression of many genes in parallel and identify groups of genes that exhibit similar expression patterns. The similar expression patterns are often a result of the fact that the genes involved are in the same metabolic pathway and have similar functions. The genetic basis of the coregulation could be the result of common promoters and regulatory regions.

To discover genes with similar gene expression patterns based on the microarray data requires partitioning the data into subsets according to similarity. To achieve this goal, hybridization signals from microarray images are organized into matrices where rows represent genes and columns represent experimental sampling conditions (such as time points or drug concentrations). Each matrix value is the Cy5/Cy3 intensity ratio representing the relative expression of a gene under a specific condition (see Box 18.1). Various classification tools are subsequently used to classify the values in the matrices for gene expression comparison.

Distance Measure

The first step towards gene classification is to define a measure of the distance or dissimilarity between genes. This requires converting a gene expression matrix in a distance matrix. The distance can be expressed as Euclidean distance or Pearson correlation coefficient. *Euclidean distance* is the square root of the sum of squared distances between expression data points. When comparing X gene expression with Y gene expression at time point i (assuming there are n time points in total), the distance score (d) can be calculated by the following formula:

$$d = \sqrt{\sum_{i=1}^{n}(x_i - y_i)^2} \qquad\qquad \text{(Eq. 18.1)}$$

Euclidean distances are widely used but suffer from the problem that when variations between genes are very small, the gene profiles can be very difficult to differentiate.

Alternatively, a Pearson correlation coefficient between two groups of data points can be used. This measures the overall similarity between the trends or shapes of the two sets of data. In this measure, a perfect positive correlation is +1 and a perfect negative correlation is −1. The distance score (d) between gene X and gene Y can be calculated using the following formula:

$$d = \frac{1}{n}\sum_{i=1}^{n}\left(\frac{x_i - \bar{x}}{sd_i}\right)\left(\frac{y_i - \bar{y}_i}{sd_i}\right) \qquad\qquad \text{(Eq. 18.2)}$$

where n is the total number of time points; x̄ and ȳ are average values for the X gene and Y gene data, respectively; and sd are standard deviation values.

The choice of the distance measures can sometimes make a big difference in the final result. Sometimes, a small change in expression data can cause a significant change in an Euclidean distance matrix. Pearson correlation coefficients are more robust than Euclidean distances in guarding against small variations and noise in the experimental data. One notable feature of the Pearson correlation coefficients is that, when the genes to be compared have exactly the same expression patterns, their gene expression profiles have identical shapes. The correlation coefficient of the gene profiles equals to +1, in which case, the relative distance between the genes is zero. When the concerned genes have absolute opposite expression patterns, the correlation coefficient becomes -1. That means that, when one gene is up-regulated, the other is down-regulated, and vice versa. In such case, the distance is converted to $+2$ (the absolute value of $|(-1) - 1|$), the maximum distance value in the matrix (see Box 18.1). The conversion to a positive distance value makes data classification more convenient.

Supervised and Unsupervised Classification

Based on the computed distances between genes in an expression profile, genes with similar expression patterns can be grouped. The classification analysis can be either supervised or unsupervised. A *supervised analysis* refers to classification of data into a set of predefined categories. For example, depending on the purpose of the experiment, the data can be classified into predefined "diseased" or "normal" categories. An *unsupervised analysis* does not assume predefined categories, but identifies data categories according to actual similarity patterns. The unsupervised analysis is also called *clustering*, which is to group patterns into clusters of genes with correlated profiles.

For microarray data, clustering analysis identifies coexpressed and coregulated genes. Genes within a category have more similarity in expression than genes from different categories. When genes are coregulated, they normally reflect related functionality. Through gene clustering, functions of previously uncharacterized genes may be discovered. Clustering methods include hierarchical clustering and partitioning clustering (e.g., k-means, self-organizing maps [SOMs]). The following discussion focuses on several of the most frequently used clustering methods.

The clustering algorithms can be further divided into two types, agglomerative and divisive (Fig. 18.6). An agglomerative method begins by clustering the two most similar data points and repeats the process to successively merge groups of data according to similarity until all groups of data are merged. This is also known as the *bottom-up approach*. A divisive method works the other way around by lumping all data points in a single cluster and successively dividing the data into smaller groups according to dissimilarity until all the hierarchical levels are resolved. This is also called the *top-down approach*.

Hierarchical Clustering. A hierarchical clustering method is in principle similar to the distance phylogenetic tree-building method (see Chapter 11). It produces a treelike structure that represents a hierarchy or relative relatedness of data groups. In the tree leaves, similar gene expression profiles are placed more closely together than dissimilar gene expression profiles. The tree-branching pattern illustrates a higher degree of relationship between related gene groups. When genes with similar expression profiles are grouped in such a way, functions for unknown genes can often be inferred.

Hierarchical clustering uses the agglomerative approach that works in much the same way as the UPGMA method (see Chapter 11), in which the most similar data pairs are joined first to form a cluster. The new cluster is treated as a single entity

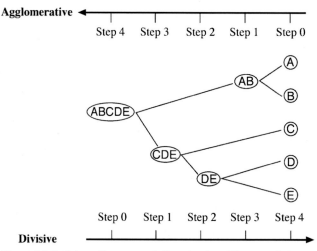

Figure 18.6: Schematic representation showing differences between agglomerative and divisive clustering methods.

creating a reduced matrix. The reduced matrix allows the next closest data point to be added to the previous cluster leading to the formation of a new cluster. By repeating the process, a dendrogram showing the clustering pattern of all data points is built.

The hierarchical clustering algorithms can be further divided in three subtypes known as single linkage, complete linkage, and average linkage. The single linkage method chooses the minimum value of a pair of distances as the cluster distance. The complete linkage method chooses the maximum value of a pair of distances, and the average linkage method chooses the mean of the two distances, which is the same as the UPGMA tree building approach. The UPGMA-based method is considered to be the most robust in discriminating expression clusters. It is important to point out that although a tree structure is produced as the final result, the resulting tree has no evolutionary meaning, but merely represents groupings of similarity patterns in gene expression.

In a tree produced by hierarchical clustering, the user has the flexibility to define a threshold for determining the boundaries of data clusters. The flexibility, however, sometimes can be a disadvantage in that it lacks objective criteria to distinguish clusters. Another potential drawback is that the hierarchical relationships of gene expression represented by the tree may not in fact exist. Some of the drawbacks can be alleviated by using alternative clustering approaches such as the k-means or self-organizing maps.

k-Means Clustering. In contrast to hierarchical clustering algorithms, k-means clustering does not produce a dendrogram, but instead classifies data through a single step partition. Thus, it is a divisive approach. In this method, data are partitioned into k-clusters, which are prespecified at the outset. The value of k is normally randomly set but can be adjusted if results are found to be unsatisfactory. In the first step, data

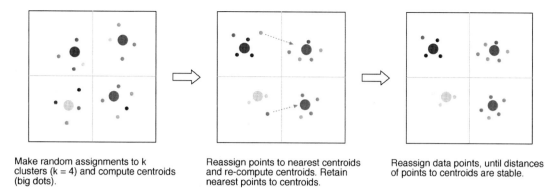

Make random assignments to k
clusters (k = 4) and compute centroids
(big dots).

Reassign points to nearest centroids
and re-compute centroids. Retain
nearest points to centroids.

Reassign data points, until distances
of points to centroids are stable.

Figure 18.7: Example of k-means clustering using four partitions. Closeness of data points is indicated by resemblance of colors (see color plate section).

points are randomly assigned to each cluster. The average of the data in a group (*centroid value*) is calculated. The distance of each data point to the centroid is also calculated. The second step is to have all the data points randomly reassigned among the k-clusters. The centroid of each cluster and distances of data points to the centroid are recomputed. Then each data point is reassigned to a different cluster. If a data point is found to be closer to the centroid of a particular cluster than to any other cluster, that data point is retained in the partition. Otherwise, it is subject to reassignment in the next iteration. This process is repeated many times, until the distances between the data points and the new centroids no longer decrease. At this point, a final clustering pattern is reached (Fig. 18.7).

As described, the number of k-clusters is specified by the user at the outset, which is either chosen randomly or determined using external information. The cluster number can be adjustable, increased or decreased to get finer or coarser data distinctions. The k-means method may not be as accurate as hierarchical clustering because it has an inherent problem of being sensitive to the selection of the initial arbitrary number of clusters. Depending on the initial position of centroids, this may lead to a different partitioning solution each time when k-means is run for the same datasets. Without searching all possible initial partitions, a suboptimal solution may be reached. However, computationally speaking, it is faster than hierarchical clustering and is still widely used.

Self-Organizing Maps. Clustering by SOMs is in principle similar to the k-means method. This pattern recognition algorithm employs neural networks. It starts by defining a number of nodes. The data points are initially assigned to the nodes at random. The distance between the input data points and the centroids are calculated. The data points are successively adjusted among the nodes, and their distances to the centroids are recalculated. After many iterations, a stabilized clustering pattern are reached with the minimum distances of the data points to the centroids.

The differences between SOM and k-means are that, in SOM, the nodes are not treated as isolated entities, but as connected to other nodes. The calculation of the centroid values in SOM takes into account not only information from within each cluster, but also information from adjacent clusters. This allows the analysis to be better at handling noisy data. Another difference is that, in SOM, some nodes are allowed to contain no data at all. Thus, at the completion of the clustering, the final number of clusters may be smaller than the initial nodes. This feature renders SOM less subjective than k-means. However, this type of algorithm is also much slower than the k-means method.

Clustering Programs. Cluster (http://rana.lbl.gov/EisenSoftware.htm) is a Windows program capable of hierarchical clustering, SOM, and k-means clustering. Outputs from hierarchical clustering are visualized with the Treeview program.

EPCLUST (www.ebi.ac.uk/EP/EPCLIST) is a web-based server that allows data to be uploaded and clustered with hierarchical clustering or k-means methods. In addition, the user can perform data selection, normalization, and database similarity searches with this program.

TIGR TM4 (www.tigr.org/tm4) is a suite of multiplatform programs for analyzing microarray data. This comprehensive package includes four interlinked programs, TIGR spot finder (for image analysis), MIDAS (for data normalization), MeV (for clustering analysis and visualization), and MADAM (for data management). The package provides different data normalization schemes and clustering options.

SOTA (Self-Organizing Tree Algorithm; www.almabioinfo.com/sota/) is a web server that uses a hybrid approach of SOM and hierarchical clustering. It builds a tree based on the divisive approach starting from the root node containing all data patterns. Instead of using the distance-based criteria to resolve a tree, the algorithm using the neural network based SOM algorithm to separate clusters of genes at each node. The homogeneity of gene clusters at each node is analyzed using SOM. The tree building stops at any point if desired homogeneity level is reached.

COMPARISON OF SAGE AND DNA MICROARRAYS

SAGE and DNA microarrays are both high throughput techniques that determine global mRNA expression levels. A number of comparative studies have indicated that the gene expression measurements from these methods are largely consistent with each other. However, the two techniques have important differences. First, SAGE does not require prior knowledge of the transcript sequence, whereas DNA microarray experiments can only detect the genes spotted on the microarray. Because SAGE is able to measure all the mRNA expressed in a sample, it has the potential to allow discovery of new, yet unknown gene transcripts. Second, SAGE measures "absolute" mRNA expression levels without arbitrary reference standards, whereas DNA microarrays indicate the relative expression levels. Therefore, SAGE expression data are more comparable across experimental conditions and platforms. This makes public SAGE databases more informative by allowing comparison of data from reference conditions with various experimental treatments. Third, the PCR amplification step involved in the SAGE procedure means that it requires only a minute quantity of sample mRNA. This compares favorably to the requirement for a much larger quantity of mRNA for microarray experiments, which may be impossible to obtain under certain circumstances. Fourth, collecting a SAGE library is very labor intensive and expensive compared with carrying out a DNA microarray experiment, however. Therefore, SAGE is not suitable for rapid screening of cells whereas the microarray analysis is. Fifth, Gene identification from SAGE data is also more cumbersome because the mRNA tags have to be extracted, compiled, and identified computationally, whereas in DNA microarrays, the identities of the probes are already known. In SAGE, comparison of gene expression profiles to discover differentially expressed genes and coexpressed genes is performed manually, whereas for microarrays, there are a large number of software algorithms to automate the process.

SUMMARY

Transcriptome analysis using ESTs, SAGE, and DNA microarrays forms the core of functional genomics and is key to understanding the interactions of genes and their regulation at the whole-genome level. EST sampling, although widely used, has a number of drawbacks in terms of error rates, efficiency, and cost. The high throughput SAGE and DNA microarray approaches provide a more quantitative measure of global gene expression. SAGE measures the "absolute" mRNA expression levels, whereas microarrays indicate relative mRNA expression levels. DNA microarrays currently enjoy greater popularity because of the relative ease of experimentation. It is also a more suitable method to probe differential gene expression between different tissue and cell samples. This requires comparing gene profiles using statistical approaches. Another goal of microarray analysis is to identify coordinated gene expression patterns, which requires clustering analysis of microarray data.

The most popular microarray data clustering techniques include hierarchical clustering, SOM, and k-means. The hierarchical approach is very similar to the phylogenetic distance tree building method. SOM and k-means normally do not generate a treelike structure as a result of clustering. Once coregulated genes are identified, upstream sequences belonging to a cluster can be retrieved and analyzed for common regulatory sequences.

In conclusion, among the three techniques for studying global gene expression, the most popular one is DNA microarrays, which has the capability to provide information that is not possible with traditional techniques. However, one should also be aware of its limitations. This technique is a multistep procedure in which errors and biases can be introduced in each step (scanning, image processing, normalization, and choice of classification method). Thus, it is a rather crude assay and may contain considerable levels of false positives and false negatives. The results from microarray analysis only provide hypotheses for gene functions based on classification of expression data. To verify the hypotheses, one has to rely on traditional biochemical and molecular biological approaches. The fundamental limitation of this method lies in the use of transcription as the sole indicator of gene expression, which may or may not correlate with expression at the protein level. The expression of proteins is what dictates the phenotypes. The last limitation is addressed in Chapter 19.

FURTHER READING

Causton, H. C., Quackenbush, J., and Brazma, A. 2003. *Microarray Gene Expression Data Analysis: A Beginner's Guide.* Malden, MA: Blackwell.

Forster, T., Roy, D., and Ghazal, P. 2003. Experiments using microarray technology: Limitations and standard operating procedure. *J. Endocrinol.* 178:195–204.

Gill, R. W., and Sanseau, P. 2000. Rapid in silico cloning of genes using expressed sequence tags (ESTs). *Biotechnol. Annu. Rev.* 5:25–44.

Quackenbush, J. 2002. Microarray data normalization and transformation. *Nat Genet.* 32(Suppl): 496–501.

Scott, H. S., and Chrast, R. 2001. Global transcript expression profiling by Serial Analysis of Gene Expression (SAGE). *Genet. Eng.* 23:201–19.

Slonim, D. K. 2002. From patterns to pathways: Gene expression data analysis comes of age. *Nat. Genet.* 32(Suppl):502–8.

Stanton, L. W. 2001. Methods to profile gene expression. *Trends Cardiovasc. Med.* 11:49–54.

Stekel, D. 2003. *Microarray Bioinformatics.* Cambridge, UK: Cambridge University Press.

Ye, S. Q., Usher, D. C., and Zhang, L. Q. 2002. Gene expression profiling of human diseases by serial analysis of gene expression. *J. Biomed. Sci.* 9:384–94.

Proteomics

Proteome refers to the entire set of expressed proteins in a cell. In other words, it is the full complement of translated product of a genome. *Proteomics* is simply the study of the proteome. More specifically, it involves simultaneous analyses of all translated proteins in a cell. It encompasses a range of activities including large-scale identification and quantification of proteins and determination of their localization, modifications, interactions, and functions. This chapter covers the major topics in proteomics such as analysis of protein expression, posttranslational modifications, protein sorting, and protein–protein interaction with an emphasis on bioinformatics applications.

Compared to transcriptional profiling in functional genomics, proteomics has clear advantages in elucidating gene functions. It provides a more direct approach to understanding cellular functions because most of the gene functions are realized by proteins. Transcriptome analysis alone does not provide clear answers to cellular functions because there is generally not a one-to-one correlation between messenger RNAs (mRNAs) and proteins in the cells. In addition, a gene in an eukaryotic genome may produce more varied translational products owing to alternative splicing, RNA editing, and so on. This means that multiple and distinct proteins may be produced from one single gene. Further complexities of protein functions can be found in posttranslational modifications, protein targeting, and protein–protein interactions. Therefore, the noncorrelation of mRNA with proteins means that studying protein expression can provide more insight on understanding of gene functions.

TECHNOLOGY OF PROTEIN EXPRESSION ANALYSIS

Characterization of protein expression at the whole proteome level involves quantitative measurement of proteins in a cell at a particular metabolic state. Unlike in DNA microarray analysis, in which the identities of the probes are known beforehand, the identities of the expressed proteins in a proteome have to determined by performing protein separation, identification, quantification, and identification procedures. The classic protein separation methods involve two-dimensional gel electrophoresis followed by gel image analysis. Further characterization involves determination of amino acid composition, peptide mass fingerprints, and sequences using mass spectrometry (MS). Finally, database searching is needed for protein identification. The outline of the procedure is shown in Figure 19.1.

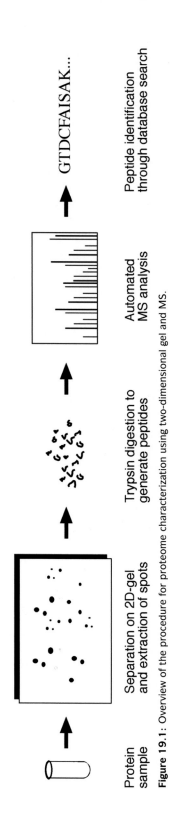

Figure 19.1: Overview of the procedure for proteome characterization using two-dimensional gel and MS.

2D-Page

Two-dimensional polyacrylamide gel electrophoresis (2D-PAGE) is a high-resolution technique that separates proteins by charge and mass. The gel is run in one direction in a pH gradient under a nondenaturing condition to separate proteins by isoelectric points (pI) and then in an orthogonal dimension under a denaturing condition to separate proteins by molecular weights (MW). This is followed by staining, usually silver staining, which is very sensitive, to reveal the position of all proteins. The result is a two-dimensional gel map; each spot on the map corresponds to a single protein being expressed. The stained gel can be further scanned and digitized for image analysis.

However, not all proteins can be separated by this method or stained properly. One of the challenges of this technique is the separation of membrane proteins, which are largely hydrophobic and not readily soulblized. They tend to aggregate in the aqueous medium of a two-dimensional gel. To overcome this problem, membrane proteins can be fractionated using specialized protocols and then electrophoresed using optimized buffers containing zwitterionic detergents. Subfractionation can be carried out to separate nuclear, cytosol, cytoskeletal, and other subcellular fractions to boost the concentrations of rare proteins and to reveal subcellular localizations of the proteins.

Gel image analysis is the next step that helps to reveal differential global protein expression patterns. This analysis includes spot determination, quantitation, and normalization. Image analysis software is used to measure the center, edges, and densities of the spots. Comparing two-dimensional gel images from various experiments can sometimes pose a challenge because the gels, unlike DNA microarrays, may shrink or warp. This requires the software programs to be able to stretch or maneuver one of the gels relative to the other to find a common geometry. When the reference spots are aligned properly, the rest of the spots can be subsequently compared automatically. There are a number of web-based tools available for this type of image analysis.

Melanie (http://us.expasy.org/melanie/) is a commercially available comprehensive software package for Windows. It carries out background subtraction, spot detection, quantitation, annotation, image manipulation and merging, and linking to 2D-PAGE databases as well as image comparison through statistical tests.

CAROL (http://gelmatching.inf.fu-berlin.de/Carol.html) is a free Java program for two-dimensional gel matching, which takes into account geometrical distortions of gel spots.

Comp2Dgel (www2.imtech.res.in/raghava/comp2dgel/) is a web server that allows the user to compare two-dimensional gel images with a two-dimensional gel database or with other gels that the user inputs. A percentage deviation of the images is obtained through superimposition of the images.

SWISS-2DPAGE (www.expasy.ch/) is a database of two-dimensional gel maps of cells of many organisms at metabolic resting conditions (control conditions), which can be used for comparison with experimental or diseased conditions. It can be searched by a spot identifier or keyword.

Mass Spectrometry Protein Identification

Once the proteins are separated on a two-dimensional gel, they can be further identified and characterized using MS. In this procedure, the proteins from a two-dimensional gel system are first digested in situ with a protease (e.g., trypsin). Protein spots of interest are excised from the two-dimensional gel. The proteolysis generates a unique pattern of peptide fragments of various MWs, which is termed a peptide fingerprint. The fragments can be analyzed with MS, a high-resolution technique for determining molecular masses. Currently, electrospray ionization MS and matrix-assisted laser desorption ionization (MALDI) MS are commonly used. These two approaches only differ in the ionization procedure used. In MALDI-MS, for example, the peptides are charged with positive ions and forced through an analyzing tube with a magnetic field. Peptides are analyzed in the gas phase. Because smaller peptides are deflected more than larger ones in a magnetic field, the peptide fragments can be separated according to molecular mass and charges. A detector generates a spectrum that displays ion intensity as a function of the mass-to-charge ratio.

As a step toward further identification, the peptides can be sequenced with successive phases of fragmentation and mass analysis. This is the technique of tandem mass spectrometry (MS/MS), in which a peptide has to pass through two analyzers for sequence determination. In the first analyzer, the peptide is fragmented by physical means generating fragments with nested sizes differing by only one amino acid. The molecular masses of these fragments are more precisely determined in the second analyzer yielding the sequence of the fragment.

Protein Identification through Database Searching

MS characterization of proteins is highly dependent on bioinformatic analysis. Once the peptide mass fingerprints or peptide sequences are determined, bioinformatics programs can be used to search for the identity of a protein in a database of theoretically digested proteins. The purpose of the database search is to find exact or nearly exact matches. However, in reality, protease digestion is rarely perfect, often generating partially digested products as a result of missed cuts at expected cutting sites. Peptides resulting from MALDI-MS are also charged, which increases their mass slightly. To increase the discriminatory ability of the database search, the search engine must allow some leeway in matching molecular masses of peptides in the cases of missed cuts and charge modifications. The user is required to provide as much information as possible as input. For example, molecular masses of peptide fingerprints, peptide sequence, MW, and pI of the intact protein, even the species names are important in obtaining unique identification of a particular protein. A basic requirement for peptide identification through database matching is the availability of all the protein sequences from an organism. Thus, this method only works well with model organisms that have completely sequenced and well-annotated genomes, but has much limitation to be applied in nonmodel organisms.

ExPASY (www.expasy.ch/tools/) is a comprehensive proteomics web server with a suite of programs for searching peptide information from the SWISS-PROT and TrEMBL databases. There are twelve database search tools in this server dedicated to protein identification based on MS data. For example, the AACompIdent program identifies proteins based on pI, MW, and amino acid composition and compares these values with theoretical compositions of all proteins in SWISS-PROT/TrEMBL. The number of candidate proteins can be further narrowed down by using species names and keywords. The TagIdent program can narrow down the candidate list by peptide sequences because of the high specificity of short sequence matches. The PeptIdent program incorporates mass fingerprinting information with information such as pI, MW, and species name. Candidate proteins are ranked by the number of matching peptides. The CombSearch tool takes advantage of the strength of multiple parameters by using combined composition, sequence tags, and peptide fingerprinting information to perform combined searches against the databases.

ProFound (http://prowl.rockefeller.edu/profound_bin/WebProFound.exe) is a web server with a set of interconnected programs. It searches a protein sequence database using MS fingerprinting information. A Bayesian algorithm ranks the database matches according to the probability of database sequences producing the peptide mass fingerprints.

Mascot (www.matrixscience.com/search_form_select.html) is another web server that identifies proteins based on peptide mass fingerprints, sequence entries, or raw MS/MS data from one or more peptides.

Differential In-Gel Electrophoresis

Differences in protein expression patterns can be detected in a similar way as in fluorescent-labeled DNA microarrays, using a technique called *differential in-gel electrophoresis* (DIGE) (Fig. 19.2). Proteins from experimental and control samples are labeled with differently colored fluorescent dyes. They are mixed together before electrophoresis on a two-dimensional gel. Differentially expressed proteins in both conditions can be coseparated and visualized in the same gel. Compared to regular 2D-PAGE, the process reduces the noise and improves the reproducibility and sensitivity of detection. In principle, it resembles the two-color DNA microarray analysis. The drawbacks of this approach are that different proteins take up fluorescent tags to different extents and that some proteins labeled with the fluorophores may become less soluble and precipitate before electrophoresis.

Protein Microarrays

Protein microarray chips are conceptually similar to DNA microarray chips (see chapter 17) and can be built to contain high-density grids with immobilized proteins for high throughput analysis. The chips contain entire immobilized proteome. However, they are not meant to be used to bind and quantitate complementary molecules as in DNA microarrays. Instead, they are used for studying protein function by

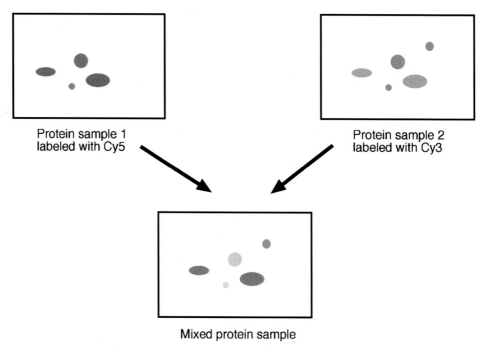

Figure 19.2: Schematic diagram showing protein differential detection using DIGE. Protein sample 1 (representing the experimental condition) is labeled with a red fluorescent dye (Cy5). Protein sample 2 (representing the control condition) is labeled with a green fluorescent dye (Cy3). The two samples are mixed together before running on a two-dimensional gel to obtain a total protein differential display map (see color plate section).

providing a solid support for assaying enzyme activity, or protein–protein interactions, protein–DNA/RNA interactions or protein–ligand interactions in an all-against-all format.

To make protein chips truly analogous to DNA chips, the solid support has to contain specific proteins or ligands that capture protein molecules by complementarity. A classical approach to this problem is to perform an immunoassay by using a spectrum of antibodies against the whole proteome. The antibodies can be fixed on a solid support for assaying thousands of proteins simultaneously. However, a major drawback of this approach is that natural antibodies are easily denatured and have a high tendency to cross-react with nonspecific antigens. In addition, producing antibodies for every single protein from an organism is prohibitively expensive.

To overcome this hurdle, a new technique is being developed that uses "protein scaffolds" to capture target molecules. The scaffolds are similar to antibodies but smaller, more stable and more specific in their binding of target proteins. They can be made in a cell-free system and attached with two fluorescence tags. This technique uses the principle of fluorescence resonance energy transfer, which is an excitation energy transfer between two fluorescent dye molecules whose excitation and absorption spectra overlap. The efficiency of the energy transfer depends on the distance of the two dyes. If one portion of the tagged protein is involved in binding to a target protein, the protein conformational changes cause the two fluorescent tags to move apart, disrupting the excitation energy transfer between the dyes such that it can be monitored on fluorescence spectra.

A technology called *Protein-Print* is in early development, which is essentially a molecular imprinting method. Chemical monomers are used to coat target proteins, which are then allow to polymerize. When polymerization is complete and the target molecules removed, a mould is formed that resembles the shape of the target protein. The moulds can then be used to capture like molecules with high specificity.

These are some of the promising technologies currently under development. Their high throughput nature means that they may eventually succeed the two-dimensional gel-based method. When the proteome chips become available, data analysis for identifying coregulated proteins should be relatively easy because it will be similar to that used for DNA microarrays. Similar image analysis and clustering algorithms can be applied to identify coregulated proteins.

POSTTRANSLATIONAL MODIFICATIONS

Another important aspect of the proteome analysis concerns posttranslational modifications. To assume biological activity, many nascent polypeptides have to be covalently modified before or after the folding process. This is especially true in eukaryotic cells where most modifications take place in the endoplasmic reticulum and the Golgi apparatus. The modifications include proteolytic cleavage; formation of disulfide bonds; addition of phosphoryl, methyl, acetyl, or other groups onto certain amino acid residues; or attachment of oligosaccharides or prosthetic groups to create mature proteins. Posttranslational modifications have a great impact on protein function by altering the size, hydrophobicity and overall conformation of the proteins. The modifications can directly influence protein–protein interactions and distribution of proteins to different subcellular locations.

It is therefore important to use bioinformatics tools to predict sites for posttranslational modifications based on specific protein sequences. However, prediction of such modifications can often be difficult because the short lengths of the sequence motifs associated with certain modifications. This often leads to many false-positive identifications. One such example is the known consensus motif for protein phosphorylation, [ST]-x-[RK]. Such a short motif can be found multiple times in almost every protein sequence. Most of the predictions based on this sequence motif alone are likely to be wrong, producing very high rates of false-positives. Similar situations can be found in other predicted modification sites. One of the reasons for the false predictions is that neighboring environment of the modification sites is not considered.

To minimize false-positive results, a statistical learning process called *support vector machine* (SVM) can be used to increase the specificity of prediction. This is a data classification method similar to the linear or quadratic discriminant analysis (see Chapter 8). In this method, the data are projected in a three-dimensional space or even a multidimensional space. A *hyperplane* – a linear or nonlinear mathematical function – is used to best separate true signals from noise. The algorithm has more environmental variables included that may be required for the enzyme modification. After training the algorithm with sufficient structural features, it is able to correctly recognize many posttranslational modification patterns.

AutoMotif (http://automotif.bioinfo.pl/) is a web server predicting protein sequence motifs using the SVM approach. In this process, the query sequence is chopped up into a number of overlapping fragments, which are fed into different kernels (similar to nodes). A hyperplane, which has been trained to recognize known protein sequence motifs, separates the kernels into different classes. Each separation is compared with known motif classes, most of which are related to posttranslational modification. The best match with a known class defines the functional motif.

Prediction of Disulfide Bridges

A disulfide bridge is a unique type of posttranslational modification in which covalent bonds are formed between cysteine residues. Disulfide bonds are important for maintaining the stability of certain types of proteins.

The disulfide prediction is the prediction of paring potential or bonding states of cysteines in a protein. Accurate prediction of disulfide bonds may also help to predict the three-dimensional structure of the protein of interest. This problem can be tackled by using either profiles constructed from multiple sequence alignment or residue contact potentials calculated based on the local sequence environment. Advanced neural networks or SVM or hidden Markov model (HMM) algorithms are often used to discern long-distance pairwise interactions among cysteine residues. The following program is one of the publicly available programs specialized in disulfide prediction.

Cysteine (http://cassandra.dsi.unifi.it/cysteines/) is a web server that predicts the disulfide bonding states of cysteine residues in a protein sequence by building profiles based on multiple sequence alignment information. A recursive neural network (see Chapter 14) ranks the candidate residues for disulfide formation.

Identification of Posttranslational Modifications in Proteomic Analysis

Posttranslational modifications can be experimentally identified based on MS fingerprinting data. Certain peptide identification tools are able to search for known posttranslational modification sites in a sequence and incorporate extra mass based on the type of modifications during database fragment matching. There are two subprograms in the ExPASY proteomics server and an independent RESID database that are related to predicting posttranslational modifications.

ExPASY (www.expasy.ch/tools) contains a number of programs to determine posttranslational modifications based on MS molecular mass data. FindMod is a subprogram that uses experimentally determined peptide fingerprint information to compare the masses of the peptide fragments with those of theoretical peptides. If a difference is found, it predicts a particular type of modification based on a set of predefined rules. It can predict twenty-eight types of modifications, including methylation, phosphorylation, lipidation, and sulfation. GlyMod is a subprogram that specializes in glycosylation determination based on the difference in mass between experimentally determined peptides and theoretical ones.

RESID (http://pir.georgetown.edu/pirwww/search/textresid.html) is an independent posttranslational modification database listing 283 types of known modifications. It can search by text or MWs.

PROTEIN SORTING

Subcellular localization is an integral part of protein functionality. Many proteins exhibit functions only after being transported to certain compartments of the cell. The study of the mechanism of protein trafficking and subcellular localization is the field of protein sorting (also known as protein targeting), which has become one of the central themes in modern cell biology. Identifying protein subcellular localization is an important aspect of functional annotation, because knowing the cellular localization of a protein often helps to narrow down its putative functions.

For many eukaryotic proteins, newly synthesized protein precursors have to be transported to specific membrane-bound compartments and be proteolytically processed to become functional. These compartments include chloroplasts, mitochondria, the nucleus, and peroxisomes. To carry out protein translocation, unique peptide signals have to be present in the nascent proteins, which function as "zip codes" that direct the proteins to each of these compartments. Once the proteins are translocated within the organelles, protease cleavage takes place to remove the signal sequences and generate mature proteins (another example of posttranslational modification). Even in prokaryotes, proteins can be targeted to the inner or outer membranes, the periplasmic space between these membranes, or the extracellular space. The sorting of these proteins is similar to that in eukaryotes and relies on the presence of signal peptides.

The signal sequences have a weak consensus but contain some specific features. They all have a hydrophobic core region preceded by one or more positively charged residues. However, the length and sequence of the signal sequences vary tremendously. Peptides targeting mitochondria, for example, are located in the N-terminal region. The sequences are typically twenty to eighty residues long, rich in positively charged residues such as arginines as well as hydroxyl residues such as serines and threonines, but devoid of negatively charged residues, and have the tendency to form amphiphilic α-helices. These targeting sequences are cleaved once the precursor proteins are inside the mitochondria. Chloroplast localization signals (also called transit peptides) are also located in the N-terminus and are about 25 to 100 residues in length, containing very few negatively charged residues but many hydroxylated residues such as serine. An interesting feature of the proteins targeted for the chloroplasts is that the transit signals are bipartite. That is, they consist of two adjacent signal peptides, one for targeting the proteins to the stroma portion of the chloroplast before being cleaved and the other for targeting the remaining portion of the proteins to the thylakoids. Localization signals targeting to the nucleus are variable in length (seven to forty-one residues) and are found in the internal region of the proteins. They typically consist of one or two stretches of basic residues with a consensus motif K(K/R)X(K/R). Nuclear signal sequences are not cleaved after protein transport.

Considerable variations in length and sequence make accurate prediction of signal peptides using computational approaches difficult. Nonetheless, various computational methods have been developed to predict the subcellular localization signals. In general, they fall within three categories. Some algorithms are signal based, depending on the knowledge of charge, hydrophobicity, or consensus motifs. Some are content based, depending on the sequence statistics such as amino acid composition. The third group of algorithm combines the virtue of both signals and content and appears to be more successful in prediction. Neural network- and HMM-based algorithms are examples of the combined approach. Here are some of the most frequently used programs for the prediction of subcellular localization and protein sorting signals with reasonable accuracy (65% to 70%).

SignalP (www.cbs.dtu.dk/services/SignalP-2.0/#submission) is a web-based program that predicts subcellular localization signals by using both neural networks and HMMs. The neural network algorithm combines two different scores, one for recognizing signal peptides and the other for protease cleavage sites. The HMM-based analysis discriminates between signal peptides and the *N*-terminal transmembrane anchor segments required for insertion of the protein into the membrane. The program is trained by three different training sets, namely, eukaryotes, Gram-negative bacteria and Gram-positive bacteria. This distinction is necessary because there are significant differences in the characteristics of the signal peptides from these organisms. Therefore, appropriate datasets need to be selected before analyzing the sequence. The program predicts both the signal peptides and the protease cleavage sites of the query sequence.

TargetP (www.cbs.dtu.dk/services/TargetP/) is a neural network-based program, similar to SignalP. It predicts the subcellular locations of eukaryotic proteins based on their *N*-terminal amino acid sequence only. It uses analysis output from SignalP and feeds it into a decision neural network, which makes a final choice regarding the target compartment.

PSORT (http://psort.nibb.ac.jp/) is a web server that uses a nearest neighbor method to make predictions of subcellular localizations. It compares the query sequence to a library of signal peptides for different cellular localizations. If the majority of the closest signal peptide matches (nearest neighbors) are for a particular cellular location, the sequence is predicted as signal peptide for that location. It is functionally similar to TargetP, but may have lower sensitivity. An iPSORT is available in the same website that predicts *N*-terminal sorting signals and is an equivalent to SignalP.

PROTEIN–PROTEIN INTERACTIONS

In general, proteins have to interact with each other to carry out biochemical functions. Thus, mapping out protein–protein interactions is another important aspect of proteomics. Interprotein interactions include strong interactions that allow formation of stable complexes and weaker ones that exist transiently. Proteins involved in forming complexes are generally more tightly coregulated in expression than those involved in transient interactions. Protein–protein interaction analysis at the proteome level helps reveal the function of previously uncharacterized proteins on the basis of the "guilt by-association" rule.

Experimental Determination

Protein interactions are commonly detected by using the classic yeast two-hybrid method that relies on the interaction of "bait" and "prey" proteins in molecular constructs in yeast. In this strategy, a two-domain transcriptional activator is employed as a helper for determining protein–protein interactions. The two domains which are a DNA-binding domain and a trans-activation domain normally interact to activate transcription. However, molecular constructs are made such that each of the two domains is covalently attached to each of the two candidate proteins (bait and prey). If the bait and prey proteins physically interact, they bring the DNA-binding and trans-activation domains in such close proximity that they reconstitute the function of the transcription activator, turning on the expression of a reporter gene as a result. If the two candidate proteins do not interact, the reporter gene expression remains switched off.

This technique is essentially a low throughput approach because each bait and prey construct has to be prepared individually to map interactions between all proteins. Nonetheless, it has been systematically applied to study interactions at the whole proteome level. Protein–protein interaction networks of yeast and a small number of other species have been subsequently determined using this method. A major flaw in this method is that it is an indirect approach to probe protein–protein interaction and has a tendency to generate false positives (spurious interactions) and false negatives (undetected interactions). It has been estimated from proteome-wide characterizations that the rate of false positives can be as high as 50%. Another weakness is that only pairwise interactions are measured, and therefore interactions that only take place when multiple proteins come together are omitted.

There are many alternative approaches to determining protein–protein interactions. One of them is to use a large-scale affinity purification technique that involves attaching fusion tags to proteins and purifying the associated protein complexes in an affinity chromatography column. The purified proteins are then analyzed by gel electrophoresis followed by MS for identification of the interacting components. The protein microarray systems mentioned above also provide a high throughput alternative for studying protein–protein interactions. Although none of the methods

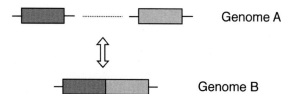

Figure 19.3: Rosetta stone method for prediction of genes encoding interacting proteins based on domain fusion patterns in different genomes. In genome A, two different domains exist in separate open reading frames. In genome B, they are fused together in one protein-encoding frame. Conversely, the two domains of the same protein encoded in genome B may become separate in genome A, but still perform the same function through physical interactions.

are guaranteed to eliminate false positives and false negatives, combining multiple approaches in theory compensates for the potential weaknesses of each technique and minimizes the artifacts.

Prediction of Protein–Protein Interactions

Decades of research on protein biochemistry and molecular biology has accumulated tremendous amount of data related to protein–protein interactions, which allow the extraction of some general rules governing these interactions. These rules have facilitated the development of algorithms for automated prediction of protein–protein interactions. The currently available tools are generally based on evolutionary studies of gene sequences, gene linkage patterns, and gene fusion patterns, which are described in detail next.

Predicting Interactions Based on Domain Fusion

One of the prediction methods is based on gene fusion events. The rationale goes like this: if A and B exist as interacting domains in a fusion protein in one proteome, the gene encoding the protein is a fusion gene. Their homologous gene sequences A′ and B′ existing separately in another genome most likely encode proteins interacting to perform a common function. Conversely, if ancestral genes A and B encode interacting proteins, they may have a tendency to be fused together in other genomes during evolution to enhance their effectiveness. This method of predicting protein–protein interactions is called the "Rosetta stone" method (Fig. 19.3) because a fused protein often reveals relationships between its domain components.

The further justification behind this method is that when two domains are fused in a single protein, they have to be in extremely close proximity to perform a common function. When the two domains are located in two different proteins, to preserve the same functionality, their close proximity and interaction have to be preserved as well. Therefore, by studying gene/protein fusion events, protein–protein interactions can be predicted. This prediction rule has been proven to be rather reliable and since successfully applied to a large number of proteins from both prokaryote and eukaryotes.

Predicting Interactions Based on Gene Neighbors

Gene orders, generally speaking, are poorly conserved among divergent prokaryotic genomes (see Chapter 16). However, if a certain gene linkage is found to be indeed conserved across divergent genomes, it can be used as a strong indicator of formation of an operon that encodes proteins that are functionally and even physically coupled. This rule of predicting protein–protein interactions holds up for most prokaryotic genomes. For eukaryotic genomes, gene order may be a less potent predictor of protein interactions than a tight coregulation for gene expression.

Predicting Interactions Based on Sequence Homology

If a pair of proteins from one proteome are known to interact, their conserved homologs in another proteome are likely to have similar interactions. The homologous pairs are referred to as *interologs*. This method relies on the correct identification of orthologs and the use of existing protein interaction databases. The method has potential to model protein quaternary structure if one pair of proteins have known structures.

InterPreTS (www.russell.embl-heidelberg.de/people/patrick/interprets/interprets.html) is a web server that has a built-in database for interacting domains based on known three-dimensional protein structures. Two protein sequences are used as query to search against the database for homologs. The alignment of the query sequences and database domains is carried out using HMMer (see Chapter 6). If the alignment scores for both sequences are above the threshold and the contact residues are found to be conserved, the two proteins are considered to be interacting proteins.

IPPRED (http://cbi.labri.fr/outils/ippred/IS_part_simple.php) is a similar web-based program that allows the user to submit multiple protein sequences. The program searches homologous sequences using BLAST in a database of known interacting protein pairs (BIND). If any two query sequences have strong enough similarity with known interacting protein pairs, they are inferred as interacting partners.

Predicting Interactions Based on Phylogenetic Information

Protein interactions can be predicted using phylogenetic profiles, which are defined as patterns of gene pairs that are concurrently present or absent across genomes. In other words, this method detects the copresence or co-absence of orthologs across a number of genomes. Genes having the same pattern of presence or absence across genomes are predicted as encoding interacting proteins. The logic behind the cooccurrence approach is that proteins normally operate as a complex. If one of the components of the complex is lost, it results in the failure of the entire complex. Under the selective pressure, the rest of the nonfunctional interacting partners in the complex are also lost during evolution because they have become functionally unnecessary. The rule based on concurrent gene loss or gene gain has proven to be less accurate than the rules based on gene fusion and gene neighbors. An example of using the phylogenetic profile method to predict interacting proteins is shown in Figure 19.4.

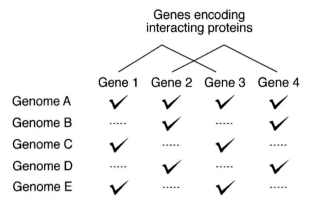

Genes encoding
interacting proteins

Figure 19.4: Phylogenetic profile method for predicting interacting proteins based on copresence and co-absence of the encoding genes across genomes. The presence is indicated by checks and absence by dashed lines. The protein pairs encoded by genes one and three as well as genes two and four are predicted as interacting partners.

A more quantitative phylogenetic method to predict protein interactions is the "mirror tree" method, which examines the resemblance between phylogenetic trees of two sequence families (Fig. 19.5). The rationale is that if two protein trees are nearly identical in topology and are highly correlated in terms of evolutionary rate, they are highly likely to interact with each other. This is because if mutations occur at

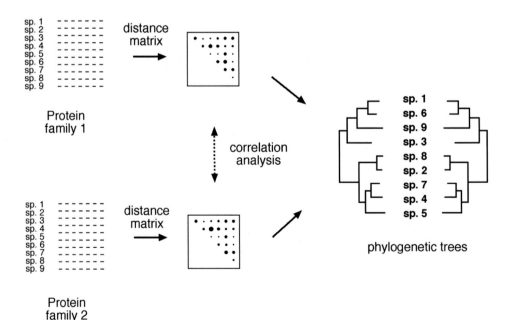

Figure 19.5: Mirror tree method for prediction of interacting proteins based on strong statistical correlation of evolutionary distance matrices used to build two phylogenetic trees for the two protein families of interest. The two trees have a near identical topology resulting in a near mirror image. The distance matrices used to construct the trees are compared using correlation analysis.

the interaction surface for one of the proteins, corresponding mutations are likely to occur in the interacting partner to sustain the interaction. As a result, the two interacting proteins should have very similar phylogenetic trees reflecting very similar evolutionary history. To analyze the extent of coevolution, correlation coefficients (r) of evolutionary distance matrices for the two groups of protein homologs used in constructing the trees are examined. It has been shown that if $r > 0.8$, there is a strong indication for protein interactions.

Matrix (http://orion.icmb.utexas.edu/cgi-bin/matrix/matrix-index.pl) is a web server that predicts interaction between two protein families. The server aligns two individual protein data sets (assuming each representing a protein family) using Clustal. It then derives distance matrices from the two alignment files and aligns the matrices to discover similar portions that may indicate interacting partners from the two protein families.

ADVICE (Automated Detection and Validation of Interaction based on the Co-Evolutions, http://advice.i2r.a-star.edu.sg/) is a similar web server providing prediction of interacting proteins using the mirror-tree approach. It performs automated BLAST searches for a given protein sequence pair to derive two sets of homologous sequences. The sequences are multiply aligned using CLUSTAL. A distance matrix for each set of alignment is then derived. The Pearson's correlation coefficient is subsequently calculated for detecting similarities between the two distance matrices. If the coefficient r > 0.8, the two query sequences are predicted to be a interacting pair.

Predicting Interactions Using Hybrid Methods

It needs to be emphasized that each of these prediction methods is based on a particular hypothesis and may exhibit a certain degree of bias associated with the hypothesis. Because it is difficult to evaluate the performance of each individual prediction method, the user of these prediction algorithms is recommended to use a combined approach that uses multiple methods to reduce bias and error rates and to yield a higher level of confidence in the protein interaction prediction. The following internet program is a good example of combining multiple lines of evidence in predicting protein-protein interactions.

STRING (Search Tool for the Retrieval of Interacting Genes/Proteins, http://www.bork.embl-heidelberg.de/STRING/) is a web server that predicts gene and protein functional associations based on combined evidence of gene linkage, gene fusion and phylogenetic profiles. The current version also includes experimental co-expression data as well as documented interactions resulted from literature mining. Functional associations include both direct and indirect protein-protein interactions. Indirect interactions can mean enzymes in the same pathway sharing a common substrate or proteins regulating each other in the genetic pathway. The server contains information for orthologous groups from 110 completely sequenced genomes. The query sequence is first classified into an orthologous group based on the COG classification (see Chapter 7) and is then used to search the database for known conserved linkage pattern, gene fusions, and phylogenetic profiles. The server uses a weighted scoring system that evaluates the significance of all three types of protein associations among the genomes. To reduce false positives and increase reliability of the prediction, the three types of genomic associations are checked against an internal reference set. A single score of pairwise interactions is given as the final output which also contains all three types of evidence plus a summary of combined protein interaction network involving multiple partners. The server returns a list of predicted protein-protein associations and a graphic representation of the association network.

SUMMARY

Protein expression analysis at the proteome level promises more accurate elucidation of cellular functions. This is an advantage over genomic analysis, which does not necessarily lead to prediction of protein functions. Traditional experimental approaches to proteomics include large-scale protein identification using 2D-PAGE and MS. The identification process requires the integration of bioinformatics tools to search databases for matching peptides. Newer protein expression profiling techniques include DIGE and protein microarrays. Protein functions can be modulated as a result of posttranslational modifications. Sequence based prediction often results in high rates of false-positives owing to limited understanding of the structural features required for the modifications. A step toward minimizing the false-positive rates in prediction is the use of SVM. Another area of proteomics is defining protein subcellular localization signals. Several web tools such as TargetP, SignalP, and PSORT are available to give reasonably successful prediction of signal peptides. Protein–protein interactions are normally determined using yeast two-hybrid experiments or other experimental methods. However, theoretical prediction of such interactions is providing a promising alternative. The current prediction methods are based on domain fusion, gene linkage pattern, sequence homology, and phylogenetic information. The ability to predict protein interactions is of tremendous value in genome annotation and in understanding the function of genes and their encoded proteins. The computational approach helps to generate hypotheses to be tested by experiments.

FURTHER READING

Aebersold, R., and Mann, M. 2003. Mass spectrometry-based proteomics. *Nature* 422:198–207.

Cutler, P. 2003. Protein arrays: The current state-of-the-art. *Proteomics* 3:3–18.

Donnes, P., and Hoglund, A. 2004. Predicting protein subcellular localization: past, present, and future. *Genomics Proteomics Bioinformatics* 2:209–15.

Droit, A., Poirier, G. G., and Hunter, J. M. 2005. Experimental and bioinformatic approaches for interrogating protein-protein interactions to determine protein function. *J. Mol. Endocrinol.* 34:263–80.

Eisenhaber, F., Eisenhaber, B., and Maurer-Stroh, S. 2003. "Prediction of post-translational modifications from amino acid sequence: Problems, pitfalls, and methodological hints." In *Bioinformatics and Genomes: Current Perspectives,* edited by M. A. Andrade, 81–105. Wymondham, UK: Horizon Scientific Press.

Emanuelsson, O. 2002. Predicting protein subcellular localisation from amino acid sequence information. *Brief. Bioinform.* 3:361–76.

Huynen, M. A., Snel, B., Mering, C., and Bork, P. 2003. Function prediction and protein networks. *Curr. Opin. Cell Biol.* 15:191–8.

Mann, M., and Jensen, O. N. 2003. Proteomic analysis of post-translational modifications. *Nature Biotechnol.* 21:255–61.

Nakai, K. 2000. Protein sorting signals and prediction of subcellular localization. *Adv. Protein Chem.* 54:277–344.

Nakai, K. 2001. Review: Prediction of in vivo fates of proteins in the era of genomics and proteomics. *J. Struct. Biol.* 134:103–16.

Phizicky, E., Bastiaens, P. I. H., Zhu, H., Snyder, M., and Fields, S. 2003. Protein analysis on a proteomic scale. *Nature* 422:208–15.

Sadygov, R. G., Cociorva, D., and Yates, J. R. III. 2004. Large-scale database searching using tandem mass spectra: Looking up the answer in the back of the book. *Nat. Methods* 1:195–202.

Tyers, M., and Mann, M. 2003. From genomics to proteomics. *Nature* 422:193–7.

Valencia, A., and Pazos, F. 2002. Computational methods for the prediction of protein interactions. *Curr. Opin. Struct. Biol.* 12:368–73.

Valencia, A., and Pazos, F. 2003. Prediction of protein-protein interactions from evolutionary information. *Methods Biochem. Anal.* 44:411–26.

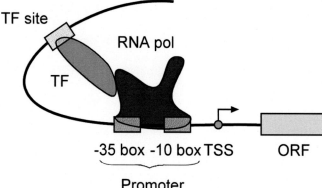

Figure 9.1. Schematic representation of elements involved in bacterial transcription initiation. RNA polymerase binds to the promoter region, which initiates transcription through interaction with transcription factors binding at different sites. *Abbreviations:* TSS, transcription start site; ORF, reading frame; pol, polymerase; TF, transcription factor (see page 114).

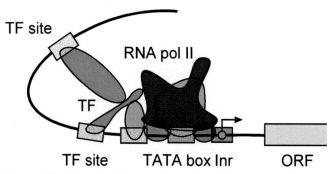

Figure 9.2. Schematic diagram of an eukaryotic promoter with transcription factors and RNA polymerase bound to the promoter. *Abbreviations:* Inr, initiator sequence; ORF, reading frame; pol, polymerase; TF, transcription factor (see page 115).

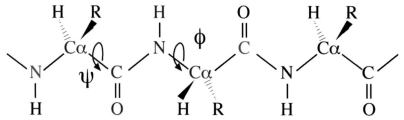

Figure 12.3. Definition of dihedral angles of ϕ and ψ. Six atoms around a peptide bond forming a peptide plane are colored in red. The ϕ angle is the rotation about the N–Cα bond, which is measured by the angle between a virtual plane formed by the C–N–Cα and the virtual plane by N–Cα–C (C in green). The ψ angle is the rotation about the Cα–C bond, which is measured by the angle between a virtual plane formed by the N–Cα–C (N in green) and the virtual plane by Cα–C–N (N in red) (see page 176).

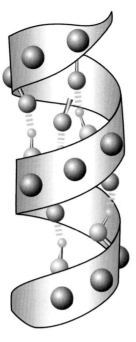

Figure 12.5. A ribbon diagram of an α-helix with main chain atoms (as gray balls) shown. Hydrogen bonds between the carbonyl oxygen (red) and the amino hydrogen (green) of two residues are shown in yellow dashed lines (see page 178).

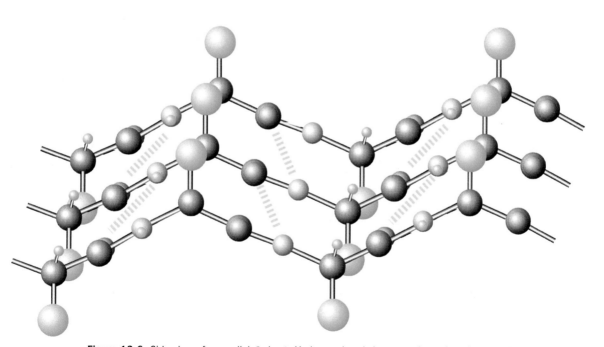

Figure 12.6. Side view of a parallel β-sheet. Hydrogen bonds between the carbonyl oxygen (red) and the amino hydrogen (green) of adjacent β-strands are shown in yellow dashed lines. R groups are shown as big balls in cyan and are positioned alternately on opposite sides of β-strands (see page 179).

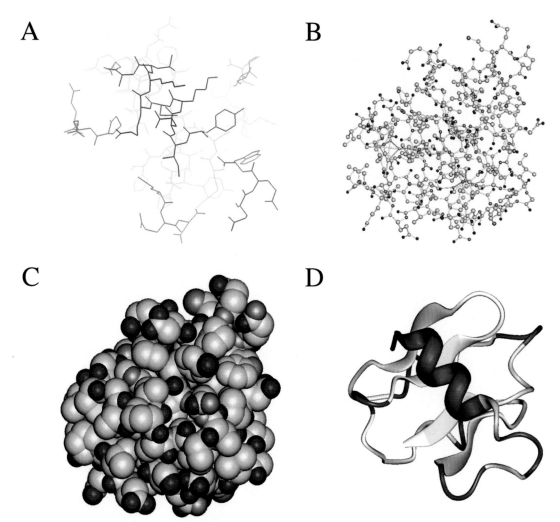

Figure 13.1. Examples of molecular structure visualization forms. **(A)** Wireframes. **(B)** Balls and sticks. **(C)** Space-filling spheres. **(D)** Ribbons (see page 188).

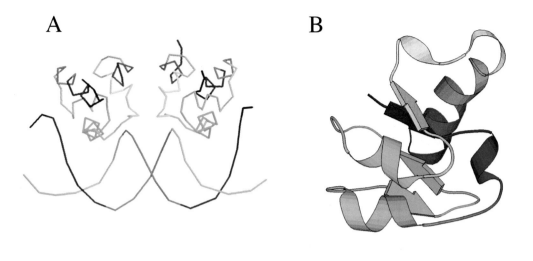

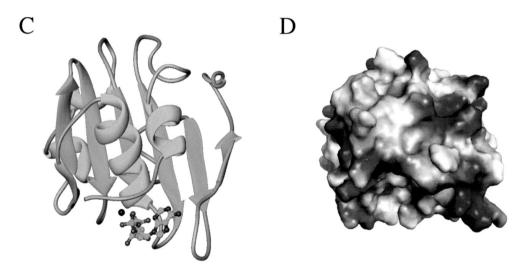

Figure 13.2. Examples of molecular graphic generated by **(A)** Rasmol, **(B)** Molscript, **(C)** Ribbons, and **(D)** Grasp (see page 189).

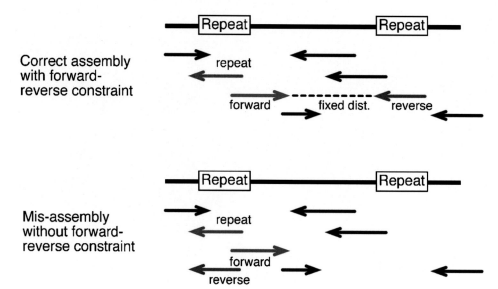

Figure 17.4. Example of sequence assembly with or without applying forward–reverse constraint, which fixes the sequence distance from both ends of a subclone. Without the restraint, the red fragment is misassembled due to matches of repetitive element in the middle of a fragment (see page 248).

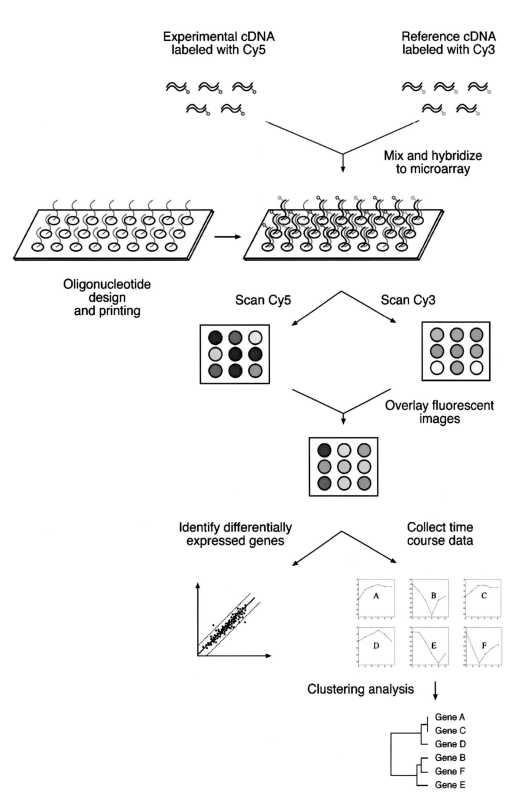

Figure 18.4. Schematic of a multistep procedure of a DNA microarray assay experiment and subsequent data analysis (see page 268).

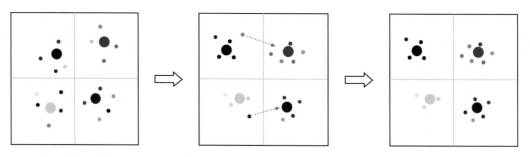

Make random assignments to k clusters (k = 4) and compute centroids (big dots).

Reassign points to nearest centroids and re-compute centroids. Retain nearest points to centroids.

Reassign data points, until distances of points to centroids are stable.

Figure 18.7. Example of k-means clustering using four partitions. Closeness of data points is indicated by resemblance of colors (see page 277).

	0 hr	1 hr	2 hr	3 hr	4 hr	5 hr
Gene A	1	4	6	8	6	6
Gene B	1	0.6	0.3	0.1	0.3	0.4
Gene C	1	2	4	4	3	3
Gene D	1	1.5	2	3	2	1
Gene E	1	1	0.5	0.2	0.1	0.2
Gene F	1	0.3	0.1	0.2	0.3	0.4

convert to false colors

log$_2$ conversion

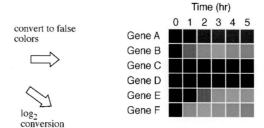

	Gene B	Gene C	Gene D	Gene E	Gene F
Gene A	-0.82	0.96	0.65	-0.68	-0.79
Gene B		-0.85	-0.86	0.66	0.67
Gene C			0.70	-0.65	-0.87
Gene D				-0.41	-0.72
Gene E					0.26

calculating Pearson correlation coefficients between genes

	0 hr	1 hr	2 hr	3 hr	4 hr	5 hr
Gene A	0	2	2.6	3	2.6	2.6
Gene B	0	-0.7	-1.7	-3.3	-1.7	-1.3
Gene C	0	1	2	2	1.6	1.6
Gene D	0	0.6	1	1.6	1	0
Gene E	0	0	-1	-2.3	-3.3	-2.3
Gene F	0	-1.7	-3.3	-2.3	-1.7	-1.3

conversion of coefficients to positive distance values

	Gene B	Gene C	Gene D	Gene E	Gene F
Gene A	1.82	0.04	0.35	1.68	1.79
Gene B		1.85	1.86	0.34	0.33
Gene C			0.30	1.65	1.87
Gene D				1.41	1.72
Gene E					0.74

hierarchical clustering

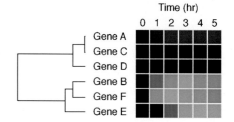

Box 18.1. Outline of the Procedure for Microarray Data Ananlysis (see page 271).

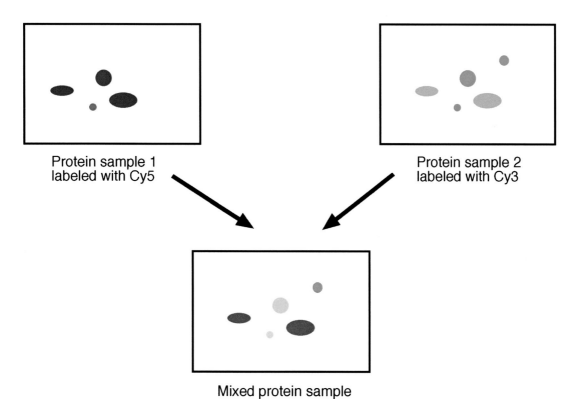

Protein sample 1
labeled with Cy5

Protein sample 2
labeled with Cy3

Mixed protein sample

Figure 19.2. Schematic diagram showing protein differential detection using DIGE. Protein sample 1 (representing experimental condition) is labeled with a red fluorescent dye (Cy5). Protein sample 2 (representing control condition) is labeled with a green fluorescent dye (Cy3). The two samples are mixed together before running on a two-dimensional gel to obtain a total protein differential display map (see page 286).

Appendix

Practical Exercises

Note: all exercises were originally designed for use on a UNIX workstation. However, with slight modifications, they can be used on any other operating systems with Internet access.

EXERCISE 1. DATABASE SEARCHES

In this exercise, you will learn how to use several biological databases to retrieve information according to certan criteria. After learning the basic search techniques, you will be given a number of problems and asked to provide answers from the databases.

1. Use a web browser to retrieve a protein sequence of lambda repressor from SWISS-PROT (http://us.expasy.org/sprot/). Choose "Full text search in Swiss-Prot and TrEMBL." In the following page, Enter "lambda repressor" (space is considered as logical operator AND) as keywords in the query window. Select "Search in Swiss-Prot only." Click on the "submit" button. Note the search result contains hypertext links taking you to references that are cited or to other related information. Spend a little time studying the annotations.

2. In the same database, search more sequences for "human MAP kinase inhibitor," "human catalase," "synechocystis cytochrome P450," "coli DNA polymerase," "HIV CCR5 receptor," and "*Cholera* dehydrogenase." Record your findings and study the annotations.

3. Go to the SRS server (http://srs6.ebi.ac.uk/) and find human genes that are larger than 200 kilobase pairs and also have poly-A signals. Click on the "Library Page" button. Select "EMBL" in the "Nucleotide sequence databases" section. Choose the "Extended" query form on the left of the page. In the following page, Select human ("hum") at the "Division" section. Enter "200000" in the "SeqLength >=" field. Enter "polya_signal" in the "AllText" field. Press the "Search" button. How many hits do you get?

4. Use your knowledge and creativity to do the following SRS exercises.
    1) Find protein sequences from *Rhizobium* submitted by Ausubel between 1991 and 2001 in the UniProt/Swiss-Prot database (hint: the date

expression can be 1-Jan-1991 and 31-Dec-2001). Study the annotations of the sequences.

2) Find full-length protein sequences of mammalian tyrosine phosphatase excluding partial or fragment sequences in the UniProt/SwissProt database (hint: the taxonic group of mammals is *mammalia*). Once you get the query result, do a Clustal multiple alignment on the first five sequences from the search result.

5. Go to the web page of NCBI Entrez (http://www.ncbi.nlm.nih.gov/) and use the advanced search options to find protein sequences for human kinase modified or added in the last 30 days in GenBank. In the Entrez "Protein" database, enter "human[ORGN] kinase", and then select "last 30 days" in the "Modification Date" field of the "Limits" section. Select "Only from" "GenBank" as database. Finally, select " Go."

6. Using Entrez, search DNA sequences for mouse fas antigen with annotated exons or introns. (Do not forget to deselect "Limits" from the above exercise.) In Entrez, select the Nucleotide database. Type mouse[ORGN] AND fas AND (exons OR introns). Click "Go."

7. For the following exercises involving the NCBI databases, design search strategies to find answers (you will need to decide which database to use first).

   1) Find gene sequences for formate dehydrogenase from *Methanobacterium*.

   2) Find gene sequences for DNA binding proteins in *Methanobacterium*.

   3) Find all human nucleotide sequences with D-loop annotations.

   4) Find protein sequences of maltoporin in Gram-negative bacteria (hint: use logic operator NOT. Gram-positive bacteria belong to Firmicutes).

   5) Find protein structures related to *Rhizobium* nodulation.

   6) Find review papers related to protein electrostatic potentials by Honig published since 1990.

   7) Find the number of exons and introns in *Arabidopsis* phytochrome A (*phyA*) gene (hint: use [GENE] to restrict search).

   8) Find two upstream neighboring genes for the hypoxanthine phosphoribosyl transferase (HPRT) gene in the *E. coli* K12 genome.

   9) Find neurologic symptoms for the human Lesch–Nyhan syndrome. What is the chromosomal location of the key gene linked to the disease? What are its two upstream neighboring genes?

   10) Find information on human gene therapy of atherosclerosis from NCBI online books.

   11) Find the number of papers that Dr. Palmer from Indiana University has published on the subject of lateral gene transfer in the past ten years.

EXERCISE 2. DATABASE SIMILARITY SEARCHES AND PAIRWISE SEQUENCE ALIGNMENT

Database Searching

In this exercise, you will learn about database sequence similarity search tools through an example: flavocytochrome b2 (PDB code 1fcb). This enzyme has been shown to be very similar to a phosphoribosylanthranilate isomerase (PDB code 1pii) by detailed three-dimensional structural analysis (Tang et al. 2003. *J. Mol. Biol.* 334:1043–62). This similarity may not be detectable by traditional similarity searches. Perform the following exercise to test the capability of various sequence searching methods to see which method has the highest sensitivity to detect the distant homologous relationship.

1. Obtain the yeast flavocytochrome b2 protein sequence from NCBI Entrez (accession number NP_013658). This is done by choosing "FASTA" in the format pull-down menu and clicking on the "Display" button. Copy the sequence into clipboard.

2. Perform a protein BLAST search (select Protein-protein BLAST at www.ncbi.nlm.nih.gov/blast/). Paste the sequence into the BLASTP query box. Choose pdb as database (this will reduce the search time). Leave all other settings as default. Click on the "BLAST!" button. To get the search result, click on the "Format!" button in the following page. Summarize the number of hits, highest and lowest bit scores in a table.

3. Change the *E*-value to 0.01 and change the word size from 3 to 2, and do the search again. Do you see any difference in the number of hits? Can you find 1pii in the search result?

4. Reset the *E*-value to 10. Change the substitution matrix from BLOSUM62 to BLOSUM45. Compare the search results again. What is your conclusion in terms of selectivity and sensitivity of your searches? Record the number of hits, and the highest and lowest scores in a table.

5. Reset the substitution matrix to BLOSUM62, run the same search with and without the low-complexity filter on. Compare the results.

6. Run the same search using FASTA (www.ebi.ac.uk/fasta33/). Choose pdb as database and leave other parameters as default. Compare the results with those from BLAST.

7. Run an exhaustive search using ScanPS (www.ebi.ac.uk/scanps/) using the default setting. This may take a few minutes. Compare results with BLAST and FASTA. Can you find 1pii in the result page?

8. Go back to the NCBI BLAST homepage, run PSI-BLAST of the above protein sequence by selecting the subprogram "PHI- and PSI-BLAST" (PHI-BLAST is pattern matching). Paste the sequence in the query box and choose pdb as database. Select "BLAST!" (for a regular query sequence, PSI-BLAST is automatically invoked). Click on "Format!" in the next page. The results will be

returned in a few minutes. Notice the symbols (New or green circle) in front of each hit.

9. By default, the hits with *E*-values below 0.005 should be selected for use in multiple sequence alignment and profile building. Click on the "Run PSI-Blast iteration 2" button. This refreshes the previous query page. Click the "Format!" button to retrieve results.

10. In the results page, notice the new hits generated from the second iteration. Perform another round of PSI-BLAST search. Record the number of hits and try to find 1pii in the result page.

11. Finally, do the same search using a hidden Markov model based approach. Access the HHPRED program (http://protevo.eb.tuebingen.mpg.de/toolkit/index.php?view=hhpred) and paste the same query sequence in the query window. Click the "Submit job" button.

12. The search may take a few minutes. When the results are returned, can you find 1pii in the search output?

13. Compare the final results with those from other methods. What is your conclusion regarding the ability of different programs to find remote homologs?

Pairwise Sequence Alignment

1. In the NCBI database, retrieve the protein sequences for mouse hypoxanthine phosphoribosyl transferase (HPRT) and the same enzyme from *E. coli* in FASTA format.

2. Perform a dot matrix alignment for the two sequences using Dothelix (www.genebee.msu.su/services/dhm/advanced.html). Paste both sequences in the query window and click on the "Run Query" button. The results are returned in the next page. Click on the diagonals on the graphic output to see the actual alignment.

3. Perform a local alignment of the two sequences using the dynamic programming based LALIGN program (www.ch.embnet.org/software/LALIGN_form.html). Make sure the two sequences are pasted separately in two different windows. Save the results in a scratch file.

4. Perform a global alignment using the same program by selecting the dial for "global." Save the results and compare with those from the local alignment.

5. Change the default gap penalty from "−14/−4" to "−4/−1". Run the local alignment and compare with previous results.

6. Do a pairwise alignment using BLAST (in the BLAST homepage, select the bl2seq program). Compare results with the previous methods.

7. Do another alignment with an exhaustive alignment program SSEARCH (http://pir.georgetown.edu/pirwww/search/pairwise.html). Compare the results.

8. Run a PRSS test to check whether there is any statistically significant similarity between the two sequences. Point your browser to the PRSS web page (http://fasta.bioch.virginia.edu/fasta/prss.htm). Paste the sequences in the

FASTA format in the two different windows. Use 1,000 shuffles and leave every-thing else as default. Click on the "Compare Sequence" button. Study the output and try to find the critical statistical parameters.

EXERCISE 3. MULTIPLE SEQUENCE ALIGNMENT AND MOTIF DETECTION

Multiple Sequence Alignment

In this exercise you will learn to use several multiple alignment programs and compare the robustness of each. The exercise is on the Rieske iron sulfur protein from a number of species. The critical functional site of this protein is a iron-sulfur center with a bound [2Fe-2S] cluster. The amino acid binding motifs are known to have consensi of C-X-H-X-G-C and C-X-X-H. Evaluate the following alignment programs for the ability to discover the conserved motifs as well as to correctly align the rest of the protein sequences. The result you obtain may aid in the understanding of the origin and evolution of the respiratory process.

1. Retrieve the following protein sequences in the FASTA format using NCBI Entrez: P08067, P20788, AAD55565, P08980, P23136, AAC84018, AAF02198.
2. Save all the sequences in a single file using a text editor such as *nedit*.
3. First, use a progressive alignment program Clustal to align the sequences. Submit the multisequence file to the ClustalW server (www.ch.embnet.org/software/ClustalW.html) for alignment using the default settings. Save the result in ClustalW format in a text file.
4. To evaluate quality of the alignment, visually inspect whether the key residues that form the iron–sulfur centers are correctly aligned and whether short gaps are scattered throughout the alignment. A more objective evaluation is to use a scoring approach. Go to a web server for alignment quality evaluation (http://igs-server.cnrs-mrs.fr/Tcoffee/tcoffee_cgi/index.cgi?action=Evaluate%20a%20Multiple%20Alignment::Regularstage1=1). Bookmark this site for later visits. Paste the alignment in Clustal format in the query box. Click on "Submit."
5. To view the result in the next page, click on the "score_html" link. The overall quality score is given in the top portion of the file. And the alignment quality is indicated by a color scheme. Record the quality score.
6. Align the same sequences using a profile-based algorithm MultAlin (http://prodes.toulouse.inra.fr/multalin/multalin.html) using the default parameters. Click the button "Start MultAlin." Save the results in the FASTA format first and convert it to the Clustal format for quality comparison.
7. Select the hyperlink for "Results as a FASTA file." Copy the alignment and open the link for Readseq (http://iubio.bio.indiana.edu/cgi-bin/readseq.cgi/). Paste

the FASTA alignment in the query box. Select the output sequence format as Clustal. Click "Submit."

8. Copy and paste the Clustal alignment in the quality evaluation server and record the score.

9. Submit the same unaligned sequences to a semi-exhaustive alignment program DCA (http://bibiserv.techfak.uni-bielefeld.de/dca/submission.html). Click on the "Submission" link on the right (in the green area) and paste the sequences in the query box. Select the output format as "FASTA." Click "Submit." Save the results for format conversion using Readseq. Do quality evaluation as above.

10. Do alignment with the same unaligned sequences using an iterative tree-based alignment program PRRN (http://prrn.ims.u-tokyo.ac.jp/). Select the output format as "FASTA." Select the "Copy and Paste" option and enter your e-mail address before submitting the alignment. Compare the quality of the alignment with other methods.

11. Finally, align the sequences using the T-Coffee server (www.ch.embnet.org/software/TCoffee.html). Score of alignment is directly presented in the HTML format. Record the score for comparison purposes.

12. Carefully compare the results from different methods. Can you identify the most reasonable alignment? Which method appears to be the best?

Hidden Markov Model Construction and Searches

This exercise is about building a hidden Markov model (HMM) profile and using it to search against a protein database.

1. Obtain the above sequence alignment from T-Coffee in the Clustal format.

2. Copy and paste the alignment file to the query box of the HMMbuild program for building an HMM profile (http://npsa-pbil.ibcp.fr/cgi-bin/npsa_automat.pl?page=/NPSA/npsa_hmmbuild.html). Click "Submit."

3. You may receive an error message "Your clustalw alignment doesn't start with the sentence : CLUSTAL W (...) multiple sequence alignment". Replace the header (beginning line) of the input file with "CLUSTAL W (...) multiple sequence alignment". Click "Submit."

4. When the HMM profile is constructed, click on the link "PROFILE" to examine the result.

5. Choose HMMSEARCH and UniProt-SwissProt database before clicking "Submit." This process takes a few minutes. Once the search is complete, the database hits that match with the HMM are returned along with multiple alignment files of the database sequences.

6. You have options to build a new HMM profile or to extract the full database sequences. Click "HMMBUILD" at the bottom of the page. The HMM profile building can be iterated as many times as desired similar to PSI-BLAST. For the interest of time, we stop here.

Protein Motif Searches

1. Align four sequences of different lengths in a file named "zf.fasta" (downloadable from www.cambridge.org/us/catalogue/catalogue.asp?isbn= 0521600820) using T-Coffee and DIALIGN2 (http://bibiserv.techfak.uni-bielefeld.de/dialign/submission.html).

2. Which of the programs is able to identify a zinc finger motif [C(X4)C(X12) H(X3)H]?

3. Verify the result with the INTERPRO motif search server (www.ebi.ac.uk/ interpro/) by cutting and pasting each of the unaligned sequences, one at a time, to the query box. Submit the query and inspect the search result.

4. Retrieve the protein sequence AAD42764 from Entrez. Do motif search of this sequence using a number of search programs listed. Pay attention to statistical scores such as *E*-values, if available, as well as the boundaries of the domains/motifs.

    a) BLOCKS Impala Searcher (http://blocks.fhcrc.org/blocks/impala.html).

    b) Reverse PSI-BLAST (http://blocks.fhcrc.org/blocks-bin/rpsblast.html).

    c) ProDom (http://prodes.toulouse.inra.fr/prodom/current/html/form.php).

    d) SMART (http://smart.embl-heidelberg.de/), select the "Normal" mode and paste the sequence in the query window.

    e) InterPro (www.ebi.ac.uk/interpro/), choose the link "Sequence Search" in the left grey area. Paste the sequence in the following query page.

    f) Scansite (http://scansite.mit.edu/), in the Motif Scan section, select "Scan a Protein by Input Sequence." Enter a name for the sequence and paste the sequence in the query window. Click "Submit Request." In the following page with the graphic representation of the domains and motifs, click "DOMAIN INFO" to get a more detailed description.

    g) eMatrix (http://fold.stanford.edu/ematrix/ematrix-search.html).

    h) Elm (http://elm.eu.org/). Use *Homo sapiens* as default organism.

Compile the results. What is your overall conclusion of the presence of domains and motifs in this protein?

DNA Motif Searches

DNA motifs are normally very subtle and can only be detected using "alignment-independent" methods such as expectation maximization (EM) and Gibbs motif sampling approaches.

1. Use the DNA sequence file "sd.fasta" (downloadable from www.cambridge. org/us/catalogue/catalogue.asp?isbn=0521600820) and generate alignment using the EM-based program Improbizer (www.cse.ucsc.edu/~kent/improbizer/ improbizer.html) with default parameters.

2. Do the same search using a Gibbs sampling-based algorithm AlignAce (http://atlas.med.harvard.edu/cgi-bin/alignace.pl) using default parameters.

3. Compare the results of best scored motifs from both methods. Are there overlaps?

4. Copy and paste the first motif derived from AlignAce to *nedit*. Remove the illegal characters (spaces and numbers).

5. Cut and paste the motif alignment into the WebLogo program (http://weblogo.berkeley.edu/logo.cgi). Click the "Create logo" button.

6. Can you identify the bacterial Shine-Dalgarno sequence motif from the sequences?

EXERCISE 4. PHYLOGENETIC ANALYSIS

In this exercise, you will reconstruct the phylogeny of HIV by building an unrooted tree for the HIV/SIV gp120 proteins using the distance neighbor joining, maximum parsimony, maximum likelihood, and Bayesian inference methods.

Constructing and Refining a Multiple Sequence Alignment

1. Open the file "gp120.fasta" (downloadable from www.cambridge.org/us/catalogue/catalogue.asp?isbn=0521600820) using nedit.

2. Go to the MultAlin alignment server (http://prodes.toulouse.inra.fr/multalin/multalin.html). Copy and paste the sequences to the query box and submit the sequences for alignment using the default parameters.

3. Visually inspect the alignment result and pay attention to the matching of cysteine residues, which roughly indicate the correctness of the alignment.

4. View the result in FASTA format by clicking the hyperlink "Results as a fasta file." Save the FASTA alignment in a new text file using nedit.

5. Refine the alignment using the Rascal program that realigns certain portion of the file. Open the Rascal web page (http://igbmc.u-strasbg.fr/PipeAlign/Rascal/rascal.html) and upload the previous alignment file in FASTA format.

6. After a minute or so, the realignment is displayed in the next window. Examine the new alignment. If you accept the refinement, save the alignment in FASTA format.

7. Next, use the Gblocks program to further eliminate poorly aligned positions and divergent regions to make the alignment more suitable for phylogenetic analysis. Go to the Gblocks web page (http://molevol.ibmb.csic.es/Gblocks_server/index.html) and upload the above refined alignment into the server.

8. By default, the program should be set for protein sequences. Check the three boxes that allow less stringent criteria for truncation. These three boxes are "Allow smaller final blocks," "Allow gap positions within the final blocks," and "Allow less strict flanking positions."

9. Click the "Get Blocks" button. After the program analyzes the alignment quality, conserved regions are indicated with blue bars.

10. If you accept the selection, click on the "Resulting alignment" hyperlink at the bottom of the page to get selected sequence blocks in the FASTA format.
11. Copy the sequence alignment and change its format using the Readseq program (http://iubio.bio.indiana.edu/cgi-bin/readseq.cgi) by pasting the sequences to the query box of the program. Select "Phylip|Phylip4" as output format. Click submit. Save the final alignment in a scratch file.

Constructing a Distance-Based Phylogenetic Tree

1. Go to the WebPhylip web page (http://biocore.unl.edu/WEBPHYLIP/).
2. Select "Distance Computation" in the left window. In the subsequent window, select "PAM Matrix" under "Protein Sequences."
3. Copy and paste the above Phylip alignment in the query box on the lower right portion of the window. Leave everything else as default. Click the "Submit" button.
4. Once the distance matrix is computed, a split window on the upper right is refreshed to give the distance matrix of the dataset.
5. To construct a tree with the matrix, select "Run" the distance methods in the lower left window.
6. In the next window, select "Run" under "Neighbor-joining and UPGMA methods."
7. This refreshes the lower right window. By default, the "Neighbor-joining tree" is selected. Select "Yes" for the question "Use previous data set?" (highlighted in red). Click the "Submit" button.
8. A crude diagram of the phylogenetic tree is displayed in the upper right window.
9. To draw a better tree, select the "Draw trees" option (in green) in the left window.
10. In the next window, select "Run" under "Draw Cladograms and phenograms."
11. In the refreshed lower right window, make sure "Yes" is selected for the question "Use tree file from last stage?" Click the "Submit" button.
12. A postscript file is returned. Save it to hard drive as "filename.ps." Convert the postscript file to the PDF format using the command "ps2pdf filename.ps". Open the PDF file using the xpdf filename.pdf command.
13. Using the same alignment file, do a phylogenetic tree using the Fitch–Margoliash method and compare the final result with the neighbor-joining tree.

Constructing a Maximum Parsimony Tree

1. In the same WebPhylip web page, click "Back to menu."
2. Select "Protein" in the "Run phylogeny methods for" section in the left window.
3. Choose "Run" "Parsimony" in the next window.
4. In the refreshed window on the right, repaste the sequence alignment in the query window and choose "Yes" for "Randomize input order of sequences?"
5. Leave everything else unchanged and click the Submit button. This initiates the calculation for parsimony tree construction, which will take a few minutes.
6. Two equally most parsimonious trees are shown in the upper right window.

7. Choose "Do consensus" on the left and "Run" "Consensus tree" in the next window.

8. In the refreshed lower right window, make sure "Yes" is selected for the question "Use tree file from last stage?" Click the "Submit" button.

9. Choose "Draw trees" and then "Run" for "Draw Cladogram" on the left.

10. Make sure "Yes" for "Use tree file from last stage?" is selected and leave everything else as default.

11. A postscript image is returned for viewing.

Constructing a Quartet Puzzling Tree

1. Access the Puzzle program web page (http://bioweb.pasteur.fr/seqanal/interfaces/Puzzle.html).

2. Copy and paste (or upload) the gp120 Phylip alignment into the query window.

3. Select "protein" for the sequence type.

4. Scroll down the window to the Protein Options section. Select "JTT model" for amino acid substitutions.

5. Leave other parameters as default. Provide your e-mail address before submitting the query.

6. The URL for the results will be sent to you by e-mail (check your e-mail in about 10 or 20 minutes).

7. Get your result by following the URL in e-mail. Select the "drawgram" option and click "Run the selected program on results.tree" button.

8. In the following Phylip page, choose "Phenogram" as "Tree Style" in the next page, and click "Run drawgram."

9. The tree is returned in a postscript file. Open the image file by clicking on the hyperlink plotfile.ps.

Constructing a Maximum Likelihood Tree Using Genetic Algorithm

1. Go to the PHYML web page (http://atgc.lirmm.fr/phyml/).

2. Select "File" next to the query window. Click "Browse" to select the sequence alignment file for uploading.

3. Select "Amino-Acids" for "Data Type."

4. Leave everything else as default and provide your name, country, and e-mail address before submitting the query.

5. The treeing result will be sent to you by e-mail (the process takes about 10 or 20 minutes).

6. One of your e-mail attachment files should contain the final tree in the Newick format which can be displayed using the Drawtree program (www.phylodiversity.net/~rick/drawtree/).

7. Copy and paste the Newick file to the query window. Leave everything else as default. Click the "Draw Tree" button.

8. The graphical tree is returned in the PDF format.

Constructing a Phylogenetic Tree Using Bayesian Inference

1. Access the BAMBE program (http://bioweb.pasteur.fr/seqanal/interfaces/bambe.html).
2. Copy and paste (or upload) the gp120 alignment in Phylip format into the query window. Leave all parameters as default (6,000 cycles with 1,000 cycles as burn-in). Provide your e-mail address before submitting the query.
3. The URL of the result is returned via e-mail (in 10 or 20 minutes).
4. Verify that the likelihood value of the final tree has reached near convergence by checking the end of the "results.par" file. (In this file, the first column represents the number of cycles and second column the lnL values of the intermediate trees.)
5. If the log likelihood of the trees is indeed stabilized, go back to the previous page and draw the final consensus tree by selecting the "drawgram" option and clicking the "Run the selected program on results.tre" button.
6. In the following page, choose "Phenogram" as Tree Style in the next page, and click "Run drawgram."
7. The tree is returned in a postscript file. Open the image file by clicking on the hyperlink plotfile.ps.
8. Compare the phylogenetic results from different methods and draw a consensus of the results and consider the evolutionary implications. What is the phylogenetic relationship of HIV-1, HIV-2, and SIV? What do the trees tell you about the origin of HIV and how many events of cross-species transmissions? (*Note:* SIVCZ is from chimpanzee and SIVM is from macaque/mangabeys.)

EXERCISE 5. PROTEIN STRUCTURE PREDICTION

Protein Secondary Structure Prediction

In this exercise, use several web programs to predict the secondary structure of a globular protein and a membrane protein, both of which have known crystal structures. The predictions are used to compare with experimentally determined structures so you can get an idea of the accuracy of the prediction programs.

1. Retrieve the protein sequence YjeE (accession number ZP_00321401) in the FASTA format from NCBI Entrez. Download the sequence into a text file.
2. Predict its secondary structure using the GOR method. Go to the web page http://fasta.bioch.virginia.edu/fasta_www/garnier.htm and paste the sequence into the query box. Click the "Predict" button.
3. Save the result in a text file.
4. The crystal structure of this protein has a PDB code 1htw (as a homotrimeric complex). The secondary structure of each monomer can be retrieved at the PDBsum database (www.ebi.ac.uk/thornton-srv/databases/pdbsum/).
5. Enter the PDB code 1htw in the query box. Click "Find."

6. In the menu bar on the top right, select the "Protein" menu. This brings up the protein secondary structure as well as the CATH classification information. If the background of the secondary structure window is black, select and download the PDF file format next to the window. Open it using the xpdf command.

7. Compare the real secondary structure with the GOR prediction. What conclusion can you draw from this?

8. Now do another prediction using the neural-network–based Predator program (http://bioweb.pasteur.fr/seqanal/interfaces/predator-simple.html). Enter your e-mail address before submitting the query.

9. The result is returned in the "predator.out" file. Compare the result with the GOR prediction and the known secondary structure. What conclusion can you draw from this?

10. Do the structure prediction again using the BRNN-based Porter program (http://distill.ucd.ie/porter//).

11. Paste sequence in the query window and enter the e-mail address. Click the "Predict" button.

12. The result is e-mailed to you in a few minutes.

13. Compare the result with the previous predictions and the known secondary structure. What can you learn from this?

14. Retrieve the human aquaporin sequence (AAH22486) from NCBI.

15. Predict the transmembrane structure using the Phobius program (http://phobius.cgb.ki.se/). Record the result.

16. The PDB code of this protein structure is 1h6i, which you can use to retrieve the experimentally determined secondary structure from PDBsum.

17. Compare the prediction result with the known structure. Do the total number of transmembrane helices and their boundaries match in all cases?

Protein Homology Modeling

In the following exercise, construct a homology model for a small protein from a cyanobacterium, *Anabaena variabilis*. The protein, which is called HetY, may be involved in nitrogen fixation but has no well-defined function. The objective of this exercise is to help provide some functional clues of the protein. The protein model is displayed and rendered using a shareware program Chimera (downloadable from www.cgl.ucsf.edu/chimera/).

1. Retrieve the protein sequence (ZP_00161818) in the FASTA format from NCBI Entrez. Save the sequence in a text file.

2. To search for structure templates, do a BLASTP search (www.ncbi.nlm.nih.gov/BLAST/) against the "pdb" database.

3. Examine the BLAST result. Select the top hit if the *E*-value of the alignment is significant enough. This sequence should correspond to the structure 1htw and can serve can as the structure template.

4. Perform more refined alignment between HetY and the template. Click on the hyperlink in the header of the template sequence to retrieve the full-length sequence in the FASTA format. Save it in a text file.

5. Align HetY and the template sequence (1htw) using T-Coffee (www.ch.embnet. org/software/TCoffee.html).

6. Convert the alignment into the FASTA format using Readseq (http://iubio. bio.indiana.edu/cgi-bin/readseq.cgi/). Save it into a text file.

7. Refine the alignment using the Rascal server (http://igbmc.u-strasbg.fr/ PipeAlign/Rascal/rascal.html) by uploading the FASTA alignment file.

8. Download the refined alignment in the FASTA format.

9. Perform comprehensive homology modeling using the GetAtoms server (www.softberry.com/berry.phtml?topic=getatoms&group=programs& subgroup=propt).

10. Paste the alignment in the query window. Select "FASTA" for format. Enter "1htw" for PDB identifier and "A" for chain identifier. Make sure the input order is the target sequence before the template sequence.

11. Select "Add H-atoms at the end of optimization" and "Process loops and insertions." Click the "PROCESS" button.

12. The coordinates of the model are returned in the next page. Save the coordinate file using the "Save As" option in File menu.

13. Open the coordinate file using nedit. Delete the dashes, trademark, and other HTML-related characters at the end of the file.

14. The raw model can be refined by energy minimization. Upload the edited coordinate file to the Hmod3DMM program (www.softberry.com/berry. phtml?topic=molmech&group=programs&subgroup=propt). Press "START."

15. A refined model is returned in a few minutes. Save the energy-minimized coordinates to the hard drive.

16. Check the quality of final structure using Verify3D (http://nihserver.mbi.ucla. edu/Verify_3D/). Upload the structure to the Verify3D server. Click "Send File."

17. In the resulting quality plot, scores above 0 indicate favorable conformations. Check to see whether if any residue scores are below 0. If the scores are significantly below 0, reminimization of the model is required.

18. Assuming the modeled protein is final, the next step is to add cofactor to the protein.

19. Assuming that the target protein has similar biochemical functions as the template protein (ATPase), important ligands from the template file that can be transferred to the target protein.

20. Download the template structure (1htw) from the PDB website (www.rcsb. org/pdb/). Click the "Download/Display File" link in the menu on the left. Download the noncompressed PDB file.

21. To extract the cofactor, open the 1HTW.pdb with nedit. Go the HETATM section near the bottom of the file. Find the coordinates for ADP 560. Copy the coordinates (make sure you include all the atoms for this cofactor).

22. Open the HetY model using nedit and paste the HETATM coordinates immediately after the ATOM section (near the end of the file). Delete the dashes, trademark, and other HTML-related characters.

23. Before using Chimera to visualize the model, you need add an alias to your .cshrc file. Open .cshrc with nedit and add a line at the end of the file: alias chimera /chem/ chimera/chimera-1.2065/bin/chimera. Save the file and quit nedit. In the UNIX window, type source .cshrc.

24. Invoke the Chimera program by typing chimera.

25. In the File menu, select "Open." For File type, select "all (ask type)." Select your model (e.g., hetY.pdb) to be opened.

26. The structure file is initially uncolored. Color the atoms by going to the menu Actions → Color → by element.

27. The structure can be rotated using the left mouse button, moved using the middle mouse button, and zoomed using right mouse button.

28. You can display a smooth solid surface showing electrostatic distribution as well as bound cofactor ADP. Go to Actions → Surface → show.

29. To select the cofactor, go to Select → Chain → het. To color it, select Actions → Color → cyan. To render it in spheres, select Actions → Atoms/ bonds → sphere.

30. To finalize selection, go to Select → Clear selection. Rotate the model around to study the protein-ligand binding.

31. To reset the rendering, go to Actions → Surface → hide; Actions → Atoms/ bonds → wire.

32. Now draw the secondary structure of the model. Select Actions → Ribbon → show; Actions → Ribbon → round. To color it, go to Tools → Graphics → Color Secondary Structure. In the pop-up window, click OK for default setting.

33. To hide the wire frames, select Actions → Atoms/bonds → hide.

34. To show the cofactor, Select → Chain → het. Then Actions → Atoms/bonds → ball & stick.

35. The publication quality image can be saved for printing purposes. To make it more printer-friendly, the background can be changed to white. Select Actions → Color → background; Actions → Color → white.

36. To save the image, go to File → Save Image. In the pop-up window, click "Save As."

37. As the program is writing an image, the model may dance around on the screen for a while. When it stabilizes, a new window pops up to prompt you for filename and file type (default format is .png). Give it a name and click "Save."

38. Quit the Chimera program. To view the image in a UNIX window, type imgview filename.png.

39. The image file can be e-mailed to yourself as attachment for printing on your own printer.

40. When you are done, close all the programs and log out.

EXERCISE 6. GENE AND PROMOTER PREDICTION AND GENE ANNOTATION

Gene Prediction

In this exercise, you do gene predictions using a bacterial sequence from *Heliobacillus mobilis* (IIm_dna.fasta) (downloadable from www.cambridge.org/us/catalogue/catalogue.asp?isbn=0521600820). This provides the foundation for operon predictions and promoter predictions. One way to verify the gene prediction result is to check the presence of Shine–Dalgarno sequence in front of each gene which is a purine-rich region with a consensus AGGAGG and is located within 20 bp upstream of the start codon.

1. Point your browser to the GeneMark web page (frame-by-frame module) (http://opal.biology.gatech.edu/GeneMark/fbf.cgi).
2. Upload the Hm_dna.fasta sequence file and choose *Bacillus subtilis* as "Species" (the closest organism).
3. Leave other options as default and start the GeneMark program.
4. Save the prediction result using nedit.
5. To confirm the prediction result, the sequence needs to have numbering.
6. Convert the original sequence file into a GenBank format using the ReadSeq server (http://iubio.bio.indiana.edu/cgi-bin/readseq.cgi). Save the results in a new file.
7. Based on the prediction by GeneMark, find the gene start sites in the sequence file. Can you find the Shine–Dalgarno sequence in each predicted frame?
8. Do another gene prediction using the Glimmer program (http://compbio.ornl.gov/GP3/pro.html). Select "Glimmer Genes." Use *B. subtilis* as the closest organism. Upload the sequence file and perform the Glimmer search.
9. When the data processing is complete, click the "Get Summary" button. In the following page, select Retrieve: → TextTable.
10. Compare the prediction result with that from GeneMark. Pay attention to the boundaries of open reading frames. For varied gene predictions, verify the presence of Shine–Dalgarno sequence in each case. Have you noticed problems of overpredictions or missed predictions with Glimmer? Can you explain why?

Operon Prediction

In this exercise, you predict operons of the above heliobacterial sequence using the 40-bp rule: if intergenic distance of a pair of unidirectionally transcribed genes is smaller than 40 bp, the gene pair can be called an *operon*. This rule was used widely before the development of the scoring method of Wang et al., which is a little too complicated for this lab.

1. Using the gene prediction result from GeneMark, calculate the intergenic distance of each pair of genes.
2. How many operons can you derive based on the 40-bp rule?

Promoter Prediction

In this exercise, perform ab initio promoter predictions based on the operon prediction from the previous exercise. Algorithms for promoter prediction are often written to predict the transcription start sites (TSS) instead. The −10 and −35 boxes can be subsequently deduced from the upstream region of this site.

1. Using the operon prediction result that you believe is correct, copy ~150-bp upstream sequence from the first operon start sites and save the sequence in a new file.
2. Convert the sequence to the FASTA format, using the Readseq program.
3. Do a promoter prediction using BPROM (www.softberry.com/berry.phtml?topic=bprom&group=programs&subgroup=gfindb). Paste the sequence in the query window and press the PROCESS button. Record the result. (*Note:* TSS predicted by this program is labeled as LDF.)
4. Do another promoter prediction using the SAK program (http://nostradamus.cs.rhul.ac.uk/%7Eleo/sak_demo/), which calculates the likelihood scores of sites being the TSS.
5. In the output page, find the position that has the highest likelihood score (listed in the second column), which is the TSS prediction.
6. Compare the results from the two sets of predictions. Are they consistent?

Gene Annotation

A major issue in genomics is gene annotation. Although a large number of genes and proteins can be assigned functions simply by sequence similarity, about 40% to 50% of the genes from newly sequenced genomes have no known functions and can only be annotated as encoding "hypothetical proteins." In this exercise, you are given one of such "difficult" protein sequences for functional annotation. This protein is YciE from *E. coli,* which has been implicated in stress response. However, its actual biochemical function has remained elusive. In this exercise, use advanced bioinformatics tools to derive functional information of the protein sequence.

1. Retrieve the protein sequence of YciE from NCBI Entrez (www.ncbi.nlm.nih.gov/Entrez/, accession P21363) in the FASTA format and study the existing annotation in the GenBank file.
2. Do domain and motif searches of this sequence using RPS-BLAST (http://blocks.fhcrc.org/blocks/rpsblast.html), SMART (http://smart.embl-heidelberg.de/) (Use the Normal mode. Check all four boxes below the query box for Outlier homologs, PFAM domains, signal peptides, and internal repeats before starting the search) and InterPro (www.ebi.ac.uk/interpro/). Compile the results. What is the consensus domains and motifs in this protein?
3. Do a functional prediction based on protein interactions by using the STRING server (http://string.embl.de/). Paste your sequence into the query box and click the "GO" button.

4. In the result page for predicted protein–protein interactions, check to see what are the predicted interacting proteins. What is the evidence for the interaction prediction?

5. Do a protein threading analysis using the HHPred server (http://protevo.eb. tuebingen.mpg.de/toolkit/index.php?view=hhpred). This program searches protein folds by combining HMMs and secondary structure prediction information.

6. Paste the sequence in the query box and submit the job using all the default settings. The query is processed. The result is returned in a minute.

7. Pick the top hit showing the most significant *E*-value and study the annotation of the structure match and visually inspect the alignment result.

8. Get more detailed information about the best matching protein structure by clicking the link with the PDB code. This brings up the PDB beta page with detailed annotation information especially the bibliographic information of the structure.

9. You can retrieve the original publication on the structure by selecting the "PubMed" link. Read the "Introduction" section of this paper. Can you get any functional description about the protein in that paper in relation to the stress response?

10. In the PDB site, retrieve the sequence of this structure (only one subunit) by selecting the menu "Summarize" → "Sequence Details." In the next page, scroll down the window to click the "Download" button for chain A in FASTA format.

11. Open the sequence in the FASTA format and save it to the hard disk.

12. Do a refined pairwise alignment of this sequence with YciE using the AliBee server (www.genebee.msu.su/services/malign_reduced.html). What is the percent identity for the best set of alignment?

13. Do a PRSS test on the two original sequences by copying and pasting the two unaligned sequences in two individual query boxes of the PRSS server (http:// fasta.bioch.virginia.edu/fasta/prss.htm). Select 1,000 shuffles with "window" setting. Are the two protein sequences significantly related? Can you designate them as homologous sequences?

14. Compile the results from each of the predictions. What is your overall conclusion of the function of this protein?

Glossary

Ab initio prediction: computational prediction based on first principles or using the most elementary information.

Accession number: unique number given to an entry in a biological database, which serves as a permanent identifier for the entry.

Agglomerative clustering: microarray data clustering method that begins by first clustering the two most similar data points and subsequently repeating the process to merge groups of data successively according to similarity until all groups of data are merged. This is in principle similar to the UPGMA phylogenetic approach.

Alternative splicing: mRNA splicing event that joins different exons from a single gene to form variable transcripts. This is one of the mechanisms of generating a large diversity of gene products in eukaryotes.

Bayesian analysis: statistical method using the Bayes theorem to describe conditional probabilities of an event. It makes inferences based on initial expectation and existing observations. Mathematically, it calculates the posterior probability (revised expectation) of two joint events (A and B) as the product of the prior probability of A event given the condition B (initial expectation) and conditional probability of B (observation) divided by the total probability of event A with and without the condition B. The method has wide applications in bioinformatics from sequence alignment and phylogenetic tree construction to microarray data analysis.

Bioinformatics: discipline of storing and analyzing biological data using computational techniques. More specifically, it is the analysis of the sequence, structure, and function of the biological macromolecules – DNA, RNA, and proteins – with the aid of computational tools that include computer hardware, software, and the Internet.

Bit score: statistical indicator in database sequence similarity searches. It is a normalized pairwise alignment score that is independent of database size. It is suitable for comparing search results from different databases. The higher the bit score, the better the match is.

BLAST (Basic Local Alignment Search Tool): commonly used sequence database search program based on sequence similarity. It has many variants, such as BLASTN, BLASTP, and BLASTX, for dealing with different types of sequences. The major feature of the algorithm is its search speed, because it is designed to rapidly detect a region of local sequence similarity in a database sequence and use it as anchor to extend to a fuller pairwise alignment.

BLOSUM matrix: amino acid substitution matrix constructed from observed frequencies of substitution in blocks of ungapped alignment of closely related protein sequences. The numbering of the BLOSUM matrices corresponds to percent identity of the protein sequences in the blocks.

Boolean expression: database retrieval method of expressing a query by connecting query words using the logical operators AND, OR, and NOT between the words.

Bootstrap analysis: statistical method for assessing the consistency of phylogenetic tree topologies based on the generation of a large number of replicates with slight modifications in input data. The trees constructed from the datasets with random modifications give a distribution of tree topologies that allow statistical assessment of each individual clade on the trees.

CASP (Critical Assessment in Structure Prediction): biannual international contest to assess protein structure prediction software programs using blind testing. This experiment attempts to serve as a rigorous test bed by providing contestants with newly solved but unpublished proteins structures to test the efficacy of new prediction algorithms. By avoiding the use of known protein structures as benchmarks, the contest is able to provide unbiased assessment of the performance of prediction programs.

Chromosome walking: experimental technique that identifies overlapping genomic DNA clones by labeling the ends of the clones with oligonucleotide probes. Through a multistep process, it is able cover an entire chromosome.

Clade: group of taxa on a phylogenetic tree that are descended from a single common ancestor. They are also referred to as being *monophyletic.*

COG (Cluster of Orthologous Groups): protein family database based on phylogenetic classification. It is constructed by comparing protein sequences encoded by completely sequenced genomes and identifying orthologous proteins shared by three or more genomes to be clustered together as orthologous groups.

Comparative genomics: subarea of genomics that focuses on comparison of whole genomes from different organisms. It includes comparison of gene number, gene location, and gene content from these genomes. The comparison provides insight into the mechanism of genome evolution and gene transfer among genomes.

Contig: contiguous stretch of DNA sequence assembled from individual overlapping DNA segments.

Cytological maps: maps showing banding patterns on a stained chromosome and observed under a microscope. The bands are often associated with the locations of genetic markers. The distance between any two bands is expressed in relative units (Dustin units).

Database: computerized archive used for storage and organization of data in such a way that information can be retrieved easily via a variety of search criteria.

Divisive clustering: microarray data clustering method that works by lumping all data points in a single cluster and successively dividing the data into smaller groups according to similarity until all the hierarchical levels are resolved. This is in principle similar to the neighbor joining phylogenetic approach.

DNA microarray: technology for high throughput gene expression profiling. Oligonucleotides representing every gene in a genome can be immobilized on tiny spots on the surface of a glass chip, which can be used for hybridization with a labeled cDNA population. By analyzing the hybridization result, levels of gene expression at the whole genome level can be revealed.

Domain: evolutionarily conserved sequence region that corresponds to a structurally independent three-dimensional unit associated with a particular functional role. It is usually much larger than a motif.

Dot plot: visual technique to perform a pairwise sequence alignment by using a two-dimensional matrix with each sequence on its dimensions and applying dots for matching residues. A contiguous line of dots in a diagonal indicates a local alignment.

Dynamic programming: algorithm to find an optimal solution by decomposing a problem into many smaller, sequentially dependent subproblems and solving them individually while storing the intermediate solutions in a table so that the highest scored solution can be chosen. To perform a pairwise sequence alignment, the method builds a two-dimensional matrix with each sequence on its dimensions and applies a scoring scheme to fill the matrix and finds the maximum scored region representing the best alignment by backtracking through the matrix.

EM (expectation maximization):   local multiple sequence alignment method for identification of shared motifs among input sequences. The motifs are discovered through random alignment of the sequences to produce a trial PSSM and successively refinement of the PSSM. A motif can be recruited after this process is repeated many times until there is no further improvement on the matrix.

EST (expressed sequence tags):   short sequences obtained from cDNA clones serving as short identifiers of full length genes. ESTs are typically in the range of 200 to 400 nucleotides in length and are generated using a high throughput approach. EST profiling can be used as a snapshot of gene expression in a particular tissue at a particular stage.

_E_-value (expectation value):   statistical significance measure of database sequence matches. It indicates the probability of a database match expected as a result of random chance. The _E_-value depends on the database size. The lower the _E_-value, the more significant the match is.

Exon shuffling: mRNA splicing event that joins exons from different genes to generate more transcripts. This is one of the mechanisms of generating a large diversity of gene products in eukaryotes.

FASTA: database sequence search program that performs the pairwise alignment by employing a heuristic method. It works by rapidly scanning a sequence to identify identical words of a certain size as the query sequence and subsequently searching for regions that contain a high density of words with high scores. The high-scoring regions are subsequently linked to form a longer gapped alignment, which is later refined using dynamic programming.

False negative: true match that fails to be recognized by an algorithm.

False positive: false match that is incorrectly identified as a true match by an algorithm.

Fingerprint: group of short, ungapped sequence segments associated with diagnostic features of a protein family. A fingerprint is a smaller unit than a motif.

Flat file: database file format that is a long text file containing database entries separated by a delimiter, a special character such as a vertical bar (|). Each field within an entry is separated by tabs.

Fold: three-dimensional topology of a protein structure described by the arrangement and connection of secondary structure elements in three dimensional space.

Fold recognition: method of protein structure prediction for the most likely protein structural fold based on structure profile similarity and sequence profile similarity. The structure profiles incorporate information of secondary structures and solvation energies. The term has been used interchangeably with *threading*.

Functional genomics: study of gene functions at the whole-genome level using high throughput approaches. This study is also termed *transcriptome analysis*, which refers to the analysis of the full set of RNA molecules produced by a cell under a given condition.

Gap penalty: part of a sequence alignment scoring system in which a penalizing score is used for producing gaps in alignment to account for the relative rarity of insertions and deletions in sequence evolution.

Gene annotation: process to identify gene locations in a newly sequenced genome and to assign functions to identified genes and gene products.

Gene ontology: annotation system for gene products using a set of structured, controlled vocabulary to indicate the biological process, molecular function, and cellular localization of a particular gene product.

Genetic algorithm: computational optimization strategy that performs iterative and randomized selection to achieve an optimal solution. It uses biological terminology as metaphor because it involves iterative "crossing," which is a mix and match of mathematical routines to generate new "offspring" routines. The offsprings are allowed to randomly "mutate." A scoring system is applied to select an offspring among many with a higher score (better "fitness") than the "parents." This offspring is allowed to "propagate" further. The iterations continue until the "fittest" offspring or an optimal solution is selected.

Genetic map: map of relative positions of genes in a genome, based on the frequency of recombinations of genetic markers through genetic crossing. The distance between two genetic markers is measured in relative units (Morgans).

Genome: complete DNA sequence of an organism that includes all the genetic information.

Genomics: study of genomes characterized by simultaneous analysis of all the genes in a genome. The topics of genomics range from genome mapping, sequencing, and functional genomic analysis to comparative genomic analysis.

Gibbs sampling: local multiple sequence alignment method for identification of shared motifs among input sequences. PSSMs are constructed iteratively from N–1 sequences and are refined with the left-out sequence. An optimal motif can be recruited after this process is repeated many times until there is no further improvement on the matrix.

Global alignment: sequence alignment strategy that matches up two or more sequences over their entire lengths. It is suitable for aligning sequences that are of similar length and suspected to have full-length similarity. If used for more divergent sequences, this strategy may miss local similar regions.

Heuristics: computational strategy to find a near-optimal solution by using rules of thumb. Essentially, this strategy takes shortcuts by reducing the search space according to certain criteria. The results are not guaranteed to be optimal, but this method is often used to save computational time.

Hidden Markov model: statistical model composed of a number of interconnected Markov chains with the capability to generate the probability value of an event by taking into account the influence from hidden variables. Mathematically, it calculates probability values of connected states among the Markov chains to find an optimal path within the network of states. It requires training to obtain the probability values of state transitions. When using a hidden Markov model to represent a multiple sequence alignment, a sequence can be generated through the model by incorporating probability values of match, insertion, and deletion states.

Hierarchical clustering: technique to classify genes from a gene expression profile. The classification is based on a gene distance matrix and groups genes of similar expression patterns to produce a dendrogram.

Hierarchical sequencing: sequencing approach that divides the genomic DNA into large fragments, each of which is cloned into a bacterial artificial chromosome (BAC). The relative order of the BAC clones are first mapped onto a chromosome. Each of the overlapping BAC clones is subsequently sequenced using the shotgun approach before they are assembled to form a contiguous genomic sequence.

Homologs: biological features that are similar owing to the fact that they are derived from a common ancestry.

Homology: biological similarity that is attributed to a common evolutionary origin.

Homology modeling: method for predicting the three-dimensional structure of a protein based on homology by assigning the structure of an unknown protein using an existing homologous protein structure as a template.

Homoplasy: observed sequence similarity that is a result of convergence or parallel evolution, but not direct evolution. This effect, which includes multiple substitutions at individual positions, often obscures the estimation of the true evolutionary distances between sequences and has to be corrected before phylogenetic tree construction.

HSP (high scoring segment pair): intermediate gapless pairwise alignment in BLAST database sequence alignment.

Identity: quantitative measure of the proportion of exact matches in a pairwise or multiple sequence alignment.

Jackknife: tree evaluation method to assess the consistency of phylogenetic tree topologies by constructing new trees using only half of the sites in an original dataset. The method is similar to bootstrapping, but its advantages are that sites are not duplicated relative to the original dataset and that computing time is much reduced because of shorter sequences.

Jukes–Cantor model: substitution model for correcting multiple substitutions in molecular sequences. For DNA sequences, the model assumes that all nucleotides are substituted with an equal rate. It is also called the *one-parameter model*.

k-Means clustering: classification technique that identifies the association of genes in an expression profile. The classification first assigns data points randomly among a number of predefined clusters and then moves the data points among the clusters while calculating the distances of the data points to the center of the cluster (centroid). The process is iterated many times until a best fit of all data points within the clusters is reached.

Kimura model: substitution model for correcting multiple substitutions in molecular sequences. For DNA sequences, the model assumes that there are two different substitution rates, one for transition and the other for transversion. It is also called the *two-parameter model*.

Lateral gene transfer: process of gene acquisition through exchange between species in a way that is incongruent with the commonly accepted vertical evolutionary scenario. It is also called *horizontal gene transfer*.

Linear discriminant analysis: statistical method that separates true signals from background noise by projecting data points in a two-dimensional graph and drawing a diagonal line that best separates signals from nonsignals based on the patterns learned from training datasets.

Local alignment: pairwise sequence alignment strategy that emphasizes matching the most similar segments between the two sequences. It can be used for aligning sequences of significant divergence and unequal lengths.

Log-odds score: score that is derived from the logarithmic conversion of an observed frequency value of an event divided by the frequency expected by random chance so that the score represents the relative likelihood of the event. For example, a positive log-odds score indicates an event happens more likely than by random chance.

Low-complexity region: sequence region that contains a high proportion of redundant residues resulting in a biased composition that significantly differs from the general sequence composition. This region often leads to spurious matches in sequence alignment and has to be masked before being used in alignment or database searching.

Machine learning: computational approach to detect patterns by progressive optimization of the internal parameters of an algorithm.

Markov process: linear chain of individual events linked together by probability values so that the occurrence of one event (or state) depends on the occurrence of the previous event(s) (or states). It can be applied to biological sequences in which each character in a sequence can be considered a state in a Markov process.

Maximum likelihood: statistical method of choosing hypotheses based on the highest likelihood values. It is most useful in molecular phylogenetic tree construction.

Maximum parsimony: principle of choosing a solution with fewest explanations or logic steps. In phylogenetic analysis, the maximum parsimony method infers a tree with the fewest mutational steps.

Minimum evolution: phylogenetic tree construction method that chooses a tree with minimum overall branch lengths. In principle, it is similar to maximum parsimony, but differs in that the minimum evolution method is distance based, whereas maximum parsimony is character based.

Molecular clock: assumption that molecular sequences evolve at a constant rate. This implies that the evolutionary time of a lineage can be estimated from its branch length in a phylogenetic tree.

Molecular phylogenetics: study of evolutionary processes and phylogenies using DNA and protein sequence data.

Monophyletic: refers to taxa on a phylogenetic tree that are descended from a single common ancestor.

Monte Carlo procedure: computer algorithm that produces random numbers based on a particular statistical distribution.

Motif: short, conserved sequence associated with a distinct function.

Needleman–Wunsch algorithm: a global pairwise alignment algorithm that applies dynamic programming in a sequence alignment.

Negative selection: evolutionary process that does not favor amino acid replacement in a protein sequence. This happens when a protein function has been stabilized. The implied function constraint deems mutations to be deleterious to the protein function. This can be detected when the synonymous substitution rate is higher than the nonsynonymous substitution rate in a protein encoding region.

Neighbor joining: phylogenetic tree-building method that constructs a tree based on phylogenetic distances between taxa. It first corrects unequal evolutionary rates of raw distances and uses the corrected distances to build a matrix. Tree construction begins from a completely unresolved tree and then decomposes the tree in a stepwise fashion until all taxa are resolved.

Neural network: machine-learning algorithm for pattern recognition. It is composed of input, hidden, and output layers. Units of information in each layer are called nodes. The nodes of different layers are interconnected to form a network analogous to a biological nervous system. Between the nodes are mathematical weight parameters that can be trained with known patterns so they can be used for later predictions. After training, the network is able to recognize correlation between an input and output.

Newick format: text representation of tree topology that uses a set of nested parentheses in which each internal node is represented by a pair of parentheses that enclose all members of a monophyletic group separated by a comma. If a tree is scaled, branch lengths are placed immediately after the name of the taxon separated by a colon.

Nonsynonymous substitutions: nucleotide changes in a protein coding region that results in alterations in the encoded amino acid sequences.

Object-oriented database: database that stores data as units that combine data and references to other records. The units are referred to as *objects*. Searching a such database involves navigating through the objects via pointers and links. The database structure is a more flexible than that of relational database but lacks the rigorous mathematical foundation of the relational databases.

OMIM (Online Mendelian Inheritance in Man)**:** database of human genetic disease, containing textual descriptions of the disorders and information about the genes associated with genetic disorders.

Orthologs: homologous sequences from different organisms or genomes derived from speciation events rather than gene duplication events.

Outgroup: taxon or a group of taxa in a phylogenetic tree known to have diverged earlier than the rest of the taxa in the tree and used to determine the position of the root.

Overfitting: phenomenon by which a machine learning algorithm overrepresents certain patterns while ignoring other possibilities. This phenomenon is a result of insufficient amounts of data in training the algorithm.

PAM matrix: amino acid substitution matrix describing the probability of one amino acid being substituted by another. It is constructed by first calculating the number of observed substitutions in a sequence dataset with 1% amino acid mutations and subsequently extrapolating the number of substitutions to more divergent sequence datasets through matrix duplication. The PAM unit is theoretically related to evolutionary time, with one PAM unit corresponding to 10 million years of evolutionary changes. Thus the higher the PAM numbering, the more divergent amino acid sequences it reflects.

Paralogs: homologous sequences from the same organism or genome, which are derived from gene duplication events rather than speciation events.

Phylogenetic footprinting: process of finding conserved DNA elements through aligning DNA sequences from multiple related species. It is widely used for identifying regulatory elements in a genome.

Phylogenetic profile: the pattern of coexistence or co-absence of gene pairs across divergent genomes. The information is useful for making inference of functionally linked genes or genes encoding interacting proteins.

Phylogeny: study of evolutionary relationships between organisms by using treelike diagrams as representations.

Physical map: map of locations of gene markers constructed by using a chromosome walking technique. The distance between gene markers is measured directly as kilobases (Kb).

Positive selection: evolutionary process that favors the replacement of amino acids in a protein sequence. This happens when the protein is adapting to a new functional role. The evidence for positive selection often comes from the observation that the nonsynonymous substitution rate is higher than the synonymous substitution rate in the DNA coding region.

Posterior probability: probability of an event estimated after taking into account a new observation. It is used in Bayesian analysis.

Profile: scoring matrix that represents a multiple sequence alignment. It contains probability or frequency values of residues for each aligned position in the alignment including gaps. A weighting scheme is often applied to correct the probability for unobserved and underobserved sequence characters. Profiles can be used to search sequence databases to detect distant homologs. This term is often used interchangeably with *position-specific scoring matrix* (PSSM).

Progressive alignment: multiple sequence alignment strategy that uses a stepwise approach to assemble an alignment. It first performs all possible pairwise alignments using the dynamic programming approach and determines the relative distances between each pair of sequences to construct a distance matrix, which is subsequently used to build a guide tree. It then realigns the two most closely related sequences using the dynamic programming approach. Other sequences are progressively added to the alignment according to the degree of similarity suggested by the guide tree. The process proceeds until all sequences are used in building a multiple alignment. The Clustal program is a good example of applying this strategy.

Protein family: group of homologous proteins with a common structure and function. A protein family is normally constructed from protein sequences with an overall identity of at least 35%.

Proteome: complete set of proteins expressed in a cell.

Proteomics: study of a proteome, which involves simultaneous analyses of all translated proteins in the entire proteome. Its topics include large-scale identification and quantification of expressed proteins and determination of their localization, modifications, interactions, and functions.

PSI-BLAST: unique version of the BLAST program that employs an iterative database searching strategy to construct multiple sequence alignments and convert them to profiles that are used to detect distant sequence homologs.

PSSM (position-specific scoring matrix): scoring table that lists the probability or frequency values of residues derived from each position in an ungapped multiple sequence alignment. A PSSM can be weighted or unweighted. In a weighted PSSM, a weighting scheme is applied to correct the probability for unobserved and underobserved sequence characters. This term is often used interchangeably with *profile*.

P-value: statistical measure representing the significance of an event based on a chance distribution. It is calculated as the probability of an event supporting the null hypothesis. The smaller the *P*-value, the more unlikely an event is due to random chance (null hypothesis) and therefore the more statistically significant it is.

Quadratic discriminant analysis: statistical method that separates true signals from background noise by projecting data points in a two dimensional graph and by drawing a curved line that best separates signals from nonsignals based on knowledge learned from a training dataset.

Quartet puzzling: phylogenetic tree construction method that relies on compiling tree topologies of all possible groups of four taxa (quartets). Individual four-taxon trees are normally derived using the exhaustive maximum likelihood method. A final tree that includes all taxa is produced by deriving a consensus from all quartet trees. The advantage of this method is computational speed.

Query: specific value used to retrieve a particular record from a database.

Ramachandran plot: two-dimensional scatter plot showing torsion angles of each amino acid residue in a protein structure. The plot delineates preferred or allowed regions of the angles as well as disallowed regions based on known protein structures. This plot helps in the evaluation of the quality of a new protein model.

Relational database: database that uses a set of separate tables to organize database entries. Each table, also called *relation*, is made up of columns and rows. Columns represent individual fields and rows represent records of data. One or more columns in a table are indexed so they can be cross-referenced in other tables. To answer a query to a relational database, the system selects linked data items from different tables and combines the information into one report.

RMSD (root mean square deviation):   measure of similarity between protein structures. It is the square root of the sum of the squared deviations of the spatial coordinates of the corresponding atoms of two protein structures that have been superimposed.

Regular expression: representation format for a sequence motif, which includes positional information for conserved and partly conserved residues.

Rotamer: preferred side chain torsion angles based on the knowledge of known protein crystal structures.

Rotamer library: collection of preferred side chain conformations that contains information about the frequency of certain conformations. Having a rotamer library reduces the computational time in a side chain conformational search.

SAGE (serial analysis of gene expression):   high throughput approach to measure global gene expression patterns. It determines the quantities of transcripts by using a large number of unique short cDNA sequence tags to represent each gene in a genome. Compared to EST analysis, SAGE analysis has a better chance of detecting weakly expressed genes.

Scaffold: continuous stretch of DNA sequence that results from merging overlapping contigs during genome assembly. Scaffolds are unidirectionally oriented along a physical map of a chromosome.

Self-organizing map: classification technique that identifies the association of genes in an expression profile. The classification is based on a neural network–like algorithm that first projects data points in a two dimensional space and subsequently carries out iterative matching of data points with a predefined number of nodes, during which the distances of the data points to the center of the cluster (centroid) are calculated. The data points stay in a particular node if the distances are small enough. The iteration continues until all data points find a best fit within the nodes.

Sensitivity: measure of ability of a classification algorithm to distinguish true positives from all possible true features. It is quantified as the ratio of true positives to the sum of true positives plus false negatives.

Sequence logo: graphical representation of a multiple sequence alignment that displays a consensus sequence with frequency information. It contains stacked letters representing the occurrence of the residues in a particular column of a multiple alignment. The overall height of a logo position reflects how conserved the position is; the height of each letter in a position reflects the relative frequency of the residue in the alignment.

Shotgun sequencing: genome sequencing approach that breaks down genomic DNA into small clones and sequences them in a random fashion. The genome sequence is subsequently assembled by joining the random fragments after identifying overlaps.

Shuffle test: statistical test for pairwise sequence alignment carried out by allowing the order of characters in one of the two sequences to be randomly altered. The shuffled sequence is subsequently used to align with the reference sequence using dynamic programming. A large number of such shuffled alignments serve to create a background alignment score distribution which is used to assess the statistical significance of the score of the original optimal pairwise alignment. A *P*-value is given to indicate the probability that the original alignment is a result of random chance.

Similarity: quantitative measure of the proportion of identical matches and conserved substitutions in a pairwise or multiple alignment.

Site: column of residues in a multiple sequence alignment.

Smith–Waterman algorithm: local pairwise alignment algorithm that applies dynamic programming in alignment.

Specificity: measure of ability of a classification algorithm to distinguish true positives from all predicted features. It is quantified as the ratio of true positives to the sum of true positives plus false positives.

Substitution matrix: two-dimensional matrix with score values describing the probability of one amino acid or nucleotide being replaced by another during sequence evolution. Commonly used substitution matrices are BLOSSUM and PAM.

Supervised classification: data analysis method that classifies data into a predefined set of categories.

Support vector machine: data classification method that projects data in a three-dimensional space. A "hyperplane" (a linear or nonlinear mathematical function) is used to separate true signals from noise. The algorithm requires training to be able to correctly recognize patterns of true features.

Synonymous substitutions: nucleotide changes in a protein coding sequence that do not result in amino acid sequence changes for the encoded protein because of redundancy in the genetic code.

Synteny: conserved gene order pattern across different genomes.

Systems biology: field of study that uses integrative approaches to model pathways and networks at the cellular level.

Taxon: each species or sequence represented at the tip of each branch of a phylogenetic tree. It is also called an *operational taxonomic unit* (OTU).

Threading: method of predicting the most likely protein structural fold based on secondary structure similarity with database structures and assessment of energies of the potential fold. The term has been used interchangeably with *fold recognition*.

Transcriptome: complete set of mRNA molecules produced by a cell under a given condition.

Transition: substitution of a purine by another purine or a pyrimidine by another pyrimidine.

Transversion: substitution of a purine by a pyrimidine or a pyrimidine by a purine.

True negative: false match that is correctly ignored by an algorithm.

True positive: true match that is correctly identified by an algorithm.

Unsupervised classification: data analysis method that does not assume predefined categories, but identifies data categories according to actual similarity patterns. It is also called *clustering*.

UPGMA (unweighted pair-group method with arithmetic means)**:**   phylogenetic tree-building method that involves clustering taxa based on phylogenetic distances. The method assumes the taxa to have equal distance from the root and starts tree building by clustering the two most closely related taxa. This produces a reduced matrix, which allows the next nearest taxa to be added. Other taxa are sequentially added using the same principle.

***Z*-score:**   statistical measure of the distance of a value from the mean of a score distribution, measured as the number of standard deviations.

Index

Printed in the United States
By Bookmasters